# Radar Polarimetry for Geoscience Applications

## *The Artech House Remote Sensing Library*

Fawwaz T. Ulaby, *Series Editor*

*Microwave Remote Sensing: Fundamentals and Radiometry, Volume I,* by F.T. Ulaby, R.K. Moore, and A.K. Fung, 1981.

*Microwave Remote Sensing: Radar Remote Sensing and Surface Scattering and Emission Theory, Volume II,* by F.T. Ulaby, R.K. Moore, and A.K. Fung, 1982.

*Microwave Remote Sensing: From Theory to Applications, Volume III,* by F.T. Ulaby, R.K. Moore, and A.K. Fung, 1986.

*Handbook of Radar Scattering Statistics for Terrain,* by F.T. Ulaby and M.C. Dobson, 1989.

*Radar Scattering Statistics for Terrain: Software and User's Manual,* by F.T. Ulaby and M.C. Dobson, 1989.

*Microwave Radiometric Systems: Design and Analysis,* by N. Skou, 1989.

*Radargrammetric Image Processing,* by F.W. Leberl, 1990.

# Radar Polarimetry for Geoscience Applications

Edited By:

**Fawwaz T. Ulaby**
University of Michigan, Ann Arbor

**Charles Elachi**
California Institute of Technology,
Jet Propulsion Laboratory, Pasadena, California

Written By:

**Charles Elachi**, Jet Propulsion Laboratory

**Yasuo Kuga**, The University of Michigan

**Kyle C. McDonald**, The University of Michigan

**Kamal Sarabandi**, The University of Michigan

**Thomas B. A. Senior**, The University of Michigan

**Fawwaz T. Ulaby**, The University of Michigan

**Jakob J. van Zyl**, Jet Propulsion Laboratory

**Michael W. Whitt**, The University of Michigan

**Howard A. Zebker**, Jet Propulsion Laboratory

**Library of Congress Cataloging-in-Publication Data**

Radar polarimetry for geoscience applications / F.T. Ulaby, C. Elachi, editors.
p. cm.
Includes bibliographical references.
ISBN 0-89006-406-7
1. Radar in earth sciences. 2. Polarimetry. I. Ulaby, Fawwaz, T. (Fawwaz Tayssir), 1943- . II. Elachi, Charles.
QE33.2.R33R32 1990 90-238
621.3848'02455--dc20 CIP

**British Library Cataloguing in Publication Data**

Radar polarimetry for geoscience applications.
1. Radar.
I. Ulaby, Fawwaz T. (Fawwaz Tayssir), 1943- II. Elachi, Charles
621.3848

ISBN 0-89006-406-7



**685 Canton Street**
**Norwood, MA 02062**



**International Standard Book Number: 0-89006-406-7**
**Library of Congress Catalog Card Number: 90-238**

**10 9 8 7 6 5 4 3 2 1**

# Contents

# The Editors

**Fawwaz T. Ulaby** earned his Ph.D. degree from the University of Texas, Austin. Currently director of the NASA Center for Space Terahertz Technology and Professor of Electrical Engineering and Computer Science at the University of Michigan, Dr. Ulaby is a Fellow of the IEEE and served as Executive Editor of the IEEE Transactions on Geoscience and Remote Sensing from 1984 to 1988. He coedited the *Manual of Remote Sensing*, published by the American Society of Photogrammetry, and coauthored Artech House's three volume series *Microwave Remote Sensing: Active and Passive* and the *Handbook of Radar Statistics for Terrain*.

**Charles Elachi** earned his Ph.D. degree from the California Institute of Technology. He has been a lecturer for Caltech for the last six years and is the Assistant Director for the Institute's Jet Propulsion Laboratory. Dr. Elachi is a fellow of the IEEE and a member of the national Academy of Engineering. His many publications include *Introduction to the Physics and Techniques of Remote Sensing* (Wiley Interscience) and *Spaceborne Radar Remote Sensing: Applications and Techniques* (IEEE Press).

# Preface

Spaceborne imaging radars are rapidly becoming one of the major tools for observing the Earth's surface and its cover. The imaging radars that were flown on Seasat in 1978 and on the Space Shuttle in 1981 and 1984 (SIR-A and SIR-B) have provided us with a wealth of data for studying the interaction of radar waves with many types of land and ocean surfaces, and have led to new insights of the scientific capability of radar for geoscientific applications. Three other spaceborne radar systems will be launched by the mid-1990s: the European ERS-1 SAR (1992), the Japanese ERS-1 SAR (1993), and the Canadian Radarsat (1994). All these instruments, which will be launched on long-duration free- flying satellites and will operate in a single-frequency single-polarization mode, are intended to emphasize scientific objectives or applications that rely mainly on qualitative tools of information extraction.

This book deals with *multifrequency polarimetric* radar systems, which represent a new generation of imaging sensors. Over the past few years, new advances in the understanding of wave-surface interaction and in the technology associated with the design, construction, and calibration of radar sensors have led to the development of new approaches that promise to offer a quantum jump in our ability to extract geoscientific information about the Earth's surface and its cover from radar data. The multifrequency and polarimetric features of these sensors make possible the separation of the backscatter contributions of the surface and its cover, thereby allowing us to extract information about each. These new capabilities have been the subject of extensive research using airborne and truck-mounted radar systems, and form the basis for future spaceborne sensors planned for the mid to late 1990s. The SIR-C mission scheduled for 1992-93 and the Eos SAR planned for the late 1990s will each carry imaging radars with total polarimetric capability, in addition to their multifrequency feature.

To publish a book that addresses the theoretical and experimental aspects of polarimetric imaging radar and illustrates some of its applications is therefore timely. This book, which is authored by researchers who are active in this field and have participated heavily in its development, consists of seven chapters. After a review of wave polarization in Chapter 1, the technique of radar polarization synthesis is introduced in Chapter 2, together with scattering matrix models for single objects. Scattering models for point and distributed targets are given in Chapters 3 and 4, respectively. Chapters 5 and 6 describe the operation and calibration techniques associated with polarimetric scatterometer systems and synthetic aperture radars. The final chapter illustrates the use of polarimetric radar observations through specific examples.

This book is addressed to graduate students, scientists, and engineers interested in conducting research in this field or in the utilization of this new tool for geoscientific research and application. Despite the fact that the book is authored by nine different individuals, a concerted effort was spent to present the material in a unified notation and consistent terminology.

The authors are indepted to many people who have played major roles in the development of radar polarimetry, in particular to the late Professor E.M. Kennaugh of Ohio State University who established the foundations of this field, and to the JPL radar team that developed the NASA/JPL polarimetric airborne radar systems which provided many of the experimental data that verified the power of this technique. Special thanks are due to Mr. Dennis Ricci of Artech House for copy-editing this manuscript, and to Ms. Bonnie Kidd and Mr. Leland Pierce for the outstanding skill and endurance they exercised in typing its many equations.

Fawwaz T. Ulaby
Charles Elachi
January, 1990

# Chapter 1

# Wave Properties and Polarization

**Fawwaz T. Ulaby and Jakob J. van Zyl**

For the treatment of problems involving absorption and scattering of electromagnetic waves by matter, Maxwell's equations provide a convenient point of departure because they define the fundamental laws governing all aspects of electromagnetism including wave generation, propagation, and interaction with matter. At any point in time $t$ and spatial location $\mathbf{r}$, where $\mathbf{r}$ is a position vector defined with respect to a specified coordinate system, the electric and magnetic fields, $\mathbf{E}(\mathbf{r}, t)$ and $\mathbf{H}(\mathbf{r}, t)$, and the magnetic flux density and electric displacement, $\mathbf{B}(\mathbf{r}, t)$ and $\mathbf{D}(\mathbf{r}, t)$, have to satisfy Maxwell's equations:

$$\nabla \cdot \mathbf{D} = \rho \tag{1.1}$$

$$\nabla \times \mathbf{E} = -\frac{\partial \mathbf{B}}{\partial t} \tag{1.2}$$

$$\nabla \cdot \mathbf{B} = 0 \tag{1.3}$$

$$\nabla \times \mathbf{H} = \mathbf{J} + \frac{\partial \mathbf{D}}{\partial t}, \tag{1.4}$$

where $\rho(\mathbf{r}, t)$ and $\mathbf{J}(\mathbf{r}, t)$ are the charge and current densities associated with free charges present at the point under consideration. Maxwell's equations are supplemented with the law of conservation of charge

$$\nabla \cdot \mathbf{J} + \frac{\partial \rho}{\partial t} = 0, \tag{1.5}$$

and the following set of constitutive relations:

$$\mathbf{D} = \epsilon' \, \mathbf{E} \tag{1.6}$$

$$\mathbf{B} = \mu \, \mathbf{H} \tag{1.7}$$

$$\mathbf{J} = \sigma \, \mathbf{E} \, , \tag{1.8}$$

where $\epsilon'$, $\mu$, and $\sigma$ are the permittivity, permeability, and conductivity of the medium under consideration.

A time dependence of the form $\exp(-i\omega t)$ will be assumed for all fields, which implies that

$$\mathbf{E}(\mathbf{r}, t) = \text{Re}\{\mathbf{E}(\mathbf{r}) e^{-i\omega t}\}, \tag{1.9}$$

where $\mathbf{E}(\mathbf{r})$ is a function of $\mathbf{r}$ only. Often we will simply write $\mathbf{E}(\mathbf{r})$ as $\mathbf{E}$, and to distinguish it from $\mathbf{E}(\mathbf{r}, t)$, we will always write the latter as a function of its arguments. In terms of this notation, substitution of the constitutive relations (1.6) to (1.8) into (1.1) to (1.4) leads to

$$\nabla \cdot \mathbf{E} = \rho / \epsilon' \tag{1.10}$$

$$\nabla \times \mathbf{E} = i \omega \, \mu \, \mathbf{H} \tag{1.11}$$

$$\nabla \cdot \mathbf{H} = 0 \tag{1.12}$$

$$\nabla \times \mathbf{H} = (\sigma - i \omega \, \epsilon') \, \mathbf{E}, \tag{1.13}$$

where $\omega$ is the angular frequency. Often convenient is to write (1.13) in the form

$$\nabla \times \mathbf{H} = -i \omega \, \epsilon_r \, \epsilon_0 \, \mathbf{E}, \tag{1.14}$$

where

$$\begin{aligned} \epsilon_r &= \frac{\epsilon'}{\epsilon_0} + i \frac{\sigma}{\omega \epsilon_0} \\ &= \epsilon_r' + i \, \epsilon_r'' \end{aligned} \tag{1.15}$$

is the relative complex dielectric constant of the medium, $\epsilon_0$ is the permittivity of free space, and $\epsilon_r'$ and $\epsilon_r''$ are the real and imaginary parts of $\epsilon_r$.

## 1-1 PLANE WAVES IN A LOSSLESS HOMOGENEOUS MEDIUM

In a source-free medium, for which $\rho = 0$ and $\mathbf{J} = 0$, Maxwell's equations can be combined to obtain the wave equation:

$$\nabla^2 \mathbf{E} + k^2 \mathbf{E} = 0 \, , \tag{1.16}$$

where

$$k = \omega\sqrt{\mu\epsilon'} \tag{1.17}$$

is the wave number. If in a Cartesian coordinate system we impose the conditions that $\mathbf{E}$ and $\mathbf{H}$ are not functions of $x$ and $y$, the solution of (1.16) is a *plane wave* given by

$$\mathbf{E}(z) = \mathbf{E}_0 e^{ikz} \; . \tag{1.18}$$

Application of (1.11) provides a similar solution for the magnetic field:

$$\mathbf{H}(z) = \mathbf{H}_0 e^{ikz} \; , \tag{1.19}$$

with

$$\mathbf{H}_0 = \frac{1}{\eta}\hat{\mathbf{z}} \times \mathbf{E}_0 \; , \tag{1.20}$$

where $\eta = \sqrt{\mu/\epsilon'}$ is the intrinsic impedance of the medium and $\hat{\mathbf{z}}$ is a unit vector in the $z$ direction.

For the general case of a plane wave propagating in the direction $\hat{\mathbf{k}}$ in a spherical coordinate system $(r, \theta, \phi)$, as depicted in Fig. 1.1, the solution of (1.16) takes the form

$$\begin{aligned} \mathbf{E}(\mathbf{r}) &= \mathbf{E}_0 e^{i\mathbf{k}\cdot\mathbf{r}} \\ &= \mathbf{E}_0 e^{ik\hat{\mathbf{k}}\cdot\mathbf{r}} \; , \end{aligned} \tag{1.21}$$

where $\mathbf{k} = k\hat{\mathbf{k}}$ is the propagation vector and $\mathbf{r}$ is the position vector to the point of interest (point $Q$ in Fig. 1.1). The unit vector $\hat{\mathbf{k}}$, which is parallel to the unit vector $\hat{\mathbf{k}}'$ in Fig. 1.1, is given by

$$\hat{\mathbf{k}} = \sin\theta\cos\phi \; \hat{\mathbf{x}} + \sin\theta\sin\phi \; \hat{\mathbf{y}} + \cos\theta \; \hat{\mathbf{z}} \; , \tag{1.22}$$

and the magnetic field is given by

$$\mathbf{H}(\mathbf{r}) = \mathbf{H}_0 e^{ik\hat{\mathbf{k}}\cdot\mathbf{r}} \; , \tag{1.23}$$

with

$$\mathbf{H}_0 = \frac{1}{\eta} \; \hat{\mathbf{k}} \times \mathbf{E}_0 \; . \tag{1.24}$$

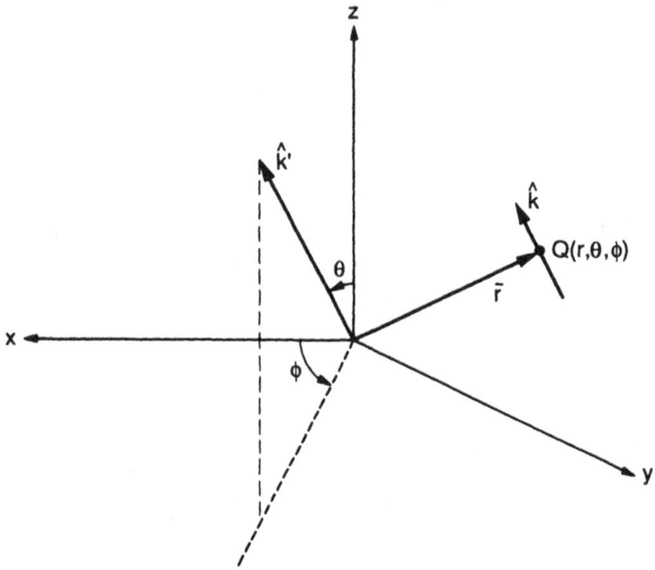


**Figure 1.1** Plane wave at point $Q$ propagating in direction $\hat{\mathbf{k}}$, which is parallel to $\hat{\mathbf{k}}'$.

# 1-2 WAVE POLARIZATION

The polarization of a plane wave describes the shape and locus of the tip of the $E$-vector (in a plane orthogonal to the direction of propagation) as a function of time. In the general case, the locus of the $E$-vector in a plane orthogonal to the direction of propagation is an ellipse, and the wave is called *elliptically polarized.* Under certain conditions, the ellipse may degenerate into a segment of a straight line or a circle, and the polarization is then called *linear* or *circular.* For problems involving radar scattering from point or distributed targets (such as terrain), the formulation usually is cast with reference to a spherical coordinate system, but to facilitate understanding of the conventions used to characterize wave polarization, we shall first examine the case of a plane wave traveling in the $z$ direction.

## 1-2.1 Cartesian Coordinate System

The electric field vector of a $z$-propagating wave lies in the $x$–$y$ plane and, therefore, consists of an $x$-component, $E_x$, and a $y$-component, $E_y$,

$$\mathbf{E}(z) = E_x(z)\,\hat{\mathbf{x}} + E_y(z)\,\hat{\mathbf{y}} , \tag{1.25}$$

with

$$\begin{aligned} E_x(z) &= A_x e^{ikz} \\ &= a_x e^{-i\delta_x} e^{ikz} , \end{aligned} \tag{1.26}$$

$$\begin{aligned} E_y(z) &= A_y e^{ikz} \\ &= a_y e^{-i\delta_y} e^{ikz} , \end{aligned} \tag{1.27}$$

where $A_x$ and $A_y$ are the amplitudes of $E_x(z)$ and $E_y(z)$, respectively, and each is composed of a magnitude ($a_x$ or $a_y$) and a phase angle ($\delta_x$ or $\delta_y$). The instantaneous value of $E_x(z,t)$ is given by

$$\begin{aligned} E_x(z,t) &= \mathrm{Re}[E_x(z)e^{-i\omega t}] \\ &= a_x \cos(kz - \omega t - \delta_x) \\ &= a_x \cos(\tau + \delta_x) , \end{aligned} \tag{1.28}$$

and, similarly,

$$\begin{aligned} E_y(z,t) &= a_y \cos(kz - \omega t - \delta_y) \\ &= a_y \cos(\tau + \delta_y) \end{aligned} \tag{1.29}$$

where $\tau = wt - kz$.

### Linear Polarization

If $\delta_x = \delta_y = \delta$, the trace of the tip of $\mathbf{E}(z,t)$ as a function of time at a specified value of $z$ forms a straight line, as illustrated in Fig. 1.2. The modulus of $\mathbf{E}(z,t)$ is given by

$$\begin{aligned} |\mathbf{E}(z,t)| &= [E_x^2(z,t) + E_y^2(z,t)]^{1/2} \\ &= [a_x^2 + a_y^2]^{1/2} \cos(\tau + \delta) \end{aligned} \tag{1.30}$$

and its inclination angle $\alpha$ is

$$\alpha = \tan^{-1}\left(\frac{E_y}{E_x}\right) = \tan^{-1}\left(\frac{a_y}{a_x}\right) \quad . \tag{1.31}$$

If $a_y = 0$, then $\alpha = 0$ and the wave is $x$-polarized, and if $a_x = 0$, then $\alpha = 90°$ and the wave is $y$-polarized.

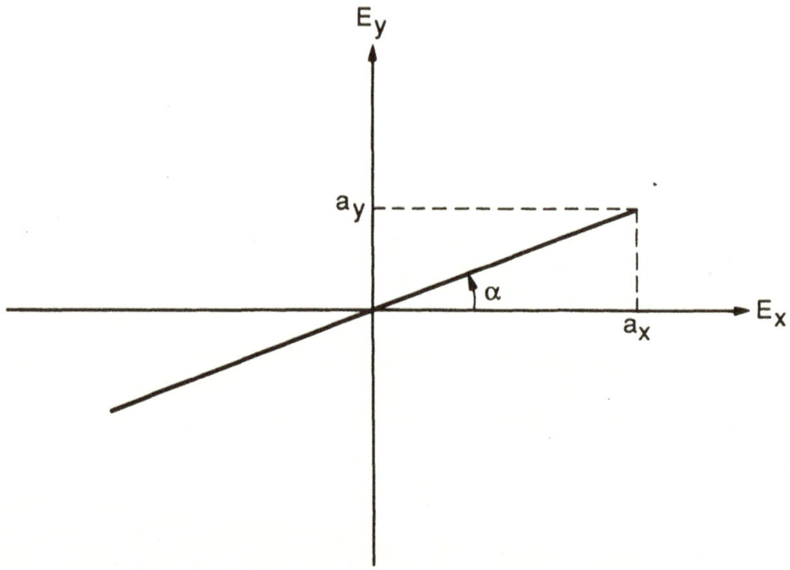


**Figure 1.2** The solid line represents the locus of the $E$-vector for linear polarization at a specified value of $z$.

### Elliptical Polarization

If $a_x \neq 0$, $a_y \neq 0$, and the phase difference $\delta = \delta_y - \delta_x \neq 0$, (1.28) and (1.29) can be combined by eliminating $\tau$ to yield the general equation for a rotated ellipse (Born and Wolf, 1965, pp. 24–27):

$$\left(\frac{E_x(z,t)}{a_x}\right)^2 + \left(\frac{E_y(z,t)}{a_y}\right)^2 - \left(\frac{2E_x(z,t)E_y(z,t)}{a_x a_y}\right)\cos\delta = \sin^2\delta \ . \tag{1.32}$$

The trace of the tip of the $E$-vector in the $x$–$y$ plane (at a fixed value of $z$) as a function of time forms an ellipse as shown in Fig. 1.3. The *sense of rotation* of the tip of the $E$-vector in the plane of polarization is called the *sense of polarization*, or ***handedness***. This sense is called *right-handed* (*left-handed*) if the direction of rotation (in a plane at a fixed $z$) is *clockwise* (*counterclockwise*) for an observer looking in the direction of propagation (IEEE Test Antenna Standard, 1979, p. 76). This definition is opposite to that used in classical optics (Born and Wolf, 1965, p. 28), which sometimes causes readers unnecessary confusion. One convenient way by which to remember the IEEE definition is in terms of the screw rule; the polarization is right-handed when the electric field rotates (in a plane) as a right-handed screw in the direction of propagation (Ishimaru, 1978, p. 30; Kraus, 1984, p. 498), as illustrated in Fig. 1.4.

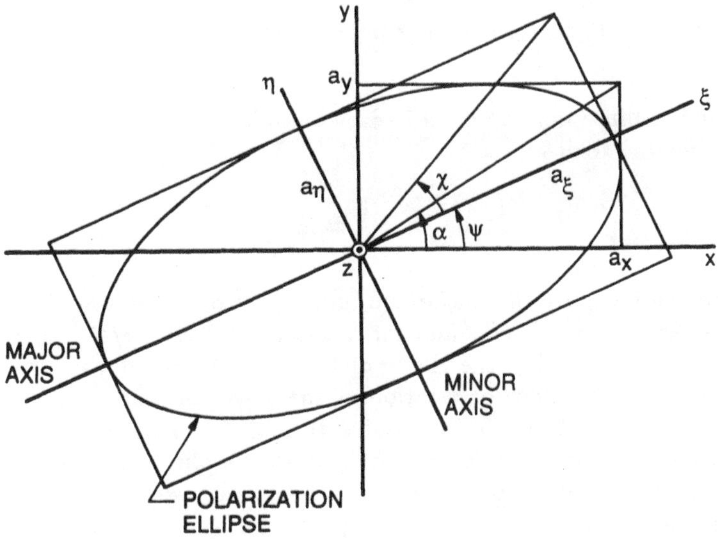


**Figure 1.3** Polarization ellipse in the $x$-$y$ plane, rotation angle $\Psi$, ellipticity angle $\chi$ and auxiliary angle $\alpha$. The wave is traveling in the $\hat{z}$ direction (out of page).

In addition to the sense of rotation, elliptical polarization is characterized by the *axial ratio* $R$, which is the ratio of the major axis to the minor axis of the ellipse, and the *rotation angle* $\psi$, which is the angle between the major axis of the ellipse and a reference direction, chosen here to be the $x$ axis. As an alternative to the axial ratio $R$, the ellipticity of the

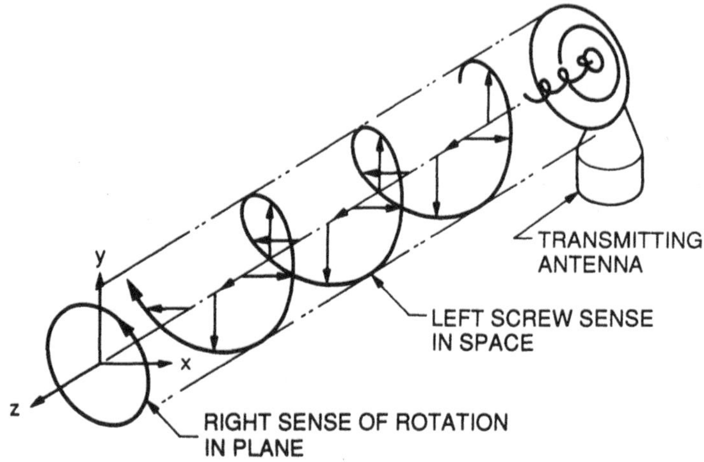


**Figure 1.4** Right-handed circular polarization.

polarization ellipse may be described by the *ellipticity angle* $\chi$ (Fig. 1.3), which is related to $R$ by

$$\tan\chi = \pm\frac{1}{R} = \pm\frac{a_\eta}{a_\xi} , \tag{1.33}$$

where $2a_\xi$ and $2a_\eta$ are the major and minor axes of the ellipse. The limits of $\psi$ are $-\pi/2 \le \psi \le \pi/2$, and for $\chi$ the limits are $-\pi/4 \le \chi \le \pi/4$. In addition to specifying the shape of the polarization ellipse, the angle $\chi$ also specifies the sense of rotation of the $E$-vector; the polarization is left-handed if $\chi > 0$ and right-handed if $\chi < 0$. If $a_x = a_y$ and $\delta = \pm\pi/2$, the ellipse degenerates to a circle, with the plus sign corresponding to left circular polarization (LCP) and the minus sign corresponding to right circular polarization (RCP).

The polarization angles $\psi$ and $\chi$ are related to the wave parameters $a_x$, $a_y$, and $\delta$ by (Born and Wolf, 1965, p. 27)

$$\tan 2\psi = (\tan 2\alpha)\cos\delta \tag{1.34}$$

$$\sin 2\chi = (\sin 2\alpha)\sin\delta , \tag{1.35}$$

where $\alpha$ is an auxiliary angle defined by

$$\tan\alpha = \frac{a_y}{a_x} . \tag{1.36}$$

## 1-2.2 Spherical Coordinate System

When considering problems involving scattering from the earth's surface, we often conveniently represent the earth's surface by the $x$–$y$ plane and represent the polarization vector of waves incident upon or scattered from the surface in terms of the spherical angles $\theta$, and $\phi$. For a plane wave traveling in the direction $\hat{\mathbf{k}}$, as shown in Fig. 1.5, we customarily characterize the vector $\mathbf{E}$ in terms of a *horizontal polarization* component, $E_h\hat{\mathbf{h}}$, and a *vertical polarization* component, $E_v\hat{\mathbf{v}}$, defined such that the coordinate system $(\hat{\mathbf{k}}, \hat{\mathbf{v}}, \hat{\mathbf{h}})$ coincides with $(\hat{\mathbf{r}}, \hat{\boldsymbol{\theta}}, \hat{\boldsymbol{\phi}})$ of a standard spherical coordinate system. Thus,

$$\mathbf{E} = (E_v\hat{\mathbf{v}} + E_h\hat{\mathbf{h}})e^{ik\hat{\mathbf{k}}\cdot\mathbf{r}} \tag{1.37}$$

with

$$\hat{\mathbf{h}} = \frac{\hat{\mathbf{z}} \times \hat{\mathbf{k}}}{|\hat{\mathbf{z}} \times \hat{\mathbf{k}}|} = -\sin\phi\ \hat{\mathbf{x}} + \cos\phi\ \hat{\mathbf{y}}\ , \tag{1.38}$$

$$\hat{\mathbf{v}} = \hat{\mathbf{h}} \times \hat{\mathbf{k}} = \cos\theta\cos\phi\ \hat{\mathbf{x}} + \cos\theta\sin\phi\ \hat{\mathbf{y}} - \sin\theta\ \hat{\mathbf{z}}\ , \tag{1.39}$$

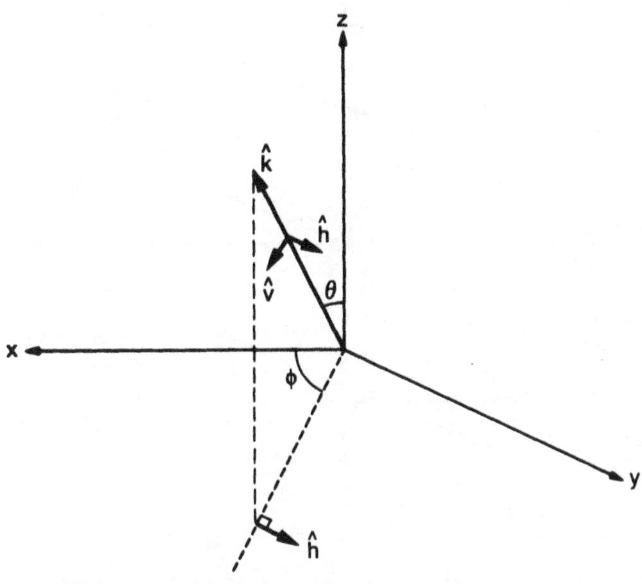


**Figure 1.5** Plane wave propagating in direction $\hat{\mathbf{k}}$, with $E$-vector components $E_v$ and $E_h$ defined such that $\hat{\mathbf{h}} = (\hat{\mathbf{k}} \times \hat{\mathbf{z}})/\mid\hat{\mathbf{k}} \times \hat{\mathbf{z}}\mid$ and $\hat{\mathbf{v}} = \hat{\mathbf{h}} \times \hat{\mathbf{k}}$.

and $\hat{\mathbf{k}}$ as defined earlier by (1.22). As should prove useful in future chapters, we adopt the matrix notation

$$\mathbf{E} = \begin{bmatrix} E_v \\ E_h \end{bmatrix} \tag{1.40}$$

wherein the phase factor of (1.37) has been suppressed for convenience.

Except for a change of subscripts from $(x, y)$ to $(v, h)$, all the expressions given in the preceding section shall hold. That is, in terms of

$$E_v = a_v e^{-i\delta_v} \ , \tag{1.41}$$

$$E_h = a_h e^{-i\delta_h} \ , \tag{1.42}$$

$$\delta = \delta_h - \delta_v \ , \tag{1.43}$$

and

$$\tan\alpha = \frac{a_h}{a_v} \ , \tag{1.44}$$

the polarization angles $\psi$ and $\chi$ are given by (1.34) and (1.35), respectively. The polarization ellipse in the $(v, h)$ plane is shown in Fig. 1.6.

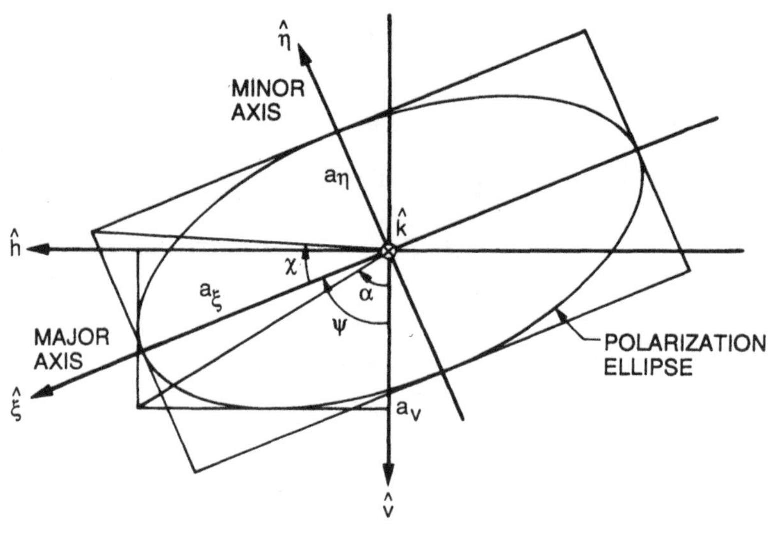


**Figure 1.6** Polarization ellipse in the $v$-$h$ plane for a wave traveling in the $\hat{\mathbf{k}}$ direction.

## 1-2.3 Stokes Vector Representation

The *polarization state* of a plane wave can be characterized by the parameters $a_v$, $a_h$, $\delta$, or alternatively by the parameters $\psi$, $\chi$, and $I_0$, where

$$I_0 = a_v^2 + a_h^2 \tag{1.45}$$

is proportional to the total intensity of the wave. In 1852, Stokes introduced a third set of parameters for characterizing polarization state with the distinction that they all have the same physical dimensions. Stokes' representation, which consists of four *Stokes parameters*, $I_0$, $Q$, $U$, and $V$, can be written in vector form as follows:

$$\mathbf{F} = \begin{bmatrix} I_0 \\ Q \\ U \\ V \end{bmatrix} = \begin{bmatrix} |E_v|^2 + |E_h|^2 \\ |E_v|^2 - |E_h|^2 \\ 2\text{Re}(E_v E_h^*) \\ 2\text{Im}(E_v E_h^*) \end{bmatrix} = \begin{bmatrix} a_v^2 + a_h^2 \\ a_v^2 - a_h^2 \\ 2a_v a_h \cos\delta \\ 2a_v a_h \sin\delta \end{bmatrix} = \begin{bmatrix} I_0 \\ I_0 \cos 2\psi \cos 2\chi \\ I_0 \sin 2\psi \cos 2\chi \\ I_0 \sin 2\chi \end{bmatrix}. \tag{1.46}$$

We shall refer to the above representation as the Stokes vector for a *completely polarized wave.*

Only three of the Stokes parameters are independent because they are related by the identity:

$$I_0^2 = Q^2 + U^2 + V^2. \tag{1.47}$$

This identity holds for completely polarized waves only.

The state of polarization of a plane monochromatic wave can be uniquely mapped to a point $P$ on the surface of a sphere of radius $I_0$, called the Poincaré sphere, by regarding the parameters $Q$, $U$, and $V$ as the Cartesian coordinates of the point $P$ (Fig. 1.7). The angles $2\chi$ and $2\psi$ define the latitude and longitude of the point $P$. Because the sign of $\chi$ determines the handedness of the polarization state, the upper hemisphere, corresponding to $\chi > 0$, displays left-handed polarizations and the lower hemisphere, corresponding to $\chi < 0$, displays right-handed polarizations. The north pole of the sphere represents left circular polarization, the south pole represents right circular polarization, and linear polarization is represented by points in the equatorial plane. Table 1.1 lists the values of $\psi$ and $\chi$ for several polarization states of common interest.

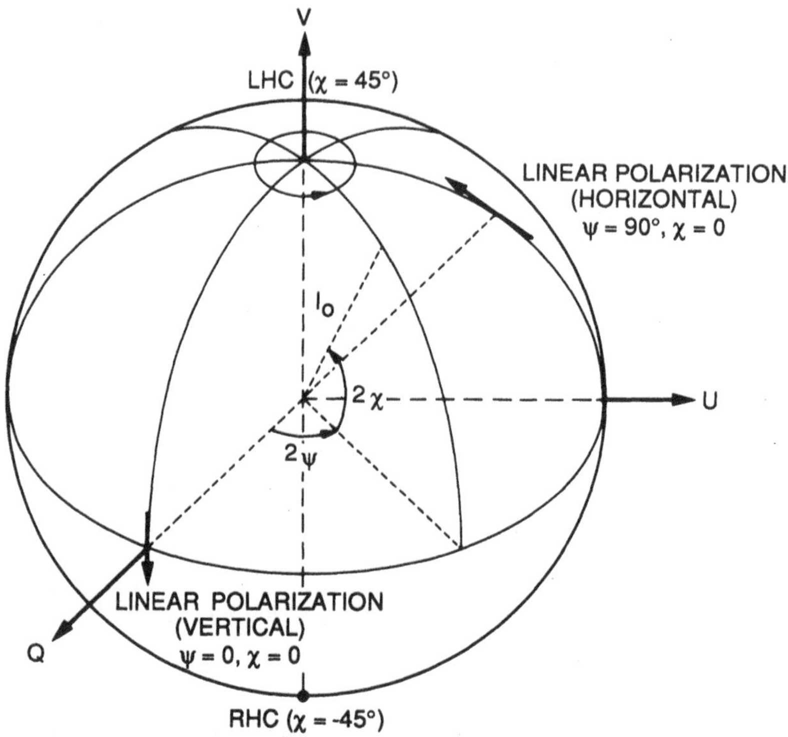


**Figure 1.7** Poincaré sphere with Stokes parameters as axes.

**Table 1.1** Values of $\psi$, $\chi$, and the normalized Stokes parameters $i_v = I_v/I_0$, $i_h = I_h/I_0$, $u = U/I_0$, and $v = V/I_0$ for some commonly used polarization states.

| Polarization State | Rotation Angle($\psi$) | Ellipticity Angle($\chi$) | Nor. Stokes Parameters | | | |
|---|---|---|---|---|---|---|
| | | | $i_v$ | $i_h$ | $u$ | $v$ |
| Vertical | 0° | 0° | 1 | 0 | 0 | 0 |
| Horizontal | 90° | 0° | 0 | 1 | 0 | 0 |
| Right circular | −90° to 90° | -45° | 1/2 | 1/2 | 0 | -1 |
| Left circular | −90° to 90° | 45° | 1/2 | 1/2 | 0 | 1 |

# 1-3 MODIFIED STOKES VECTOR

When Stokes constructed his four polarization parameters ($I_0$, $Q$, $U$, and $V$), among his objectives was to have one of the parameters represent the total intensity of the wave ($I_0$), another to represent the difference between the vertically polarized intensity and the horizontally polarized intensity ($Q$), and the last two ($U$ and $V$) to represent jointly the phase difference between the vertically polarized and horizontally polarized components of the wave. From the definition of $I_0$ and $Q$ given by (1.46), we can obtain the expressions:

$$I_v = |E_v|^2 = (I_0 + Q)/2 \tag{1.48}$$

$$I_h = |E_h|^2 = (I_0 - Q)/2 \tag{1.49}$$

for the individual intensities of the horizontally and vertically polarized components. Alternatively, we can define a *modified Stokes vector* $\mathbf{F}_m$ that consists of the set $I_v$, $I_h$, $U$, and $V$. Thus,

$$\mathbf{F}_m = \begin{bmatrix} I_v \\ I_h \\ U \\ V \end{bmatrix} = \begin{bmatrix} \frac{1}{2}(1 + \cos 2\psi \cos 2\chi) \\ \frac{1}{2}(1 - \cos 2\psi \cos 2\chi) \\ \sin 2\psi \cos 2\chi \\ \sin 2\chi \end{bmatrix} I_0 \,. \tag{1.50}$$

As noted by Chandrasekhar (1960, p. 34), the set of polarization parameters defined by the modified Stokes vector is more convenient to use in solving radiative transfer problems than the original set introduced by Stokes. The vectors $\mathbf{F}$ and $\mathbf{F}_m$ are related by [1]

$$\mathbf{F} = \mathcal{U}\mathbf{F}_m \,, \tag{1.51}$$

where

$$\mathcal{U} = \begin{bmatrix} 1 & 1 & 0 & 0 \\ 1 & -1 & 0 & 0 \\ 0 & 0 & 1 & 0 \\ 0 & 0 & 0 & 1 \end{bmatrix} . \tag{1.52}$$

[1] To avoid confusion between vectors, as in (1.50), and matrices, as in (1.52), vectors are denoted by boldface letters, whereas matrices are denoted by boldface script letters.

# 1-4 PARTIAL POLARIZATION

The previous sections dealt with *completely polarized*, monochromatic waves for which $\omega$, $a_v$, $a_h$, and $\delta$ are constants. However, the radiation from many natural and constructed (man-made) sources may extend over a wide frequency range and the wave parameters $a_v$, $a_h$, and $\delta$ may be functions of time or spatial position. In the case of radar waves, the frequency bandwidth is sufficiently small in comparison with the carrier frequency to allow us to treat the wave as quasimonochromatic as far as the polarization state is concerned. The wave transmitted by a radar system to illuminate a terrain surface can be regarded as a monochromatic, completely polarized, plane wave provided that (a) the radar is sufficiently far from the surface to satisfy the far-field condition and (b) the intervening atmosphere is electromagnetically homogeneous. Both conditions are almost always satisfied. In contrast with the transmitted signal, the signal received by the radar is seldom completely polarized when observed as a function of time or spatial position. This is because the received signal, which consists of the superposition of a large number of waves of a variety of polarizations, is the result of backscattering by a statistically random surface. If we model a given terrain surface as a statistically homogeneous target comprising randomly distributed scattering centers or facets, the electric field of the total signal backscattered from a given resolution cell (Fig. 1.8) is the vector sum of the electric fields of the waves scattered from all the scattering centers contained in that cell (Ulaby and Dobson, 1989, p. 38). If we move the resolution cell to another part of the same random surface, we will have a different set of vectors — and hence polarizations — contributing to the total field. Hence, for radar scattering from a statistically homogeneous medium, the electric field components $E_v$ and $E_h$ are, in general, random variables that are a function of the position vector $\boldsymbol{\xi}$ shown in Fig. 1.8:

$$E_v(\boldsymbol{\xi}) = a_v(\boldsymbol{\xi})e^{-i\delta_v(\boldsymbol{\xi})} \tag{1.53}$$

$$E_h(\boldsymbol{\xi}) = a_h(\boldsymbol{\xi})e^{-i\delta_h(\boldsymbol{\xi})} \ . \tag{1.54}$$

Accordingly, the Stokes parameters for a *partially polarized* wave are defined as

$$\mathbf{F} = \begin{bmatrix} I_0 \\ Q \\ U \\ V \end{bmatrix} = \begin{bmatrix} \langle |E_v|^2\rangle + \langle |E_h|^2\rangle \\ \langle |E_v|^2\rangle - \langle |E_h|^2\rangle \\ \langle 2\mathrm{Re}(E_v E_h^*)\rangle \\ \langle 2\mathrm{Im}(E_v E_h^*)\rangle \end{bmatrix} = \begin{bmatrix} \langle a_v^2\rangle + \langle a_h^2\rangle \\ \langle a_v^2\rangle - \langle a_h^2\rangle \\ \langle 2a_v a_h \cos\delta\rangle \\ \langle 2a_v a_h \sin\delta\rangle \end{bmatrix} , \tag{1.55}$$

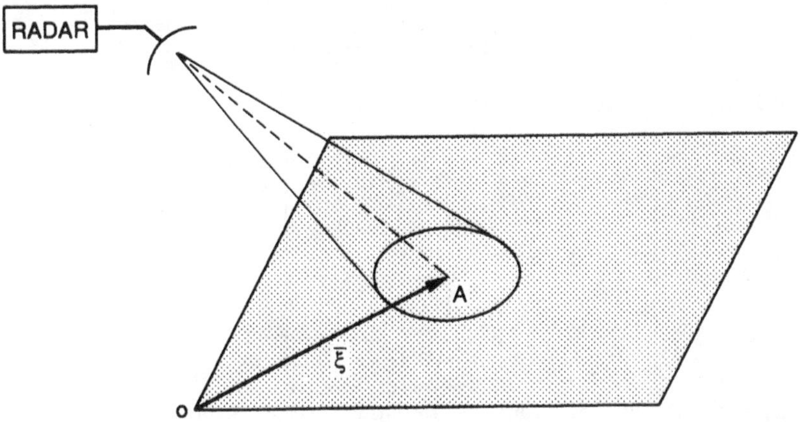


**Figure 1.8** Radar beam illuminating resolution cell of area A containing randomly distributed scatterers. The cell's center is located at $\xi$ relative to a defined coordinate system.

where $\langle\ \rangle$ indicates the ensemble average over the variable $\boldsymbol{\xi}$. For a *completely unpolarized wave*, $\langle a_v^2 \rangle = \langle a_h^2 \rangle$ and $E_v(\xi)$ and $E_h(\xi)$ are uncorrelated. This leads to the following definition for the Stokes vector of a completely unpolarized wave:

$$\mathbf{F}_{un} = \begin{bmatrix} 1 \\ 0 \\ 0 \\ 0 \end{bmatrix} I_0 \ . \tag{1.56}$$

The identity given by (1.47) holds only if the wave is completely polarized. For the general case of a partially polarized wave,

$$I_0^2 \geq Q^2 + U^2 + V^2 \ , \tag{1.57}$$

and the degree of polarization of a wave is defined as

$$\begin{aligned} m &= \frac{\text{polarized power}}{\text{total power}} \\ &= \frac{\sqrt{Q^2+U^2+V^2}}{I_0} \ . \end{aligned} \tag{1.58}$$

A partially polarized wave may be regarded as the sum of a completely unpolarized wave represented by the Stokes vector $\mathbf{F}_{un}$ and a completely polarized wave represented by the Stokes vector $\mathbf{F}_p$:

$$\mathbf{F} = (1 - m)\mathbf{F}_{un} + m\, \mathbf{F}_p \ , \tag{1.59}$$

where $\mathbf{F}_{un}$ is as given by (1.56) and

$$\mathbf{F}_p = \begin{bmatrix} 1 \\ \cos 2\psi_p \cos 2\chi_p \\ \sin 2\psi_p \cos 2\chi_p \\ \sin 2\chi_p \end{bmatrix} I_0 \ . \tag{1.60}$$

The polarization angles $\psi_p$ and $\chi_p$ pertain to the completely polarized component of the wave.

## References

Born, M., and E. Wolf, *Principles of Optics*, Pergamon Press, New York, 1965.

Chandrasekhar, S., *Radiative Transfer*, Dover, New York, 1960.

*IEEE Standard Test Procedures for Antennas*, IEEE Press, New York, 1979.

Ishimaru, A., *Wave Propagation and Scattering in Random Media*, Academic Press, New York, 1978.

Kraus, J.D., *Electromagnetics*, McGraw-Hill, New York, 1984.

Ulaby, F.T., and M.C. Dobson, *Radar Scattering Statistics for Terrain*, Artech House, Norwood, MA, 1989.

# Chapter 2

# Scattering Matrix Representation for Simple Targets

**Jakob J. van Zyl and Fawwaz T. Ulaby**

In this chapter we introduce the different mathematical representations used to describe scattering properties. In addition, we derive the concept of polarization synthesis, as described by (van Zyl *et al.*, 1987a) and (Zebker *et al.*, 1987). Results for selected simple targets are presented in terms of the various matrices as well as plots of their respective polarization responses.

## 2-1 COORDINATE SYSTEM AND SCATTERING MATRIX

Scattering calculations are performed in two major types of coordinate systems, the *forward scatter alignment* (FSA) convention and the *backscatter alignment* (BSA) convention. In both cases, the electric fields of the incident and scattered waves are expressed in local coordinate systems centered on the transmitting and receiving antennas, respectively. All coordinate systems are defined in terms of a global coordinate system centered inside the scatterer as shown in Figs. 2.1 and 2.2 for the FSA and BSA conventions, respectively.

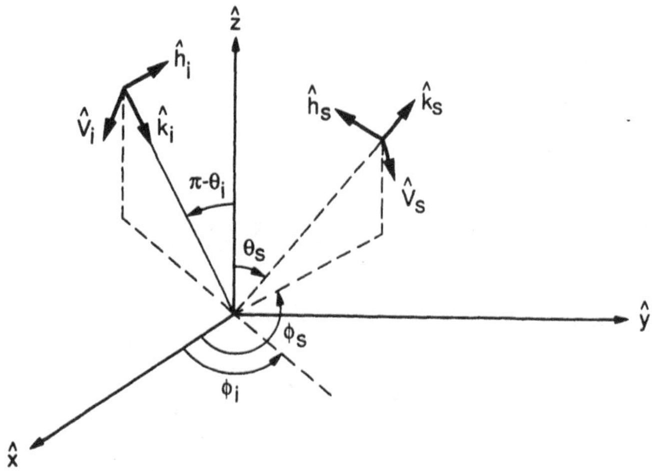

**Figure 2.1** Coordinate systems and scattering geometry for the forward scatter alignment (FSA) convention.

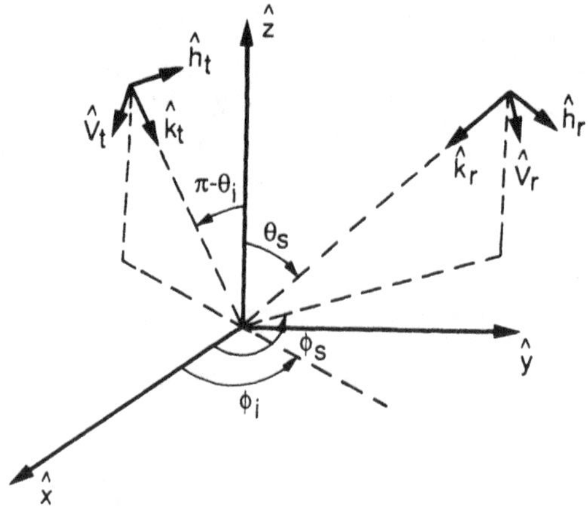

**Figure 2.2** Coordinate systems and scattering geometry for the backscatter alignment (BSA) convention.

Both conventions are used in the literature, with the FSA being the preferred system for problems involving bistatic scattering from particles and inhomogeneous media and the BSA being the preferred system for calculating the radar backscatter from a given target or medium. Because both conventions are used in this book, we shall define and interrelate their individual coordinate systems in this section. Then we will use only the BSA convention to define the various matrices and relationships associated with synthesizing the polarimetric responses of point targets. The connection between the two conventions is provided in Section 2-5 in the form of transformation matrices.

## 2-1.1 Forward Scattering Alignment (FSA) Convention

The FSA convention is commonly used in connection with problems involving *wave* scattering by particles and *wave* propagation in inhomogeneous media. FSA is a "wave-oriented" convention in that the directions of the vertical and horizontal unit vectors, $\hat{\mathbf{v}}$ and $\hat{\mathbf{h}}$, are always defined with respect to the direction of propagation of the wave, $\hat{\mathbf{k}}$. The coordinate system $(\hat{\mathbf{k}}, \hat{\mathbf{v}}, \hat{\mathbf{h}})$ is chosen to coincide with $(\hat{\mathbf{r}}, \hat{\boldsymbol{\theta}}, \hat{\boldsymbol{\phi}})$ in a standard spherical coordinate system, as shown earlier in Section 1-2.2. Thus, for a wave incident in the direction $\hat{\mathbf{k}}_i$ and scattered in the direction $\hat{\mathbf{k}}_s$, as shown in Fig. 2.1, the unit vectors are

$$\hat{\mathbf{k}}_i = \cos\phi_i \sin\theta_i\ \hat{\mathbf{x}} + \sin\phi_i \sin\theta_i\ \hat{\mathbf{y}} + \cos\theta_i\ \hat{\mathbf{z}} \tag{2.1}$$

$$\hat{\mathbf{h}}_i = \frac{\hat{\mathbf{z}} \times \hat{\mathbf{k}}_i}{|\hat{\mathbf{z}} \times \hat{\mathbf{k}}_i|} = -\sin\phi_i\ \hat{\mathbf{x}} + \cos\phi_i\ \hat{\mathbf{y}} \tag{2.2}$$

$$\hat{\mathbf{v}}_i = \hat{\mathbf{h}}_i \times \hat{\mathbf{k}}_i = \cos\phi_i \cos\theta_i\ \hat{\mathbf{x}} + \sin\phi_i \cos\theta_i\ \hat{\mathbf{y}} - \sin\theta_i\ \hat{\mathbf{z}} \tag{2.3}$$

$$\hat{\mathbf{k}}_s = \cos\phi_s \sin\theta_s\ \hat{\mathbf{x}} + \sin\phi_s \sin\theta_s\ \hat{\mathbf{y}} + \cos\theta_s\ \hat{\mathbf{z}} \tag{2.4}$$

$$\hat{\mathbf{h}}_s = \frac{\hat{\mathbf{z}} \times \hat{\mathbf{k}}_s}{|\hat{\mathbf{z}} \times \hat{\mathbf{k}}_s|} = -\sin\phi_s\ \hat{\mathbf{x}} + \cos\phi_s\ \hat{\mathbf{y}} \tag{2.5}$$

$$\hat{\mathbf{v}}_s = \hat{\mathbf{h}}_s \times \hat{\mathbf{k}}_s = \cos\phi_s \cos\theta_s\ \hat{\mathbf{x}} + \sin\phi_s \cos\theta_s\ \hat{\mathbf{y}} - \sin\theta_s\ \hat{\mathbf{z}}. \tag{2.6}$$

For forward scattering ($\theta_s = \theta_i$ and $\phi_s = \phi_i$), we have $\hat{\mathbf{k}}_s = \hat{\mathbf{k}}_i$, $\hat{\mathbf{v}}_s = \hat{\mathbf{v}}_i$, and $\hat{\mathbf{h}}_s = \hat{\mathbf{h}}_i$. In contrast, for backscattering ($\theta_s = \pi - \theta_i$ and $\phi_s = \pi + \phi_i$), the relations are $\hat{\mathbf{k}}_s = -\hat{\mathbf{k}}_i$, $\hat{\mathbf{v}}_s = \hat{\mathbf{v}}_i$, and $\hat{\mathbf{h}}_s = -\hat{\mathbf{h}}_i$. The direction of propagation $\hat{\mathbf{k}}_s$ is opposite to $\hat{\mathbf{k}}_i$, as expected, but only one of the two polarization vectors reverses direction upon backscattering.

## 2-1.2 Backscatter Alignment (BSA) Convention

Whereas the polarization unit vectors in the FSA convention are defined relative to the propagating wave, in the BSA convention they are defined with respect to the radar antennas in accordance with the IEEE standard (IEEE, 1983), which defines the polarization state of an antenna to be the polarization of the wave radiated by the antenna, even when it is used as a receiving antenna. Consequently, for the backscatter case under the BSA convention, the vertical and horizontal unit polarization vectors of the scattered wave are identical with their counterparts of the incident wave. The local coordinate system unit vectors are (Fig. 2.2):

$$\hat{\mathbf{k}}_t = \hat{\mathbf{k}}_i = \cos\phi_i \sin\theta_i\ \hat{\mathbf{x}} + \sin\phi_i \sin\theta_i\ \hat{\mathbf{y}} + \cos\theta_i\ \hat{\mathbf{z}} \tag{2.7}$$

$$\hat{\mathbf{h}}_t = \hat{\mathbf{h}}_i = -\sin\phi_i\ \hat{\mathbf{x}} + \cos\phi_i\ \hat{\mathbf{y}} \tag{2.8}$$

$$\hat{\mathbf{v}}_t = \hat{\mathbf{v}}_i = \cos\phi_i \cos\theta_i\, \hat{\mathbf{x}} + \sin\phi_i \cos\theta_i\, \hat{\mathbf{y}} - \sin\theta_i\, \hat{\mathbf{z}} \tag{2.9}$$

$$\hat{\mathbf{k}}_r = -\hat{\mathbf{k}}_s = -[\cos(\phi_s)\sin(\theta_s)\hat{\mathbf{x}} + \sin\phi_s \sin\theta_s\, \hat{\mathbf{y}} + \cos\theta_s\, \hat{\mathbf{z}}] \tag{2.10}$$

$$\hat{\mathbf{h}}_r = -\hat{\mathbf{h}}_s = \sin\phi_s\ \hat{\mathbf{x}} - \cos\phi_s\ \hat{\mathbf{y}} \tag{2.11}$$

$$\hat{\mathbf{v}}_r = \hat{\mathbf{v}}_s = \cos\phi_s \cos\theta_s\, \hat{\mathbf{x}} + \sin\phi_s \cos\theta_s\, \hat{\mathbf{y}} - \sin\theta_s\, \hat{\mathbf{z}}, \tag{2.12}$$

where the subscripts $t$ and $r$ refer to the transmitting (or incident) and receiving (or opposite of scattered) antenna orientations. Because the two coordinate systems, $(\hat{\mathbf{k}}_t,\ \hat{\mathbf{h}}_t,\ \hat{\mathbf{v}}_t)$ and $(\hat{\mathbf{k}}_r,\ \hat{\mathbf{h}}_r,\ \hat{\mathbf{v}}_r)$, are coincident when the two antennas are collocated, the BSA convention has been the preferred system in the area of radar polarimetry (van Zyl *et al.*, 1987a ; Zebker *et al.*, 1987).

To distinguish easily between the coordinate systems of the two conventions, we choose the subscripts $i$ and $s$ (which denote incident and scattered) for the FSA and the subscripts $t$ and $r$ (which denote transmitted and received) for the BSA convention. We must keep in mind, however, that the $i$ coordinate system and the $t$ coordinate systems are identical. To denote whether a given symbol represents a quantity that has been defined according to the FSA or BSA convention, we shall adopt the rule that the quantity is defined according to the BSA (or FSA) convention if the symbol has (or does not have) a solid bar over it. Thus, the scattering matrix $\overline{\boldsymbol{S}}$, for example, is defined in the BSA convention, whereas $\boldsymbol{S}$ is defined in the FSA convention. More specifically, let us consider a scatterer illuminated by an electromagnetic plane wave with incident electric field $\mathbf{E}^i$:

$$\mathbf{E}^i = E_v^i \hat{\mathbf{v}}_i + E_h^i \hat{\mathbf{h}}_i, \tag{2.13}$$

with $(\hat{\mathbf{v}}_i,\ \hat{\mathbf{h}}_i)$ according to (2.3) and (2.2). The incident wave sets up currents in the scatterer, which in turn reradiate a "scattered" wave. In the far zone of the scatterer, the scattered wave is an outgoing, spherical wave, which can be approximated by a plane wave over the relatively small area occupied by the receiving antenna. In the FSA coordinate system, the transverse components of the electric field of the scattered wave, $\mathbf{E}^s$, can be written as

$$\mathbf{E}^s = E_v^s \hat{\mathbf{v}}_s + E_h^s \hat{\mathbf{h}}_s. \tag{2.14}$$

The components of $\mathbf{E}^i$ and $\mathbf{E}^s$ are related by the complex $2 \times 2$ *scattering matrix* of the scatterer, $\boldsymbol{S}$, defined according to (Sinclair, 1950; Kennaugh, 1951; van de Hulst, 1981):

$$\begin{pmatrix} E_v^s \\ E_h^s \end{pmatrix} = \frac{e^{ik_0 r}}{r} \begin{pmatrix} S_{vv} & S_{vh} \\ S_{hv} & S_{hh} \end{pmatrix} \begin{pmatrix} E_v^i \\ E_h^i \end{pmatrix} \tag{2.15}$$

or

$$\mathbf{E}^s = \frac{e^{ik_0 r}}{r} \boldsymbol{S}\mathbf{E}^i. \tag{2.16}$$

Here, $r$ is the distance between the scatterer and the receiving antenna and $k_0$ is the wavenumber of the illuminating wave.

We note here that the scattering matrix as introduced in (2.15) is often called the *Jones matrix* in optics texts after the physicist R. Clark Jones, who in 1941 invented the $2 \times 2$ complex matrix representation for use in characterization of optical systems and materials (Jones, 1941a; Hurwitz and Jones, 1941; Jones, 1941b; Jones, 1942). We shall use the more general term *scattering matrix* in this text.

The elements of the complex scattering matrix are known as the complex scattering amplitudes. Each scattering amplitude may be a function of frequency, the scattering and illuminating angles, and the orientation of the scatterer relative to the coordinate system (the target aspect angle). This means that each element of the scattering matrix should actually be written as

$$S_{pq} = S_{pq}(\theta_s, \phi_s; \theta_i, \phi_i; \theta_j, \phi_j); \quad p, q = h, v, \tag{2.17}$$

where $(\theta_j, \phi_j)$ are the orientation angles of the scatterer. For example, for a disk-shaped scatterer, such as a leaf, $(\theta_j, \phi_j)$ defines the direction of its surface normal. For a cylindrically shaped scatterer, such as a branch or a needle, $(\theta_j, \phi_j)$ defines the direction of the cylinder axis.

In the BSA convention, the transmitted and received fields are expressed in the form:

$$\mathbf{E}^t = E_v^t \hat{\mathbf{v}}_t + E_h^t \hat{\mathbf{h}}_t, \tag{2.18}$$

$$\mathbf{E}^r = E_v^r \hat{\mathbf{v}}_r + E_h^r \hat{\mathbf{h}}_r, \tag{2.19}$$

with the unit vectors defined by (2.8), (2.9), (2.11), and (2.12). Upon comparing the expressions for the FSA and BSA local coordinate systems, the following becomes evident:

$$\mathbf{E}^i = \mathbf{E}^t \tag{2.20}$$

$$\mathbf{E}^s = \begin{pmatrix} 1 & 0 \\ 0 & -1 \end{pmatrix} \mathbf{E}^r. \tag{2.21}$$

Therefore, the fields $\mathbf{E}^t$ and $\mathbf{E}^r$ are related by

$$\mathbf{E}^r = \frac{e^{ik_0 r}}{r} \;\overline{\mathcal{S}}\mathbf{E}^t, \tag{2.22}$$

where

$$\overline{\mathcal{S}} = \begin{pmatrix} \overline{S}_{vv} & \overline{S}_{vh} \\ \overline{S}_{hv} & \overline{S}_{hh} \end{pmatrix} \tag{2.23}$$

is the scattering matrix of the object, defined according to the BSA convention. If we use (2.20) and (2.21), we obtain the relationship:

$$\mathcal{S} = \begin{pmatrix} 1 & 0 \\ 0 & -1 \end{pmatrix} \overline{\mathcal{S}}. \tag{2.24}$$

Throughout this book, we shall use the overbar symbol, as in $\overline{\mathcal{S}}$, to denote that the quantity in question has been defined according to the BSA convention. The absence of the overbar implies that the FSA convention is applicable.

### 2-1.3 Reciprocity Relation

Consider an arbitrary particle illuminated by a linearly polarized plane wave (Fig. 2.3) with electric field $\mathbf{E}^i = E_0\hat{\mathbf{q}}$. The direction of propagation of the incident wave is $\hat{\mathbf{k}}_i$ and the observation point is in the far zone at a range $r$ from the particle and corresponds to a scattering direction $\hat{\mathbf{k}}_s$ defined in the FSA convention, as shown in Figure 2.3a. Let us denote by $\mathbf{E}^s(\hat{\mathbf{q}};\ \hat{\mathbf{k}}_s,\ \hat{\mathbf{k}}_i)$ the amplitude of the scattered wave at the observation point due to a $\hat{\mathbf{q}}$-polarized wave incident in the direction $\hat{\mathbf{k}}_i$. Next, suppose we are interested in the component of $\mathbf{E}^s(\hat{\mathbf{q}};\ \hat{\mathbf{k}}_s,\ \hat{\mathbf{k}}_i)$ along any direction $\hat{\mathbf{q}}'$ such that $\hat{\mathbf{q}}'\cdot\hat{\mathbf{k}}_s = 0$; namely,

$$E^s_{q'}(\hat{\mathbf{q}};\ \hat{\mathbf{k}}_s,\ \hat{\mathbf{k}}_i) = \hat{\mathbf{q}}'\cdot\mathbf{E}^s(\hat{\mathbf{q}};\ \hat{\mathbf{k}}_s,\ \hat{\mathbf{k}}_i). \tag{2.25}$$

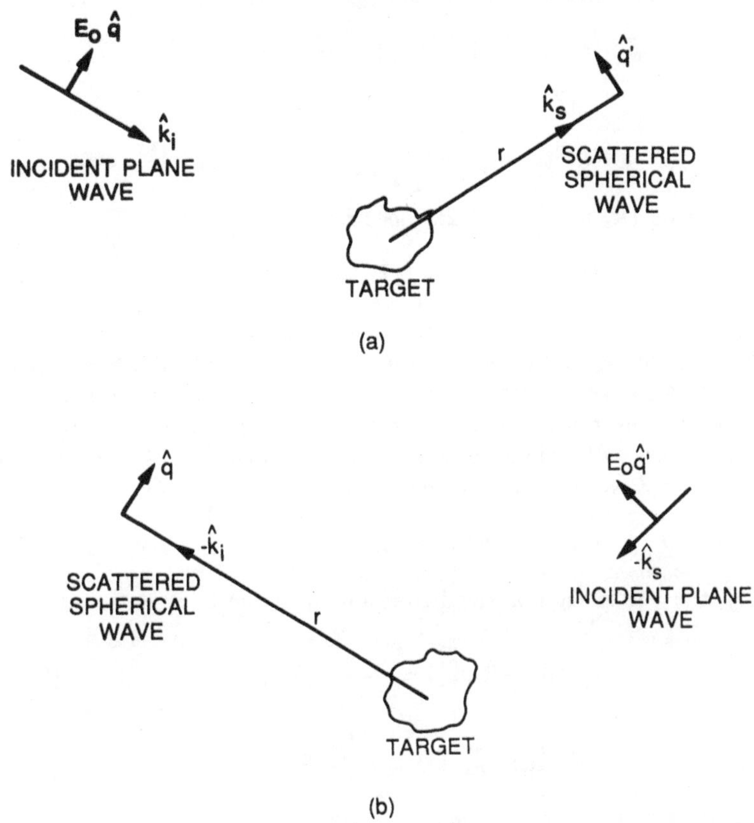


**Figure 2.3** In case (a) the target is illuminated with a $\hat{\mathbf{q}}$-polarized wave incident in the direction $\hat{\mathbf{k}}_i$ and the $\hat{\mathbf{q}}'$ component of the electric field of the scattered wave is observed from the $\hat{\mathbf{k}}_s$ direction. In case (b) the incident and scattered propagation directions are reversed and interchanged.

Now, let us interchange (a) the incident and scattered directions and (b) the direction of polarization of the incident wave and that of the observed scattered wave, and then repeat the experiment (Figure 2.3b). Thus, the particle remains unchanged, the incident wave is $\mathbf{E}^i = E_0\hat{\mathbf{q}}'$ and its direction of propagation is $-\hat{\mathbf{k}}_s$, the observation point is at range $r$, the direction of propagation of the scattered wave is $-\hat{\mathbf{k}}_i$, and the quantity of interest is the component of the scattered wave along the direction $\hat{\mathbf{q}}$. This component is

$$E^s_q(\hat{\mathbf{q}}';\ -\hat{\mathbf{k}}_i,\ -\hat{\mathbf{k}}_s) = \hat{\mathbf{q}}\cdot\mathbf{E}^s(\hat{\mathbf{q}}';\ -\hat{\mathbf{k}}_i,\ -\hat{\mathbf{k}}_s). \tag{2.26}$$

The reciprocity theorem states that (Tsang *et al.*, 1985)

$$E^s_{q'}(\hat{\mathbf{q}};\ \hat{\mathbf{k}}_s,\ \hat{\mathbf{k}}_i) = E^s_q(\hat{\mathbf{q}}';\ -\hat{\mathbf{k}}_i,\ -\hat{\mathbf{k}}_s), \tag{2.27}$$

or, equivalently, that

$$\hat{\mathbf{q}}'\cdot\mathbf{E}^s(\hat{\mathbf{q}};\ \hat{\mathbf{k}}_s,\ \hat{\mathbf{k}}_i) = \hat{\mathbf{q}}\cdot\mathbf{E}^s(\hat{\mathbf{q}}';\ -\hat{\mathbf{k}}_i,\ -\hat{\mathbf{k}}_s). \tag{2.28}$$

For backscattering ($\hat{\mathbf{k}}_s = -\hat{\mathbf{k}}_i$), (2.28) becomes

$$\hat{\mathbf{q}}'\cdot\mathbf{E}^s(\hat{\mathbf{q}};\ -\hat{\mathbf{k}}_i,\ \hat{\mathbf{k}}_i) = \hat{\mathbf{q}}\cdot\mathbf{E}^s(\hat{\mathbf{q}}';\ -\hat{\mathbf{k}}_i,\ \hat{\mathbf{k}}_i). \tag{2.29}$$

We are interested in examining the relationship between the scattering amplitude, $S_{vh}$, corresponding to illumination by a horizontally polarized wave and observation of the vertically polarized component of the scattered wave, and $S_{hv}$, corresponding to the reverse order of incident and observed wave polarizations. We shall do so by illuminating the particle with a $h$-polarized plane wave with amplitude $E_0$; thus,

$$\mathbf{E}^i = E_0\hat{\mathbf{h}}_i. \tag{2.30}$$

From (2.15), the vertically polarized component of the backscattered wave is

$$E^s_{v_s}(\hat{\mathbf{h}}_i; -\hat{\mathbf{k}}_i, \hat{\mathbf{k}}_i) = \frac{e^{ik_0r}}{r}S_{vh}E_0, \tag{2.31}$$

or

$$\hat{\mathbf{v}}_s\cdot\mathbf{E}^s(\hat{\mathbf{h}}_i; -\hat{\mathbf{k}}_i, \hat{\mathbf{k}}_i) = \frac{e^{ik_0r}}{r}S_{vh}E_0. \tag{2.32}$$

Now, let us reverse the polarizations by setting

$$\mathbf{E}^i = E_0\hat{\mathbf{v}}_i, \tag{2.33}$$

and selecting the $h$-polarized component of the scattered wave:

$$E^s_{h_s}(\hat{\mathbf{v}}_i; -\hat{\mathbf{k}}_i, \hat{\mathbf{k}}_i) = \frac{e^{ik_0r}}{r}S_{hv}E_0, \tag{2.34}$$

or

$$\hat{\mathbf{h}}_s\cdot\mathbf{E}^s(\hat{\mathbf{v}}_i; -\hat{\mathbf{k}}_i, \hat{\mathbf{k}}_i) = \frac{e^{ik_0r}}{r}S_{hv}E_0. \tag{2.35}$$

From the expressions given by (2.1) to (2.6), we can easily show that for the backscattering case, $\hat{\mathbf{v}}_s = \hat{\mathbf{v}}_i$ and $\hat{\mathbf{h}}_s = -\hat{\mathbf{h}}_i$. If we make these substitutions in (2.35) and then insert (2.32) and (2.35) into (2.29) (after having set $\hat{\mathbf{q}} = \hat{\mathbf{h}}_i$ and $\hat{\mathbf{q}}' = \hat{\mathbf{v}}_s$), we obtain the result:

$$S_{vh} = -S_{hv}, \tag{2.36}$$

derived for the backscattering condition in the FSA convention. This result is valid for a single particle as well as a volume of many particles. A repetition of the preceding steps by using the coordinate system appropriate for the BSA convention leads to

$$\overline{S}_{vh} = \overline{S}_{hv}. \tag{2.37}$$

In general, the bistatic scattering matrix may contain up to seven independent numbers (four amplitudes and three relative phases as discussed later in Section 2-3). Because of the relations given by (2.36) or (2.37), the backscattering matrix contains only up to five independent quantities (three amplitudes and two relative phases).

In the next three sections, we shall introduce the matrices involved in polarization synthesis using the BSA convention, and then in Section 2-5 we shall discuss the transformations needed to translate a matrix in the BSA convention to one in the FSA convention.

## 2-2 MUELLER AND MODIFIED MUELLER MATRICES

If we use (2.22) and the definition of the Stokes vector given by (1.46) in Chapter 1, we can easily show that the scattered-wave Stokes vector is related to the incident-wave Stokes vector through the *Mueller matrix* $\overline{\mathcal{L}}$ (Jones, 1947; van de Hulst, 1981, van Zyl, 1989):

$$\overline{\mathbf{F}}^r = \frac{1}{r^2}\,\overline{\mathcal{L}}(\theta_s, \phi_s; \theta_i, \phi_i; \theta_j, \phi_j)\overline{\mathbf{F}}^t. \tag{2.38}$$

The overbar symbol denotes that these matrices are defined according to the BSA convention. The factor $1/r^2$ is needed because the scattered wave is a spherical wave (see (2.15)).

We note here that a number of different names are used to describe the matrix $\overline{\mathcal{L}}$ in (2.38). Following (Deirmendjian, 1969; Ishimaru, 1978; and Tsang *et al.*, 1985,) we call the same matrix the *Stokes matrix.* When the Stokes matrix is averaged over all scatterer orientations and sizes, the resulting matrix is the *phase matrix* frequently used in radiative transfer calculations (Chandrasekhar, 1960). The name most often used in optics literature is the Mueller matrix (Jones, 1947; Schurcliff, 1962; Clarke and Grainger, 1971; Azzam and Bashara, 1976; Hauge, 1978) after Hans Mueller, who in 1943 started to use this matrix to characterize depolarizing optical systems. We shall use the name Mueller matrix in this text.

The expressions describing the elements of the *Mueller matrix* in terms of the elements of the scattering matrix are derived as follows. First, note that the Stokes vector can be written as

$$\overline{\mathbf{F}} = \mathcal{R}\overline{\mathbf{G}}, \tag{2.39}$$

where

$$\mathcal{R} = \begin{pmatrix} 1 & 1 & 0 & 0 \\ 1 & -1 & 0 & 0 \\ 0 & 0 & 1 & 1 \\ 0 & 0 & -i & i \end{pmatrix} \tag{2.40}$$

and

$$\overline{\mathbf{G}} = \begin{pmatrix} \overline{E}_v \overline{E}_v^* \\ \overline{E}_h \overline{E}_h^* \\ \overline{E}_v \overline{E}_h^* \\ \overline{E}_h \overline{E}_v^* \end{pmatrix}. \tag{2.41}$$

Through some straightforward but tedious algebra, we can show that

$$\overline{\mathbf{G}}^r = \frac{1}{r^2} \overline{\mathcal{W}} \overline{\mathbf{G}}^t, \tag{2.42}$$

where

$$\overline{\mathcal{W}} = \begin{pmatrix} \overline{S}_{vv}\overline{S}_{vv}^* & \overline{S}_{vh}\overline{S}_{vh}^* & \overline{S}_{vh}\overline{S}_{vv}^* & \overline{S}_{vv}\overline{S}_{vh}^* \\ \overline{S}_{hv}\overline{S}_{hv}^* & \overline{S}_{hh}\overline{S}_{hh}^* & \overline{S}_{hh}\overline{S}_{hv}^* & \overline{S}_{hv}\overline{S}_{hh}^* \\ \overline{S}_{hv}\overline{S}_{vv}^* & \overline{S}_{hh}\overline{S}_{vh}^* & \overline{S}_{hh}\overline{S}_{vv}^* & \overline{S}_{hv}\overline{S}_{vh}^* \\ \overline{S}_{vv}\overline{S}_{hv}^* & \overline{S}_{vh}\overline{S}_{hh}^* & \overline{S}_{vh}\overline{S}_{hv}^* & \overline{S}_{vv}\overline{S}_{hh}^* \end{pmatrix}. \tag{2.43}$$

Note that in calculating the elements of $\overline{\mathcal{W}}$ any information about an absolute phase multiplying $\overline{\mathcal{S}}$ will be lost. From (2.38) to (2.43), the Mueller matrix can be written as

$$\overline{\mathcal{L}} = \mathcal{R}\overline{\mathcal{W}}\mathcal{R}^{-1}. \tag{2.44}$$

Analogous to (2.38), we can relate the scattered-wave modified Stokes parameters to the modified Stokes parameters of the incident wave through a *modified Mueller matrix*:

$$\overline{\mathbf{F}}_m^r = \frac{1}{r^2} \overline{\mathcal{L}}_m(\theta_s, \phi_s; \theta_i, \phi_i; \theta_j, \phi_j)\overline{\mathbf{F}}_m^t. \tag{2.45}$$

Note that we can write the modified Stokes vector as

$$\overline{\mathbf{F}}_m = \mathcal{V}\overline{\mathbf{G}}, \tag{2.46}$$

where

$$\mathcal{V} = \begin{pmatrix} 1 & 0 & 0 & 0 \\ 0 & 1 & 0 & 0 \\ 0 & 0 & 1 & 1 \\ 0 & 0 & -i & i \end{pmatrix}. \tag{2.47}$$

We may easily show that

$$\overline{\mathcal{L}}_m = \mathcal{V}\overline{\mathcal{W}}\mathcal{V}^{-1}. \tag{2.48}$$

Even if the scattering matrix is symmetrical ($\overline{S}_{vh} = \overline{S}_{hv}$), the Mueller and modified Mueller matrices will not in general be symmetrical. We further note that the Mueller matrix and modified Mueller matrix are simply related. This follows from the relationship between the Stokes vector and modified Stokes vector. From (1.46), and (1.48) to (1.50) of the previous chapter, we note that

$$\overline{\mathbf{F}} = \mathcal{U}\overline{\mathbf{F}}_m, \tag{2.49}$$

where

$$\mathcal{U} = \begin{pmatrix} 1 & 1 & 0 & 0 \\ 1 & -1 & 0 & 0 \\ 0 & 0 & 1 & 0 \\ 0 & 0 & 0 & 1 \end{pmatrix}. \tag{2.50}$$

Thus, starting with (2.38) and using (2.49), we may easily show that

$$\begin{aligned} \overline{\mathbf{F}}^r_m &= \frac{1}{r^2}\,\overline{\mathcal{L}}_m\overline{\mathbf{F}}^t_m \\ \mathcal{U}\overline{\mathbf{F}}^r &= \frac{1}{r^2}\,\overline{\mathcal{L}}_m\mathcal{U}\overline{\mathbf{F}}^t, \end{aligned}$$

from which follows

$$\overline{\mathcal{L}} = \mathcal{U}^{-1}\overline{\mathcal{L}}_m\mathcal{U}, \tag{2.51}$$

and

$$\overline{\mathcal{L}}_m = \mathcal{U}\overline{\mathcal{L}}\mathcal{U}^{-1}. \tag{2.52}$$

## 2-3 POLARIZATION SYNTHESIS

Knowledge of the scattering matrix $\overline{\mathcal{S}}$ permits calculation of the received power for any possible combination of transmitting and receiving antennas; this process is called *polarization synthesis* (van Zyl *et al.*, 1987a; Zebker *et al.*, 1987). This important technique is what gives polarimetry its great advantage over conventional fixed-polarization radars — more information may be inferred about the surface if the full polarimetric properties are known.

The procedure for calculating the scattering cross section for any synthesized antenna pair is as follows. The power $P$ observed if antennas with electric field vectors $\overline{\mathbf{E}}^t$ and $\overline{\mathbf{E}}^r$ were utilized for transmitting and receiving, respectively, is found by evaluating the matrix equation (van Zyl, 1985; van Zyl *et al.*, 1987a, 1987b):

$$P = K(\lambda, \theta, \phi)|\overline{\mathbf{E}}^r \cdot \overline{\mathcal{S}}\,\overline{\mathbf{E}}^t|^2, \tag{2.53}$$

where

$$K(\lambda, \theta, \phi) = \frac{1}{2}\frac{\lambda^2}{4\pi}\sqrt{\frac{\epsilon_0}{\mu_0}}\frac{g(\theta, \phi)}{|\overline{\mathbf{E}}^r|^2}, \tag{2.54}$$

and $g(\theta, \phi)$ is the antenna gain function, $(\lambda^2/4\pi)g(\theta, \phi)$ is the effective area of the antenna and $\epsilon_0$ and $\mu_0$ are the permittivity and permeability of free space, respectively. The bistatic scattering cross section of the scatterer is defined as (Sinclair, 1948; Ruck *et al.*, 1970):

$$\sigma_{rt}(\psi_r, \chi_r, \psi_t, \chi_t) = \lim_{r \to \infty} 4\pi r^2 \left(\frac{P_{\text{rec }rt}}{P_{\text{trans}}}\right), \tag{2.55}$$

where $P_{\text{rec }rt}$ is the power received using polarization $t$ to transmit and polarization $r$ to receive the waves, $P_{\text{trans}}$ is the transmitted power, and the angles $(\psi_r, \chi_r)$ and $(\psi_t, \chi_t)$ are the orientation and ellipticity angles of the receiving and transmitting antenna polarization ellipses, respectively. If we use the definition (2.55) with (2.53) and assume that the factor $K$ is independent of polarization, the scattering cross section can be expressed as

$$\sigma_{rt}(\psi_r, \chi_r, \psi_t, \chi_t) = 4\pi\,|\overline{\mathbf{p}}^r \cdot \overline{\mathcal{S}}\,\overline{\mathbf{p}}^t|^2, \tag{2.56}$$

where $\overline{\mathbf{p}}^m$, the field polarization vector (Papas, 1988) is given by

$$\overline{\mathbf{p}}^m = \frac{\overline{\mathbf{E}}^m}{|\overline{\mathbf{E}}^m|}, \quad m = r, t. \tag{2.57}$$

Note that in calculating the square of the magnitude in (2.56), any absolute phase multiplying the scattering matrix $\overline{\mathcal{S}}$ will be lost.

Equations (2.22) and (2.56) are the fundamental equations of radar polarimetry (Kostinski and Boerner, 1986). An imaging polarimeter is usually implemented by configuring the hardware to measure the four scattering matrix elements directly for every resolution element in a scene. For a detailed description of typical implementations, see Chapters 5 and 6.

Polarization synthesis can also be expresed in terms of either the *Stokes scattering operator* $\overline{\mathcal{M}}$ (van Zyl et al., 1987a, 1987b), or the modified

Stokes scattering operator $\overline{\mathcal{M}}_m$. The polarization synthesis equation for the Stokes parameters is easily derived from (2.56). We can show that (Kennaugh, 1951; van Zyl, 1985)

$$\begin{aligned}\sigma_{rt}(\psi_r, \chi_r, \psi_t, \chi_t) &= 4\pi \ |\overline{\mathbf{p}}^r \cdot \overline{\mathcal{S}}\, \overline{\mathbf{p}}^t|^2 \\ &= 4\pi \ \overline{\mathbf{g}}^r \cdot \overline{\mathcal{W}}\, \overline{\mathbf{g}}^t \ , \end{aligned} \tag{2.58}$$

where

$$\overline{\mathbf{g}} = \begin{pmatrix} \overline{p}_v \overline{p}_v^* \\ \overline{p}_h \overline{p}_h^* \\ \overline{p}_v \overline{p}_h^* \\ \overline{p}_h \overline{p}_v^* \end{pmatrix} . \tag{2.59}$$

Writing the relation (2.39) for the normalized Stokes vector $\overline{\mathbf{A}}$,

$$\overline{\mathbf{A}} = \mathcal{R}\overline{\mathbf{g}} \tag{2.60}$$

and using the result in (2.58), we find the polarization synthesis equation for the Stokes parameters:

$$\sigma_{rt}(\psi_r, \chi_r, \psi_t, \chi_t) \quad = \quad 4\pi \ \overline{\mathbf{A}}^r \cdot \overline{\mathcal{M}}\, \overline{\mathbf{A}}^t \tag{2.61}$$

where $\overline{\mathcal{M}}$ is the Stokes scattering operator, given by (Kennaugh, 1951; van Zyl *et al.*, 1987a, 1987b)

$$\overline{\mathcal{M}} = \tilde{\mathcal{R}}^{-1} \overline{\mathcal{W}} \mathcal{R}^{-1}. \tag{2.62}$$

Here, $\tilde{\mathcal{R}}$ means the transpose of the matrix $\mathcal{R}$.

Notice that the matrix $\overline{\mathcal{M}}$, which we call the Stokes scattering operator, is different from, and ought not be confused with, the Mueller matrix $\overline{\mathcal{L}}$. Unfortunately, both $\overline{\mathcal{M}}$ and $\overline{\mathcal{L}}$ have been called the Stokes matrix in the past (Ishimaru, 1978; Tsang *et al.*, 1985; van Zyl *et al.*, 1987a). To avoid any confusion, we shall use the names Mueller matrix for $\overline{\mathcal{L}}$, which relates the scattered Stokes vector to the incident Stokes vector, and the name Stokes scattering operator (Kennaugh, 1951; Ioannidis and Hammers, 1978; van Zyl, 1985; van Zyl *et al.*, 1987b) for $\overline{\mathcal{M}}$, the matrix used in the Stokes parameter polarization synthesis equation. We note here that the Mueller matrix and the Stokes scattering operator are related. From (2.44) and (2.62), we may easily show that

$$\overline{\mathcal{L}} = \mathcal{R}\tilde{\mathcal{R}}\overline{\mathcal{M}}. \tag{2.63}$$

Alternatively, the polarization synthesis equation for the modified Stokes parameters is given by

$$\sigma_{rt}(\psi_r, \chi_r, \psi_t, \chi_t) \quad = \quad 4\pi \ \overline{\mathbf{A}}_m^r \cdot \overline{\mathcal{M}}_m \overline{\mathbf{A}}_m^t, \tag{2.64}$$

where $\overline{\mathbf{A}}_m = \mathcal{V}\overline{\mathbf{g}}$ and $\overline{\mathcal{M}}_m$ is the modified Stokes scattering operator given by

$$\overline{\mathcal{M}}_m = \widetilde{\mathcal{V}}^{-1}\overline{\mathcal{W}}\mathcal{V}^{-1}. \tag{2.65}$$

The modified Mueller matrix and the modified Stokes scattering operator are also related:

$$\overline{\mathcal{L}}_m = \mathcal{V}\widetilde{\mathcal{V}}\overline{\mathcal{M}}_m. \tag{2.66}$$

Finally, the modified Stokes scattering operator can be calculated from the Stokes scattering operator by using

$$\overline{\mathcal{M}}_m = \widetilde{\mathcal{U}}\overline{\mathcal{M}}\mathcal{U}. \tag{2.67}$$

When the scattering matrix is symmetrical, such as in the backscatter case when reciprocity holds (see (2.37)), $\overline{\mathcal{W}}$ is also symmetrical, i.e., $\overline{\mathcal{W}} = \widetilde{\overline{\mathcal{W}}}$. In this case, we may easily see from (2.62) and (2.65) that the Stokes scattering operator and the modified Stokes scattering operator are both symmetrical matrices. This property of the backscatter Stokes scattering operator is used to reduce the data volume of the NASA-JPL imaging radar polarimeter (see Chapter 6). Even if (2.37) is true, however, the Mueller matrix and modified Mueller matrix are not symmetrical.

An alternative method of polarization synthesis utilizes the covariance matrix $\overline{\mathcal{C}}$ (Swartz *et al.*, 1988). To derive the polarization synthesis equation, (2.56) is written as

$$\begin{aligned}\sigma_{rt}(\psi_r, \chi_r, \psi_t, \chi_t) &= 4\pi\ |\overline{\mathbf{p}}^r \cdot \overline{\mathcal{S}}\,\overline{\mathbf{p}}^t|^2 \\ &= 4\pi\ |\overline{\mathbf{B}}^* \cdot \overline{\mathbf{X}}|^2\ ,\end{aligned} \tag{2.68}$$

where $\overline{\mathbf{B}}^*$ is an antenna vector given by

$$\overline{\mathbf{B}}^* = \begin{pmatrix} \overline{p}_v^r\overline{p}_v^t \\ \overline{p}_v^r\overline{p}_h^t \\ \overline{p}_h^r\overline{p}_v^t \\ \overline{p}_h^r\overline{p}_h^t \end{pmatrix}, \tag{2.69}$$

and $\overline{\mathbf{X}}$ is a target vector:

$$\overline{\mathbf{X}} = \begin{pmatrix} \overline{S}_{vv} \\ \overline{S}_{vh} \\ \overline{S}_{hv} \\ \overline{S}_{hh} \end{pmatrix}. \tag{2.70}$$

Rewriting the second part of (2.68) leads to the polarization synthesis equation for the covariance matrix:

$$\sigma_{rt}(\psi_r, \chi_r, \psi_t, \chi_t) = 4\pi\ \overline{\mathbf{B}}^* \cdot \overline{\mathcal{C}}\,\overline{\mathbf{B}}, \tag{2.71}$$

with the covariance matrix defined as

$$\overline{\mathcal{C}} = \overline{\mathbf{X}}\widetilde{\overline{\mathbf{X}}}^{*}. \tag{2.72}$$

Performing the multiplications, we find that

$$\overline{\mathcal{C}} = \begin{pmatrix} \overline{S}_{vv}\overline{S}_{vv}^{*} & \overline{S}_{vv}\overline{S}_{vh}^{*} & \overline{S}_{vv}\overline{S}_{hv}^{*} & \overline{S}_{vv}\overline{S}_{hh}^{*} \\ \overline{S}_{vh}\overline{S}_{vv}^{*} & \overline{S}_{vh}\overline{S}_{vh}^{*} & \overline{S}_{vh}\overline{S}_{hv}^{*} & \overline{S}_{vh}\overline{S}_{hh}^{*} \\ \overline{S}_{hv}\overline{S}_{vv}^{*} & \overline{S}_{hv}\overline{S}_{vh}^{*} & \overline{S}_{hv}\overline{S}_{hv}^{*} & \overline{S}_{hv}\overline{S}_{hh}^{*} \\ \overline{S}_{hh}\overline{S}_{vv}^{*} & \overline{S}_{hh}\overline{S}_{vh}^{*} & \overline{S}_{hh}\overline{S}_{hv}^{*} & \overline{S}_{hh}\overline{S}_{hh}^{*} \end{pmatrix}. \tag{2.73}$$

We note that when the scattering matrix is symmetrical, the covariance matrix is a Hermitian matrix.

For imaging radars, the individual power measurements for each radar resolution element are statistically related only; therefore, in a typical radar system, several power measurements are added to reduce statistical variation at the expense of loss of spatial resolution. The average scattering cross section of a set of $N$ measurements then can be expressed as (van Zyl *et al.*, 1987):

$$\sigma_{rt}^{0}(\psi_r, \chi_r, \psi_t, \chi_t) = \frac{1}{A}\,\langle\sigma_{rt}^{(n)}(\psi_r, \chi_r, \psi_t, \chi_t)\rangle, \tag{2.74}$$

where $\sigma_{rt}^{(n)}$ is the scattering cross section of the $n$th individual measurement, $A$ is the illuminated area, and the symbol $\langle\ \rangle$ denotes ensemble averaging. Thus, if the same antennas are used for each observation and the multiplicative factor $K$ is assumed constant over the averaging area, we have

$$\sigma_{rt}^{0}(\psi_r, \chi_r, \psi_t, \chi_t) = \frac{4\pi}{A}\,\overline{\mathbf{A}}^{r}\cdot\langle\overline{\mathcal{M}}\rangle\overline{\mathbf{A}}^{t}, \tag{2.75}$$

$$\sigma_{rt}^{0}(\psi_r, \chi_r, \psi_t, \chi_t) = \frac{4\pi}{A}\,\overline{\mathbf{A}}_{m}^{r}\cdot\langle\overline{\mathcal{M}}_{m}\rangle\overline{\mathbf{A}}_{m}^{t}, \tag{2.76}$$

or

$$\sigma_{rt}^{0}(\psi_r, \chi_r, \psi_t, \chi_t) = \frac{4\pi}{A}\,\overline{\mathbf{B}}^{*}\cdot\langle\overline{\mathcal{C}}\rangle\overline{\mathbf{B}}. \tag{2.77}$$

Note that, in terms of the scattering matrix, (2.74) can be written as

$$\sigma_{rt}^{0}(\psi_r, \chi_r, \psi_t, \chi_t) = \frac{4\pi}{A}\langle|\overline{\mathbf{p}}^{r}\cdot\overline{\mathcal{S}}\,\overline{\mathbf{p}}^{t}|^{2}\rangle. \tag{2.78}$$

Equations (2.75) to (2.78) show the advantage of using the Stokes matrix or covariance matrix representations for polarization synthesis in the case of imaging radar polarimeters. Although the same information would be conveyed by (2.75), (2.76), (2.77), or (2.78), the Stokes scattering operator, modified Stokes scattering operator, and covariance matrix have the additional advantage that the collective properties of a group of resolution elements can be expressed using a single matrix rather than requiring the individual scattering matrices for each element. In addition, the four different approaches also require different amounts of computation to synthesize the scattering cross section for a collection of $N$ pixels. For example, in the monostatic case, the scattering matrix is symmetrical, which means that the Stokes and modified Stokes scattering operators are real, symmetrical matrices and the covariance matrix can be written as a $3 \times 3$ Hermitian matrix (Swartz *et al.*, 1988). We can then show that, in the backscatter case, the Stokes scattering operator and modified Stokes scattering operator approaches require $10N$ real-number additions to calculate the average Stokes or modified Stokes scattering operator, followed by 10 real-number multiplications to synthesize the scattering cross section from this composite area for an arbitrary polarization combination. The covariance matrix approach requires $12N$ real additions ($6N$ complex additions) and 24 real multiplications (6 complex multiplications) for the same synthesis procedure, while the scattering matrix approach requires $14N$ real multiplications and $7N$ real additions. For large values of $N$, the Stokes and modified Stokes scattering operator methods will therefore require significantly fewer computations than the scattering matrix and covariance matrix methods. The number of computations needed for each approach is summarized in Table 2.1.

**Table 2.1** Number of real number computations needed to synthesize the average scattering cross section for $N$ measurements for an arbitrary transmitting and receiving polarization combination.

| Synthesis Approach (eqn.) | Multiplications | Additions |
|---|---|---|
| Scattering matrix (2.78) | $14N$ | $7N$ |
| Stokes scattering operator (2.75) | 10 | $10N$ |
| Modified Stokes Scatt. op. (2.76) | 10 | $10N$ |
| Covariance matrix (2.77) | 24 | $12N$ |

# 2-4 POLARIZATION RESPONSE

A particular graphical representation of the variation of scattering cross section as a function of polarization, known as the *polarization signature* (van Zyl, 1985; van Zyl *et al.*, 1987a; Zebker *et al.*, 1987; Agrawal and Boerner, 1989) or *polarization response,* is quite useful for describing polarization properties of point and distributed targets. The response consists of a plot of synthesized scattering cross section as a function of the ellipticity and orientation angles of the transmitted wave. We note here that the term polarization signature has been used to describe the polarization response in earlier papers (van Zyl *et al.*, 1987; Zebker *et al.*, 1987). The word "signature" implies uniqueness, but the polarization response is not unique, as will be shown in this section, and so we shall use the term polarization response rather than polarization signature in this text. Two kinds of response will be presented here: in the case of co-polarized responses, the transmitting and receiving antennas have the same polarization; in the case of the cross-polarized responses, the receiving antenna is polarized orthogonally to the transmitting antenna. In both cases, the synthesized RCS is plotted as a function of the transmitting antenna polarization, i.e., $(\psi_t, \chi_t)$ in (2.74) to (2.78). The polarization responses of many different scatterers are described by (van Zyl *et al.*, 1987a; Zebker *et al.*, 1987; and Agrawal and Boerner, 1989). In this section, the polarization responses of a number of different simple scatterers will be presented and discussed in detail. In each case, the scattering, Stokes, modified Stokes, and covariance matrices will be listed and the co- and cross-polarized responses will be presented for the backscatter direction only. For more complex point targets, the reader is referred to Chapter 3, and for distributed targets to Chapter 4.

## 2-4.1 Large Conducting Sphere

The rigorous solution of scattering by spheres was first derived by Mie (1908). For a large conducting sphere, the scattering matrix in the BSA convention can be written as

$$\overline{S} = \frac{a}{2}\begin{pmatrix} 1 & 0 \\ 0 & 1 \end{pmatrix}, \tag{2.79}$$

where $a$ is the radius of the sphere. The corresponding Stokes scattering operator is

$$\overline{\mathcal{M}} = \frac{a^2}{8} \begin{pmatrix} 1 & 0 & 0 & 0 \\ 0 & 1 & 0 & 0 \\ 0 & 0 & 1 & 0 \\ 0 & 0 & 0 & -1 \end{pmatrix}, \tag{2.80}$$

the modified Stokes scattering operator is

$$\overline{\mathcal{M}}_m = \frac{a^2}{8} \begin{pmatrix} 2 & 0 & 0 & 0 \\ 0 & 2 & 0 & 0 \\ 0 & 0 & 1 & 0 \\ 0 & 0 & 0 & -1 \end{pmatrix}, \tag{2.81}$$

and the covariance matrix is

$$\overline{\mathcal{C}} = \frac{a^2}{4} \begin{pmatrix} 1 & 0 & 0 & 1 \\ 0 & 0 & 0 & 0 \\ 0 & 0 & 0 & 0 \\ 1 & 0 & 0 & 1 \end{pmatrix}. \tag{2.82}$$

Figure 2.4 shows the polarization responses of a large conducting sphere. The co-polarized response shows that the maximum measured scattering cross section (which is the same as the geometrical cross section if the radius of the sphere is large compared to the radar wavelength) occurs for linear polarizations and the scattering cross section is independent of the linear polarization orientation angle. This independence is a consequence of the lack of preferred orientation for spheres. For the cross-polarized response, the measured scattering cross section is greatest for the circular polarizations and smallest for linear polarizations. The maxima occur at the circular polarizations because the scattered waves for circularly polarized incident waves are also circularly polarized, but with the opposite sense of rotation.

Note that a large (meaning large enough so that the edges are not illuminated by the transmitting antenna) flat surface at normal incidence would have the same polarization responses as a large sphere.

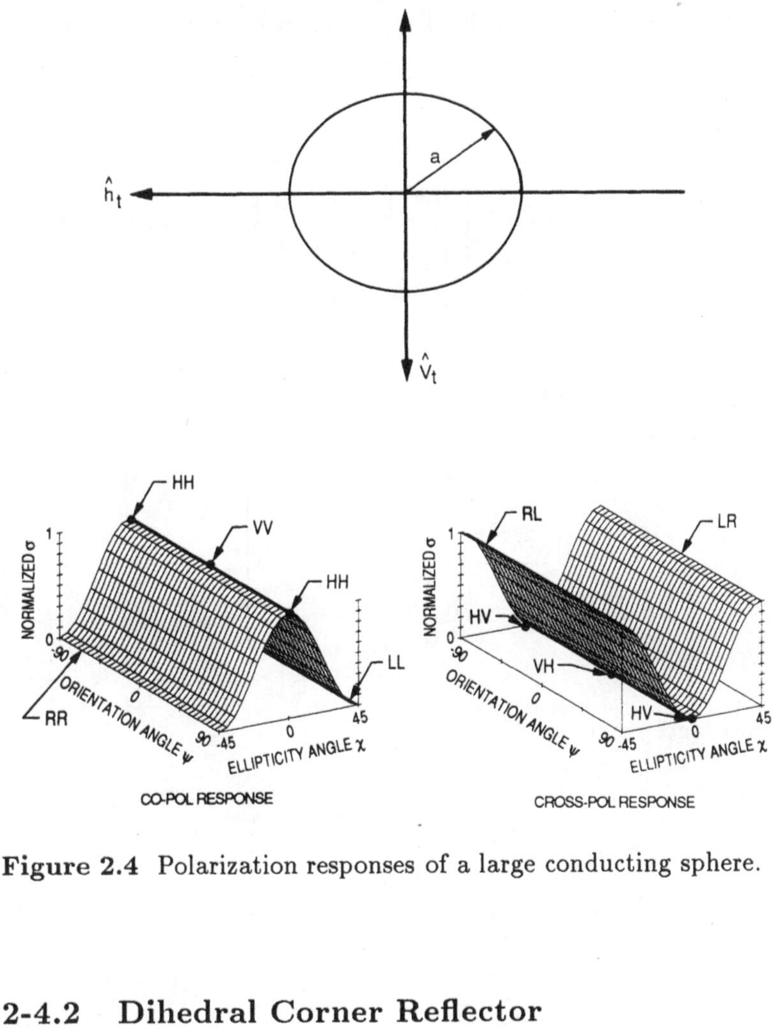


**Figure 2.4** Polarization responses of a large conducting sphere.

## 2-4.2 Dihedral Corner Reflector

A wave scattered from a dihedral corner reflector will undergo a reflection at each surface and will return in the direction from which it came (Jasik, 1961). The scattering matrix of a dihedral corner reflector in the BSA convention is

$$\overline{\mathcal{S}} = \frac{k_0 ab}{\pi} \begin{pmatrix} -1 & 0 \\ 0 & 1 \end{pmatrix}, \tag{2.83}$$

where $a$ and $b$ are the dimensions of the corner reflector (see Fig. 2.5.) The scattering matrix given in (2.83) is valid for a conducting corner reflector being illuminated along its boresight (Ruck *et al.*, 1970). The corresponding Stokes scattering operator is

$$\overline{\mathcal{M}} = \frac{k_0^2 a^2 b^2}{2\pi^2} \begin{pmatrix} 1 & 0 & 0 & 0 \\ 0 & 1 & 0 & 0 \\ 0 & 0 & -1 & 0 \\ 0 & 0 & 0 & 1 \end{pmatrix}, \tag{2.84}$$

the modified Stokes scattering operator is

$$\overline{\mathcal{M}}_m = \frac{k_0^2 a^2 b^2}{2\pi^2} \begin{pmatrix} 2 & 0 & 0 & 0 \\ 0 & 2 & 0 & 0 \\ 0 & 0 & -1 & 0 \\ 0 & 0 & 0 & 1 \end{pmatrix}, \tag{2.85}$$

and the covariance matrix is

$$\overline{\mathcal{C}} = \frac{k_0^2 a^2 b^2}{\pi^2} \begin{pmatrix} 1 & 0 & 0 & -1 \\ 0 & 0 & 0 & 0 \\ 0 & 0 & 0 & 0 \\ -1 & 0 & 0 & 1 \end{pmatrix}. \tag{2.86}$$

The co- and cross-polarized responses of the dihedral corner reflector shown in Figure 2.5 are very different from those of the sphere. The co-polarized response possesses two minima at linear polarizations offset by ±45° from the horizontal and vertical polarizations, and the cross-polarized response has maxima at these locations. The striking difference between the responses of the sphere and that of the dihedral corner reflector is due to the extra reflection that the waves undergo in the case of the corner reflector. This extra reflection introduces an additional 180° phase shift between $\overline{S}_{hh}$ and $\overline{S}_{vv}$, leading to the different shape of the responses.

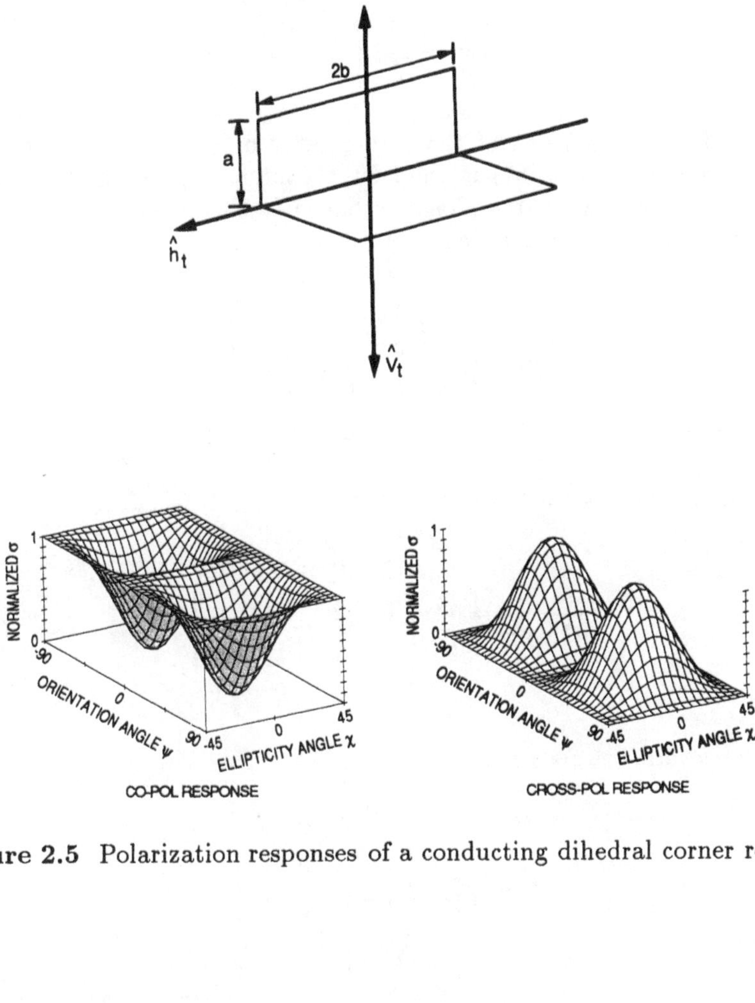


**Figure 2.5** Polarization responses of a conducting dihedral corner reflector.

## 2-4.3 Trihedral Corner Reflector

In the case of a trihedral corner reflector, the waves are reflected three times before returning to the radar. Each of the extra two reflections, compared to the sphere, introduces a 180° phase shift between $\overline{S}_{hh}$ and $\overline{S}_{vv}$, resulting in a total of 360°. Therefore, the scattering matrix of a trihedral corner reflector, when illuminated along its boresight, is identical to that of a large sphere, except for the multiplicative constant:

$$\overline{\mathcal{S}} = \frac{k_0 l^2}{\sqrt{12}\pi} \begin{pmatrix} 1 & 0 \\ 0 & 1 \end{pmatrix}, \tag{2.87}$$

where $l$ is the dimension of the corner reflector (see Fig. 2.6) and the scattering matrix in (2.87) is valid for a conducting triangular trihedral corner reflector illuminated along its boresight (Ruck *et al.*, 1970). The corresponding Stokes scattering operator is

$$\overline{\mathcal{M}} = \frac{k_0^2 l^4}{24\pi^2} \begin{pmatrix} 1 & 0 & 0 & 0 \\ 0 & 1 & 0 & 0 \\ 0 & 0 & 1 & 0 \\ 0 & 0 & 0 & -1 \end{pmatrix}, \tag{2.88}$$

the modified Stokes scattering operator is

$$\overline{\mathcal{M}}_m = \frac{k_0^2 l^4}{24\pi^2} \begin{pmatrix} 2 & 0 & 0 & 0 \\ 0 & 2 & 0 & 0 \\ 0 & 0 & 1 & 0 \\ 0 & 0 & 0 & -1 \end{pmatrix}, \tag{2.89}$$

and the covariance matrix is

$$\overline{\mathcal{C}} = \frac{k_0^2 l^4}{12\pi^2} \begin{pmatrix} 1 & 0 & 0 & 1 \\ 0 & 0 & 0 & 0 \\ 0 & 0 & 0 & 0 \\ 1 & 0 & 0 & 1 \end{pmatrix}. \tag{2.90}$$

The corresponding trihedral corner reflector polarization responses that are shown in Figure 2.6 are therefore also identical to those of the sphere. Notice as well that the polarization responses, like those of the sphere, are independent of the polarization ellipse orientation angles. This lack of preferred orientation means that the corner reflector polarization response is not sensitive to rotations of the trihedral corner reflector around the line connecting the radar antenna and corner reflector. This shape of the trihedral corner reflector's response makes it a useful target for calibration of polarimetric radar images (van Zyl, 1990).

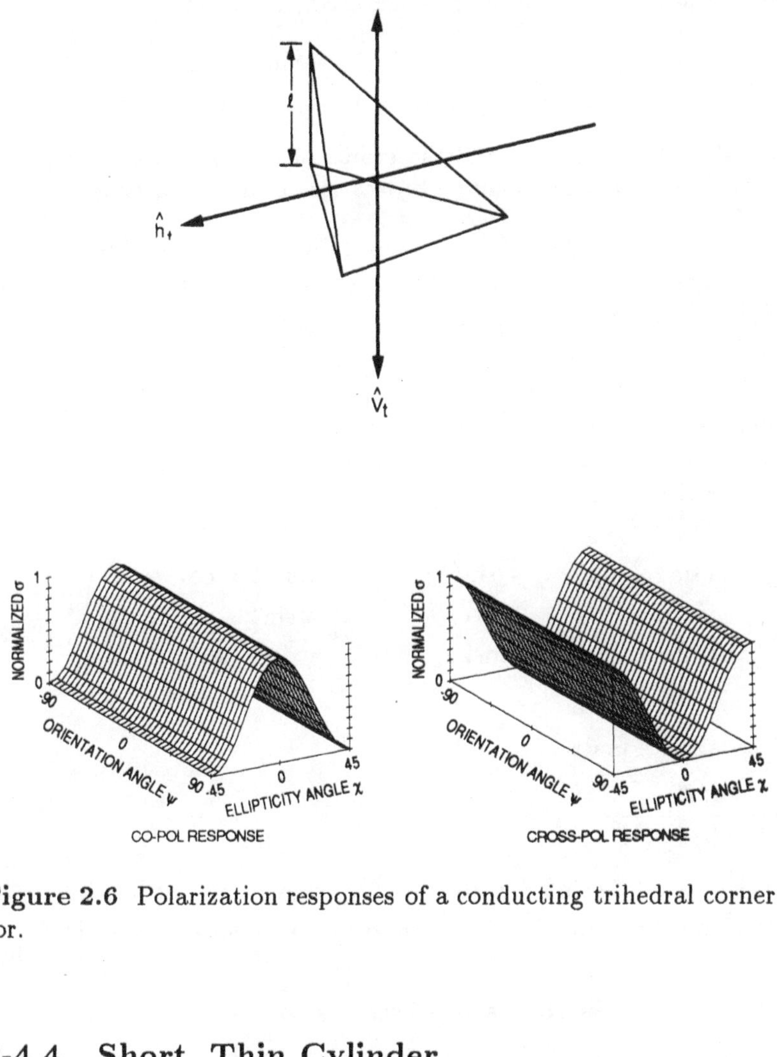


**Figure 2.6** Polarization responses of a conducting trihedral corner reflector.

## 2-4.4 Short, Thin Cylinder

Consider a short, thin conducting cylinder oriented such that its axis lies in a plane orthogonal to the direction in which the radar wave propagates. Let the angle between the axis of the cylinder and the basis vector $\hat{\mathbf{h}}_i$ (measured in the clockwise direction) be denoted by $\alpha$. If the cylinder radius and length are both much less than the wavelength, and the cylinder length is large compared to its radius, the scattering matrix for the backscatter direction in the BSA convention is (Ruck *et al.*, 1970):

$$\overline{\mathcal{S}} = \frac{k_0^2 l^3}{3[\ln(4l/a) - 1]} \begin{pmatrix} \sin^2\alpha & -\sin\alpha\cos\alpha \\ -\sin\alpha\cos\alpha & \cos^2\alpha \end{pmatrix}, \tag{2.91}$$

where $l$ and $a$ are the length and the radius of the cylinder, respectively (see Figure 2.7). The corresponding Stokes scattering operator is

$$\overline{\mathcal{M}} = \frac{k_0^4 l^6}{36[\ln(4l/a) - 1]^2} \begin{pmatrix} 1 & -\cos 2\alpha & -\sin 2\alpha & 0 \\ -\cos 2\alpha & \cos^2 2\alpha & -\sin 2\alpha\cos 2\alpha & 0 \\ -\sin 2\alpha & -\sin 2\alpha\cos 2\alpha & \sin^2 2\alpha & 0 \\ 0 & 0 & 0 & 0 \end{pmatrix}, \tag{2.92}$$

the modified Stokes scattering operator is

$$\overline{\mathcal{M}}_m = \frac{k_0^4 l^6}{36[\ln(4l/a) - 1]^2} \cdot \begin{pmatrix} 4\sin^4\alpha & \sin^2 2\alpha & -2\sin 2\alpha\ \cos^2\alpha & 0 \\ \sin^2 2\alpha & 4\cos^4\alpha & -2\sin 2\alpha\ \sin^2\alpha & 0 \\ -2\sin 2\alpha\ \cos^2\alpha & -2\sin 2\alpha\ \sin^2\alpha & \sin^2 2\alpha & 0 \\ 0 & 0 & 0 & 0 \end{pmatrix}, \tag{2.93}$$

and the covariance matrix is

$$\overline{\mathcal{C}} = \frac{k_0^4 l^6}{9[\ln(4l/a] - 1)^2} \cdot \begin{pmatrix} \sin^4\alpha & -\sin^3\alpha\ \cos\alpha & -\sin^3\alpha\ \cos\alpha & \sin^2\alpha\cos^2\alpha \\ -\sin^3\alpha\ \cos\alpha & \sin^2\alpha\cos^2\alpha & \sin^2\alpha\cos^2\alpha & -\sin\alpha\ \cos^3\alpha \\ -\sin^3\alpha\ \cos\alpha & \sin^2\alpha\cos^2\alpha & \sin^2\alpha\cos^2\alpha & -\sin\alpha\ \cos^3\alpha \\ \sin^2\alpha\cos^2\alpha & -\sin\alpha\ \cos^3\alpha & -\sin\alpha\ \cos^3\alpha & \cos^4\alpha \end{pmatrix}. \tag{2.94}$$

Figure 2.7 shows the polarization responses for the cylinder oriented horizontally, Fig. 2.8 for the cylinder oriented at 45°, and Fig. 2.9 for the cylinder oriented vertically. In all cases, the maximum in the co-polarized response and one of the minima in the cross-polarized response occur at the linear polarization with the same orientation as the cylinder, illustrating the possibility that radar polarimetry can be used to infer the orientation of scatterers. Notice that both the co-polarized and cross-polarized responses also have minima at the linear polarization orthogonal to the cylinder orientation. When this polarization is transmitted, the incident wave is polarized orthogonally to the cylinder, with the result that no current is induced in the cylinder by the incident wave, and no scattered energy is radiated. Finally, note that the cross-polarized responses of the vertically and horizontally oriented scatterers are identical. This means, whereas the co-polarized response clearly shows the orientation of the cylinder, the orientation of the cylinder cannot be estimated uniquely from the cross-polarized response alone.

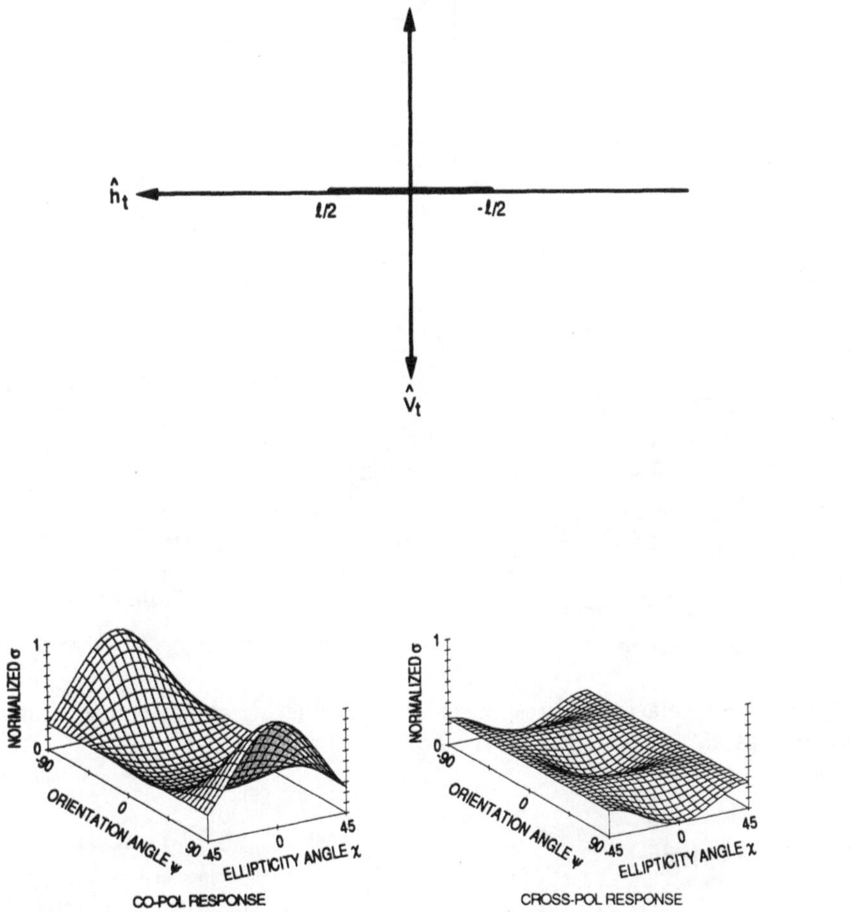


**Figure 2.7** Polarization responses of a conducting short, thin cylinder, oriented horizontally.

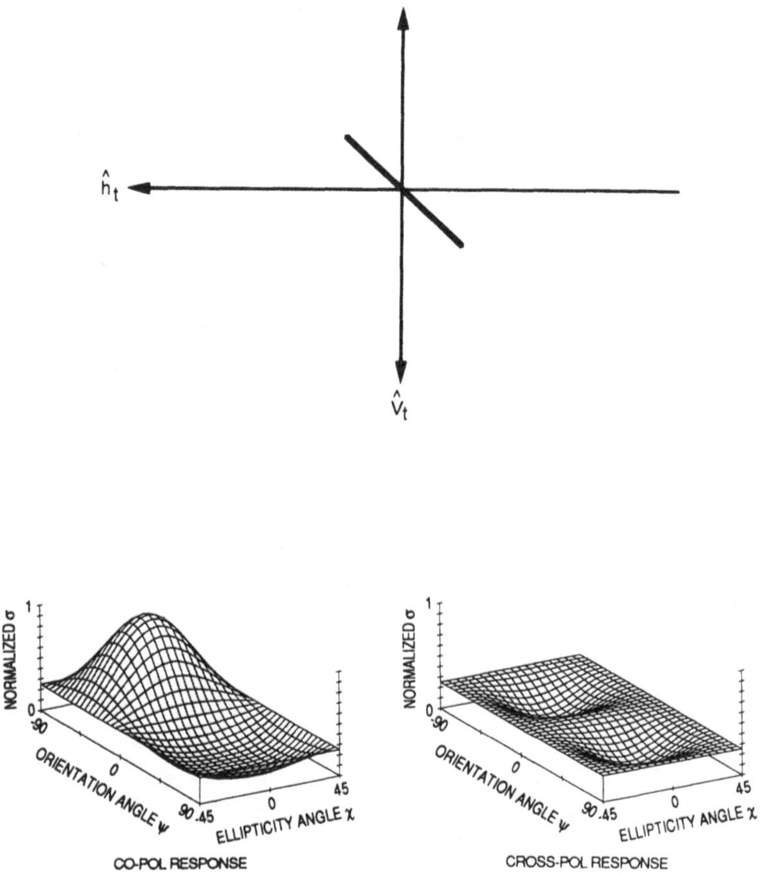


**Figure 2.8** Polarization responses of a conducting short, thin cylinder, oriented at 45° with respect to the horizontal direction.

## 2-4.5 Helix

Huynen (1970) gives the scattering matrix of a left-handed helix in the BSA convention as

$$\overline{\mathcal{S}}_l = \frac{1}{2}\begin{pmatrix} -1 & i \\ i & 1 \end{pmatrix}, \tag{2.95}$$

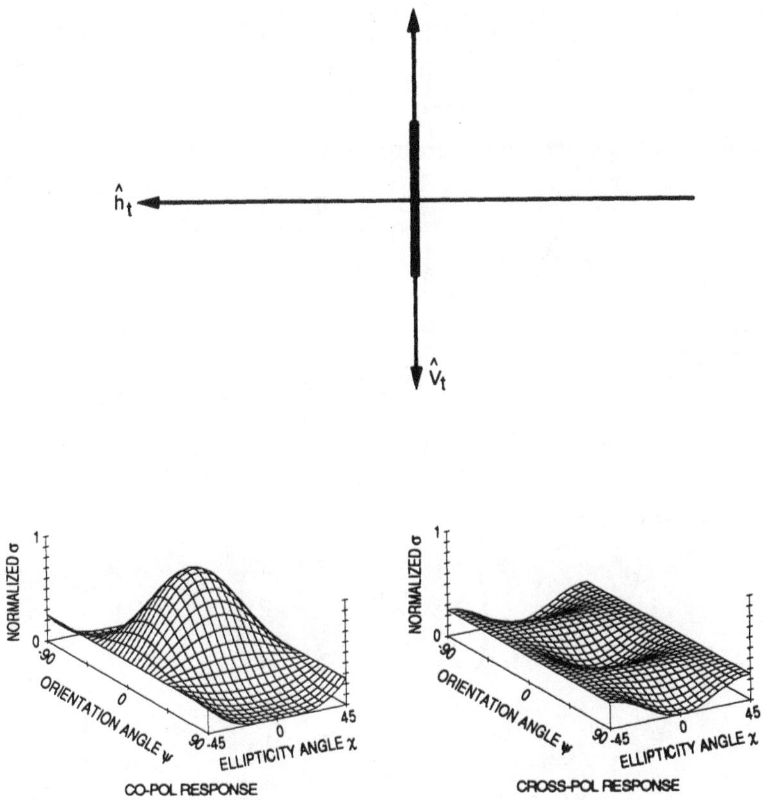


**Figure 2.9** Polarization resonses of a conducting short, thin cylinder, oriented vertically.

and that of a right-handed helix as

$$\overline{\mathcal{S}}_r = \frac{1}{2}\begin{pmatrix} -1 & -i \\ -i & 1 \end{pmatrix}. \tag{2.96}$$

The corresponding Stokes scattering operators are

$$\overline{\mathcal{M}}_l = \frac{1}{4}\begin{pmatrix} 1 & 0 & 0 & 1 \\ 0 & 0 & 0 & 0 \\ 0 & 0 & 0 & 0 \\ 1 & 0 & 0 & 1 \end{pmatrix} \tag{2.97}$$

and

$$\overline{\mathcal{M}}_r = \frac{1}{4}\begin{pmatrix} 1 & 0 & 0 & -1 \\ 0 & 0 & 0 & 0 \\ 0 & 0 & 0 & 0 \\ -1 & 0 & 0 & 1 \end{pmatrix}, \tag{2.98}$$

the modified Stokes scattering operators are

$$\overline{\mathcal{M}}_{ml} = \frac{1}{4}\begin{pmatrix} 1 & 1 & 0 & 1 \\ 1 & 1 & 0 & 1 \\ 0 & 0 & 1 & 0 \\ 1 & 1 & 0 & 1 \end{pmatrix} \tag{2.99}$$

and

$$\overline{\mathcal{M}}_{mr} = \frac{1}{4}\begin{pmatrix} 1 & 1 & 0 & -1 \\ 1 & 1 & 0 & -1 \\ 0 & 0 & 1 & 0 \\ -1 & -1 & 0 & 1 \end{pmatrix}, \tag{2.100}$$

and the covariance matrices are

$$\overline{\mathcal{C}}_l = \frac{1}{4}\begin{pmatrix} 1 & i & i & -1 \\ -i & 1 & 1 & i \\ -i & 1 & 1 & i \\ -1 & -i & -i & 1 \end{pmatrix} \tag{2.101}$$

and

$$\overline{\mathcal{C}}_r = \frac{1}{4}\begin{pmatrix} 1 & -i & -i & -1 \\ i & 1 & 1 & -i \\ i & 1 & 1 & -i \\ -1 & i & i & 1 \end{pmatrix}. \tag{2.102}$$

The polarization responses of the two kinds of helices are shown in Figs. 2.10 and 2.11, respectively. In each case, the maximum in the co-polarized response occurs at the circular polarization with the same handedness as the helix. For this particular polarization, the coupling between the incident wave electric field and the helix is maximum. The scattered wave is circularly polarized with the same handedness as the incident wave, which means that the scattered wave and the receiving antenna have the same polarization in the co-polarized case, leading to the observed maximum in the response. Because the scattered wave has the same polarization as the incident wave, no energy will be absorbed by an orthogonally polarized receiving antenna, which is evident in the minimum observed in the cross-polarized response. When a circularly polarized wave with the opposite handedness to that of the helix is transmitted, no coupling occurs between the helix and the incident wave, resulting in no scattered energy.

This explains why both the co- and cross-polarized responses have minima at the circular polarization with opposite handedness to that of the helix. Also notice, similar to the case of the vertically and horizontally oriented thin cylinders, that the cross-polarized responses of the two kinds of helices are identical. This means that the handedness of the helix cannot be deduced from the cross-polarized response, but is easily deduced from the co-polarized response.

# 2-5 COORDINATE SYSTEM TRANSFORMATIONS

In this section we shall discuss the transformations needed to calculate a matrix in the FSA convention, given a matrix in the BSA convention. Because the JPL Stokes matrix data format is defined by using a different definition of the Stokes vector, we shall also discuss the transformations needed to put the JPL data in the BSA format.

## 2-5.1 FSA to BSA Transformation

Using (2.20) and (2.21) in the definition of the Stokes vector given by (1.46), we have

$$\mathbf{F}^i = \overline{\mathbf{F}}^t \tag{2.103}$$

and

$$\mathbf{F}^s = \mathcal{T}\,\overline{\mathbf{F}}^r, \tag{2.104}$$

where $\mathbf{F}$ is defined in the FSA convention, $\overline{\mathbf{F}}$ is defined in the BSA convention, and

$$\mathcal{T} = \begin{pmatrix} 1 & 0 & 0 & 0 \\ 0 & 1 & 0 & 0 \\ 0 & 0 & -1 & 0 \\ 0 & 0 & 0 & -1 \end{pmatrix}. \tag{2.105}$$

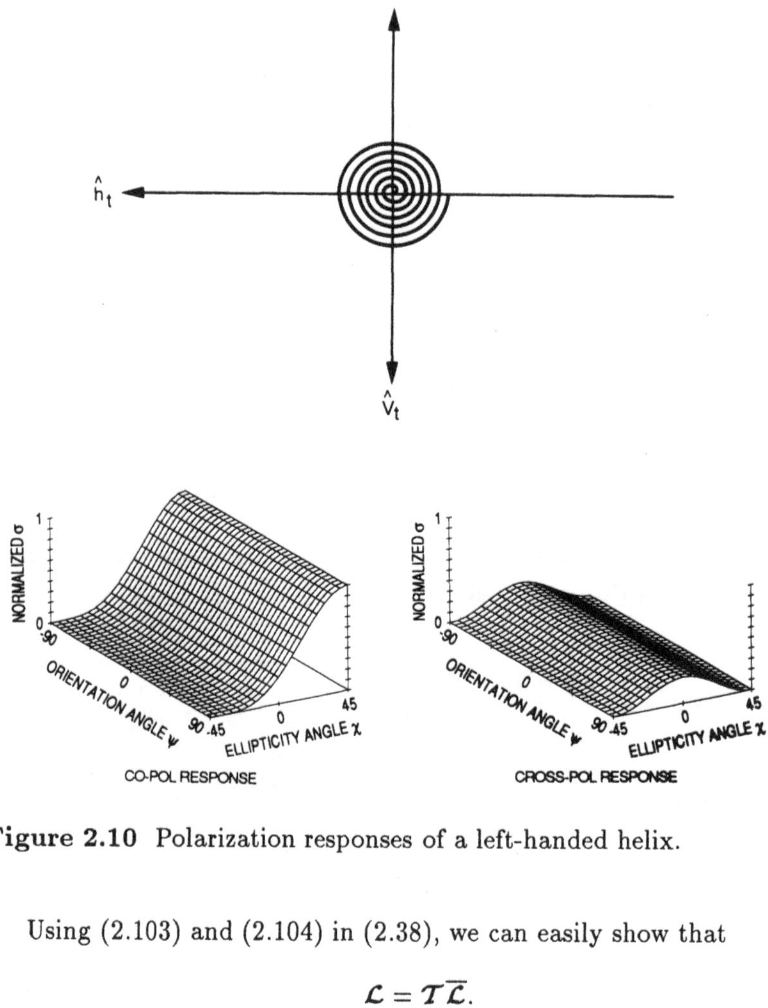


**Figure 2.10** Polarization responses of a left-handed helix.

Using (2.103) and (2.104) in (2.38), we can easily show that

$$\mathcal{L} = \mathcal{T}\overline{\mathcal{L}}. \tag{2.106}$$

From (2.20), (2.21), and (2.42), we can similarly show that

$$\mathbf{G}^i = \overline{\mathbf{G}}^t, \tag{2.107}$$

$$\mathbf{G}^s = \mathcal{T}\,\overline{\mathbf{G}}^r, \tag{2.108}$$

and therefore

$$\mathcal{W} = \mathcal{T}\overline{\mathcal{W}}. \tag{2.109}$$

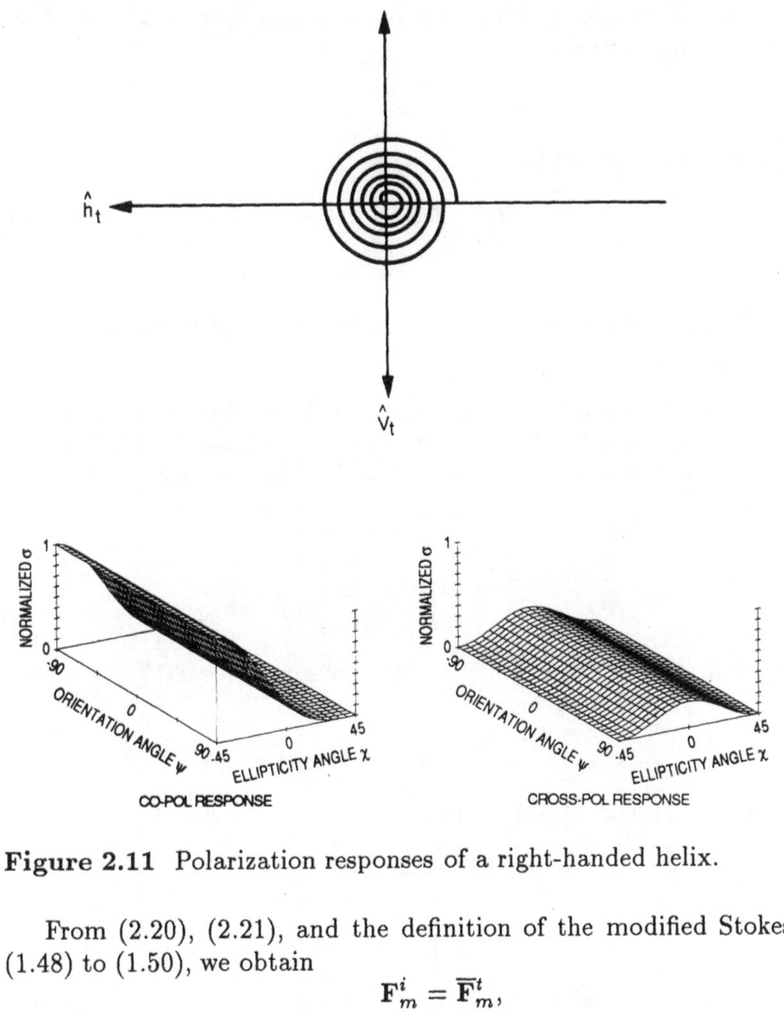


**Figure 2.11** Polarization responses of a right-handed helix.

From (2.20), (2.21), and the definition of the modified Stokes vector (1.48) to (1.50), we obtain

$$\mathbf{F}^i_m = \overline{\mathbf{F}}^t_m, \tag{2.110}$$

$$\mathbf{F}^s_m = \mathcal{T}\,\overline{\mathbf{F}}^r_m, \tag{2.111}$$

which means that

$$\mathcal{L}_m = \mathcal{T}\overline{\mathcal{L}}_m. \tag{2.112}$$

To show that the Stokes scattering operator and the modified Stokes scattering operator in the FSA and BSA conventions are related is straightforward as follows:

$$\mathcal{M} = \mathcal{T}\overline{\mathcal{M}} \tag{2.113}$$

and

$$\mathcal{M}_m = \mathcal{T}\overline{\mathcal{M}}_m. \tag{2.114}$$

To derive the transformations for the covariance matrix, we note from (2.20), (2.21), and (2.24) that

$$\mathbf{X} = \mathcal{T}\,\overline{\mathbf{X}}. \tag{2.115}$$

Thus, by using (2.115) and (2.72), we write

$$\mathcal{C} = \mathcal{T}\,\overline{\mathcal{C}}\,\mathcal{T}. \tag{2.116}$$

## 2-5.2 Transformation for JPL Stokes Matrix Data

The data recorded by the JPL polarimetric imager are formatted according to the JPL Stokes matrix, which is defined on the basis of the following definition of the Stokes vector, expressed in the BSA convention:

$$\overline{\mathbf{F}}_{\text{JPL}} = \begin{pmatrix} \langle |E_h|^2\rangle + \langle |E_v|^2\rangle \\ \langle |E_h|^2\rangle - \langle |E_v|^2\rangle \\ \langle -2\text{Re}(E_v^* E_h)\rangle \\ \langle -2\text{Im}(E_v^* E_h)\rangle \end{pmatrix}. \tag{2.117}$$

Using this definition of the Stokes matrix, a polarization synthesis equation similar to (2.61) can be derived:

$$\sigma_{rt}(\psi_r, \chi_r, \psi_t, \chi_t) \;=\; 4\pi\, \overline{\mathbf{A}}^r_{\text{JPL}} \cdot \overline{\mathcal{M}}_{\text{JPL}} \overline{\mathbf{A}}^t_{\text{JPL}}. \tag{2.118}$$

Comparing (2.117) with (1.53), clearly

$$\overline{\mathbf{F}}_{\text{JPL}} = \mathcal{Q}\,\overline{\mathbf{F}}, \tag{2.119}$$

with

$$\mathcal{Q} = \begin{pmatrix} 1 & 0 & 0 & 0 \\ 0 & -1 & 0 & 0 \\ 0 & 0 & -1 & 0 \\ 0 & 0 & 0 & 1 \end{pmatrix}. \tag{2.120}$$

Thus, normalizing the Stokes vectors, we find from (2.119) that

$$\overline{\mathbf{A}}_{\text{JPL}} = \mathcal{Q}\,\overline{\mathbf{A}} \tag{2.121}$$

and

$$\overline{\mathbf{A}} = \mathcal{Q}\,\overline{\mathbf{A}}_{\text{JPL}}. \tag{2.122}$$

Using (2.122) in (2.61), we find that

$$\overline{\mathbf{A}}^r \cdot \overline{\mathcal{M}}\,\overline{\mathbf{A}}^t = \overline{\mathbf{A}}^r_{\text{JPL}} \cdot \tilde{\mathcal{Q}}\,\overline{\mathcal{M}}\,\mathcal{Q}\,\overline{\mathbf{A}}^t_{\text{JPL}}, \tag{2.123}$$

from which, comparing (2.123) to (2.118), follows

$$\overline{\mathcal{M}} = \tilde{\mathcal{Q}}\,\overline{\mathcal{M}}_{\mathrm{JPL}}\mathcal{Q}, \tag{2.124}$$

and

$$\mathcal{M} = \mathcal{T}\,\tilde{\mathcal{Q}}\,\overline{\mathcal{M}}_{\mathrm{JPL}}\mathcal{Q}. \tag{2.125}$$

Thus, given the Stokes scattering operator in the JPL data format, all other matrices can be calculated in either the BSA or FSA conventions.

# 2-6 CONCLUSIONS

The different mathematical representations used to describe the scattering properties of simple scatterers were introduced in this chapter. The equations for polarization synthesis using four different representations were derived, and the polarization response, a graphical representation of the variation of the scattering cross-section with polarization, was introduced. Examples of polarization responses of simple scatterers were discussed. The examples show that polarization responses give useful information about the orientation of scatterers such as thin cylinders. For scatterers with some helicity, the polarization responses provide an easy way to measure the handedness of the helix, although the scattering matrix is measured by using a linearly polarized basis. By comparing the responses of the dihedral and trihedral corner reflectors, the polarization response clearly provides a simple way to discriminate between scattering with two reflections and scattering with three reflections. All these examples confirm that the polarization response may be a valuable tool in identifying the dominant scattering mechanisms that are present in a particular set of pixels in a radar image.

A word of caution about the use of the polarization response in identifying scattering mechanisms is in order at this point. Unfortunately, polarization responses are not unique. The examples show that a large sphere has the same polarization responses as a trihedral corner reflector. Conceivably, different combinations of scattering mechanisms may produce the same resultant polarization response. Nevertheless, if proper care is taken when scattering results are interpreted, the full use of polarization information increases the present capability to solve for geometrical structure and electric properties of scatterers.

# References


Agrawal, A.P., and W.M. Boerner, "Redevelopment of Kennaugh's Target Characteristic Polarization State Theory Using the Polarization Transformation Ratio for the Coherent Case," *IEEE Trans. on Geoscience and Remote Sensing,* Vol. GE-27, pp. 2–14, 1988.

Azzam, R.M.A., and N.M. Bashara, *Ellipsometry and Polarized Light,* North-Holland, Amsterdam, 1976.

Chandrasekhar, S., *Radiative Transfer,* Dover, New York, 1960.

Clarke, D., and J.F. Grainger, *Polarized Light and Optical Measurement,* Pergamon, New York, 1971.

Deirmendjian, D., *Electromagnetic Scattering on Spherical Polydispersions,* American Elsevier, New York, 1969.

Hauge, P.S., "Mueller Matrix Ellipsometry with Imperfect Compensators," *J. Optical Society of America,* Vol. 68, pp. 1519–1528, 1978.

Hurwitz, H., and R.C. Jones, "A New Calculus for the Treatment of Optical Systems II. Proof of Three General Equivalence Theorems," *J. Optical Society of America,* Vol. 31, pp. 493–499, 1941.

Huynen, J.R., "Measurement of the Target Scattering Matrix," *Proceedings of the IEEE,* Vol. 53, pp. 936–946, 1965.

Huynen, J.R., Phenomenological Theory of Radar Targets, Ph.D. Dissertation, Drukkerij Bronder-Offset, N.V., Rotterdam, 1970.

IEEE Standard Definitions of Terms for Antennas, *IEEE Trans. on Antennas and Propagation,* Vol. AP-31, No. 6, 1983.

Jasik, H. (ed.), *Antenna Engineering Handbook,* McGraw-Hill, New York p. 13-9, 1961.

Jones, R.C., "A New Calculus for the Treatment of Optical Systems I. Description and Discussion of the Calculus," *J. Optical Society of America,* Vol. 31, pp. 488–493, 1941.

Jones, R.C., "A New Calculus for the Treatment of Optical Systems III. The Sohncke Theory of Optical Activity," *J. Optical Society of America,* Vol. 31, pp. 500–503, 1941.

Jones, R.C., "A New Calculus for the Treatment of Optical Systems IV," *J. Optical Society of America,* Vol. 32, pp. 486–493, 1942.

Jones, R.C., "A New Calculus for the Treatment of Optical Systems V. A More General Formulation, and Description of Another Calculus," *J. Optical Society of America,* Vol. 37, pp. 107–110, 1947.

Kennaugh, E.M., "Effects of the type of polarization on echo characteristics," Report 389-9, Antenna Laboratory, Ohio State University, 1951.

Kostinski, A.B., and W.M. Boerner, "On Foundations of Radar Polarimetry," *IEEE Trans. on Antennas and Propagation,* Vol. AP-34, pp. 1395–1404, 1986.

Mie, G., "Beiträge zur Optik trüber Medien, speziell kolloidaler Metallösungen," *Ann. d. Physik,* Vol. 25, p. 377, 1908.

Ruck, G.T., D.E. Barrick, W.D. Stuart, and C.K. Krichbaum, *Radar Cross Section Handbook,* Volume I, pp. 7–17, Plenum Press, New York, 1970.

Shurcliff, W.A., *Polarized Light,* Harvard University Press, Cambridge, MA, 1962.

Sinclair, G., "The Transmission and Reception of Elliptically Polarized Waves," *Proceedings of the IRE,* Vol. 38, pp. 148–151, 1950.

Swartz, A.A., H.A. Yueh, J.A. Kong, L.M. Novak, and R.T. Shin, "Optimal Polarizations for Achieving Maximum Contrast in Radar Images," *J. Geophysical Research,* Vol. 93, No. B12, pp. 15252–15260, 1988.

Tsang, L., J.A. Kong, and R.T. Shin, *Theory of Microwave Remote Sensing,* Wiley and Sons, New York, 1985.

Van de Hulst, H.C., *Light Scattering by Small Particles,* pp. 28–42, Dover, New York, 1981.

van Zyl, J.J., On the Importance of Polarization in Radar Scattering Problems, Ph.D. Thesis, Caltech Antenna Laboratory Report No. 120, 152 pp., California Institute of Technology, Pasadena, CA, 1985.

van Zyl, J.J., H.A. Zebker, and C. Elachi, "Imaging Radar Polarization Signatures: Theory and Observation," *Radio Science,* Vol. 22, pp. 529–543, 1987.

van Zyl, J.J., C.H. Papas, and C. Elachi, "On the Optimum Polarizations of Incoherently Reflected Waves," *IEEE Trans. on Antennas and Propagation,* Vol. AP-35, pp. 818–825, 1987.

van Zyl, J.J., "Unsupervised Classification of Scattering Behavior Using Imaging Radar Polarimetry Data," *IEEE Trans. on Geoscience and Remote Sensing,* Vol. GE-27, pp. 36–45,1989.

van Zyl, J.J., "Calibration of Polarimetric Radar Images Using Only Image Parameters and Trihedral Corner Reflector Responses," *IEEE Trans. on Geoscience and Remote Sensing,* Vol. GE-28, 1990.

Zebker, H.A., J.J. van Zyl, and D.N. Held, "Imaging Radar Polarimetry From Wave Synthesis," *J. Geophysical Research,* Vol. 92, No. B1, pp. 683–701, 1987.

# Chapter 3

# Scattering Models for Point Targets

**Thomas B.A. Senior and Kamal Sarabandi**

To predict the scattering from a complicated target such as a tree, we need to determine the scattering from such individual parts as the leaves, twigs, branches, and trunk, each of which is modeled as a simple structure amenable to rigorous analysis. This leads to a consideration of canonical bodies such as plates, cylinders, spheroids, and spheres, the scattering properties of which can be determined. Each body is finite in extent and because, in the far zone (defined below), the scattered field appears to originate at a point, the body is described as a point target, as opposed to a distributed target such as terrain.

Vegetation materials are lossy dielectrics with a dielectric constant that is strongly influenced by the moisture content. We therefore consider nonmagnetic lossy dielectric bodies having relative permeability unity and relative permittivity $\epsilon_r = \epsilon_r' + i\epsilon_r''$. The limiting case as the conductivity $\sigma$ (and hence $\epsilon_r''$) $\to \infty$ is a perfectly conducting body, which is useful for reference purposes.

# 3-1 INTRODUCTION

The scattered field $\mathbf{E}^s$, $\mathbf{H}^s$ is a perturbation arising from the introduction of the body into the primary (or incident) field $\mathbf{E}^i$, $\mathbf{H}^i$ that formerly existed throughout all space. Thus,

$$\mathbf{E}^s = \mathbf{E} - \mathbf{E}^i \ , \quad \mathbf{H}^s = \mathbf{H} - \mathbf{H}^i \ , \tag{3.1}$$

where $\mathbf{E}$, $\mathbf{H}$ form the total field in the presence of the body. The scattering is attributable to electric currents created inside and on the surface of the body, but to allow magnetic currents as well is mathematically convenient. The scattered field at any point $\mathbf{r}$ in space can then be determined from the electric and magnetic Hertz vectors:

$$\begin{aligned} \mathbf{\Pi}_e(\mathbf{r}) &= \frac{i\eta_0}{4\pi k_0}\int_V \mathbf{J}_e(\mathbf{r}')\frac{e^{ik_0|\mathbf{r}-\mathbf{r}'|}}{|\mathbf{r}-\mathbf{r}'|}dV' \\ \mathbf{\Pi}_m(\mathbf{r}) &= \frac{i\eta_0^{-1}}{4\pi k_0}\int_V \mathbf{J}_m(\mathbf{r}')\frac{e^{ik_0|\mathbf{r}-\mathbf{r}'|}}{|\mathbf{r}-\mathbf{r}'|}dV' \ , \end{aligned} \tag{3.2}$$

where $\mathbf{J}_e$, $\mathbf{J}_m$ are the volume electric and magnetic current densities, respectively, the index 0 refers to free space, and the integration is done over the volume $V$ occupied by the body. Two special cases are of interest. If

$$\mathbf{J}_e(\mathbf{r}') = -i\omega\mathbf{p}\delta(\mathbf{r}') \ , \tag{3.3}$$

where $\delta(\mathbf{r}')$ is the three-dimensional impulse or delta function, then

$$\mathbf{\Pi}_e(\mathbf{r}) = \frac{\mathbf{p}}{4\pi\epsilon_0}\frac{e^{ik_0r}}{r} \ , \tag{3.4}$$

corresponding to an electric dipole of moment $\mathbf{p}$ located at the origin in free space. Similarly, if

$$\mathbf{J}_m(\mathbf{r}') = -i\omega\mu_0\mathbf{m}\delta(\mathbf{r}') \ , \tag{3.5}$$

we have

$$\mathbf{\Pi}_m(\mathbf{r}) = \frac{\mathbf{m}}{4\pi}\frac{e^{ik_0r}}{r} \ , \tag{3.6}$$

which represents a magnetic dipole of moment $\mathbf{m}$. Although in principle associated with a magnetic current, a magnetic dipole is produced by an electric current flowing in a loop of infinitesimal radius in a plane perpendicular to $\mathbf{m}$.

In terms of the Hertz vectors $\mathbf{\Pi}_e, \mathbf{\Pi}_m$, the scattered field is expressed as

$$\begin{aligned} \mathbf{E}^s(\mathbf{r}) &= \nabla \times \nabla \times \mathbf{\Pi}_e(\mathbf{r}) + ik_0\eta_0 \nabla \times \mathbf{\Pi}_m(\mathbf{r}) \\ \mathbf{H}^s(\mathbf{r}) &= \nabla \times \nabla \times \mathbf{\Pi}_m(\mathbf{r}) - ik_0\eta_0^{-1} \nabla \times \mathbf{\Pi}_e(\mathbf{r}) \ . \end{aligned} \tag{3.7}$$

At large distances from the body, $\nabla \times () \approx ik_0\hat{\mathbf{r}} \times ()$ and hence

$$\begin{aligned} \mathbf{E}^s(\mathbf{r}) &= -k_0^2\{\hat{\mathbf{r}} \times \hat{\mathbf{r}} \times \mathbf{\Pi}_e + \hat{\mathbf{r}} \times \eta_0 \mathbf{\Pi}_m\} \\ \mathbf{H}^s(\mathbf{r}) &= -k_0^2\{\hat{\mathbf{r}} \times \hat{\mathbf{r}} \times \mathbf{\Pi}_m - \hat{\mathbf{r}} \times \eta_0^{-1} \mathbf{\Pi}_e\} \ , \end{aligned} \tag{3.8}$$

which demonstrate the fact that $\mathbf{E}^s$, $\mathbf{H}^s$ are perpendicular to $\hat{\mathbf{r}}$. Moreover,

$$|\mathbf{r} - \mathbf{r}'| \approx r - \hat{\mathbf{r}}\cdot\mathbf{r}' + \frac{1}{2}\frac{r'^2}{r}\{1 - (\hat{\mathbf{r}}\cdot\hat{\mathbf{r}}')^2\} \ , \tag{3.9}$$

and if $r$ is so large that over the entire volume occupied by the body the third term on the right-hand side of (3.9) is negligible, even when in the phase factor,

$$\int_V \mathbf{J}_{e,m}(\mathbf{r}') \frac{e^{ik_0|\mathbf{r}-\mathbf{r}'|}}{|\mathbf{r} - \mathbf{r}'|} dV' = \frac{e^{ik_0 r}}{r} \int_V \mathbf{J}_{e,m}(\mathbf{r}') e^{-ik_0\hat{\mathbf{r}}\cdot\mathbf{r}'} dV' \ . \tag{3.10}$$

The accepted requirement for the validity of (3.10) is

$$r > \frac{2D^2}{\lambda_0} \tag{3.11}$$

where $\lambda_0$ is the free-space wavelength and $D$ is the maximum dimension (diameter) of the body, which implies that phase differences of up to 22.5° between the contributions from individual parts of the body are deemed negligible. Equation (3.11) defines the far zone of the body, and (3.8) and (3.10) then show that in the far zone the scattered field looks like a spherically outgoing wave attributable to electric and magnetic sources located at the origin. Therefore, a calculation of the far zone scattered field requires only the determination of the equivalent sources.

Unfortunately, this is not a trivial task for even the simplest body, and the one finite shape for which an exact analytical expression for the scattered field is known is the sphere, either homogeneous or spherically stratified, including the perfectly conducting sphere as a special case. In practice, we need to rely on approximate techniques, the accuracy of which can be established by comparison with experimental or computed data. There are several approximate techniques available, and their regions of validity are specified by the body's size in comparison with the wavelength. At high frequencies, when all significant dimensions of the body are large compared with the wavelength, the most useful method is that of *physical optics,* and this is applied to a sequence of simple shapes in Section 3-2. The other extreme is at low frequencies where all dimensions are small compared with the wavelength. The body now scatters as a whole, and, as shown in Section 3-3, there is less dependence on the details of the body's shape. At intermediate frequencies, the scatterings from the various parts of the body interact with each other, giving rise to resonance effects that are difficult to predict. Although some progress has been made in penetrating the resonance region from above by using more sophisticated high frequency techniques such as the *geometrical theory of diffraction* (James, 1976), to employ numerical techniques is now customary, for example, based on a numerical solution of the integral equations for the induced currents (Poggio and Miller, 1987, pp. 168–170). Nevertheless, the exceptional cases where exact analytical results are available continue to be important — if nothing else, for comparison purposes — and these results are presented in Section 3-4.

# 3-2 HIGH FREQUENCY SCATTERING

## 3-2.1 Physical Optics Approximation

This is based on the surface field of *geometrical optics* (Knott *et al.,* 1985, pp. 57–59), and eliminates the spatial discontinuities in the field that geometrical optics predicts, as well as the infinities at ray caustics. To understand the nature of the approximation, consider a small region about the point $\mathbf{r}'$ on the surface of a perfectly conducting body as shown in Fig. 3.1. If the surface has a unique normal $\hat{\mathbf{n}}'$ at $\mathbf{r}'$, the principal radii of curvature $R_1$, $R_2$ of the surface are nonzero there, and as the wavelength tends to zero, $R_1/\lambda_0$, $R_2/\lambda_0$, $\to \infty$. Thus, at high frequencies, this portion of the surface appears locally flat, and if it is regarded as part of an infinite plane, the local boundary value problem is that of a plane wave incident on a perfectly conducting plane.

The solution is known. The scattered field is simply a reflected plane

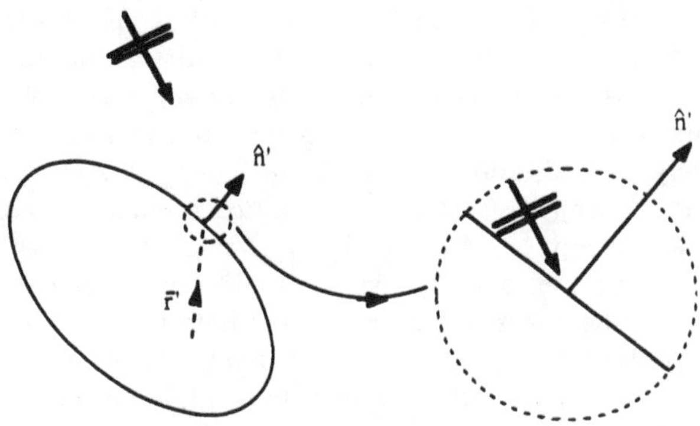


**Figure 3.1** The origin of the physical optics approximation.

wave such that

$$E^r_{\tan} = -E^i_{\tan} , \quad H^r_{\tan} = H^i_{\tan} ,$$

and hence

$$\hat{\mathbf{n}}' \times \mathbf{E} = \hat{\mathbf{n}}' \times (\mathbf{E}^i + \mathbf{E}^r) = 0 , \tag{3.12}$$

in accordance with the boundary condition, and

$$\hat{\mathbf{n}}' \times \mathbf{H} = \hat{\mathbf{n}}' \times (\mathbf{H}^i + \mathbf{H}^r) = 2\hat{\mathbf{n}}' \times \mathbf{H}^i . \tag{3.13}$$

In contrast, at a point on the rear (or, indeed, any geometrically shadowed) portion of the surface, the incident field is blocked by the body, and because there is no incident field there, $\hat{\mathbf{n}} \times \mathbf{H} = 0$. The physical optics approximation (or *tangent plane approximation* as it is sometimes called) therefore specifies the surface field on a perfectly conducting body as follows:

$$\hat{\mathbf{n}}' \times \mathbf{H} = \begin{cases} 2\hat{\mathbf{n}}' \times \mathbf{H}^i, & \text{in the lit region} \\ 0, & \text{in the shadow} \end{cases}$$

leading to the Hertz vectors:

$$\begin{aligned} \mathbf{\Pi}_e(\mathbf{r}) &= \frac{i\eta_0}{2\pi k_0} \int_{\text{lit}} \hat{\mathbf{n}}' \times \mathbf{H}^i \frac{e^{ik_0|\mathbf{r}-\mathbf{r}'|}}{|\mathbf{r}-\mathbf{r}'|} dS' \\ \mathbf{\Pi}_m(\mathbf{r}) &= 0 , \end{aligned} \tag{3.14}$$

from which the scattered field can be determined.

The shortcomings of this approach are many. It violates the reciprocity condition concerning the interchange of transmitter and receiver; it provides an erroneous contribution from the boundary of the lit and shadowed regions of the surface, and from any edges where the unit vector normal $\hat{\mathbf{n}}'$ is not uniquely defined; and because the surface field is everywhere a function of the local properties alone, it does not predict the contribution of any surface waves supported by the body. Nevertheless, this approach has its virtues. It is simple to apply, even to bodies of complicated shape and, under most circumstances, it accurately predicts the dominant features of a scattering pattern and is at least qualitatively correct with the rest.

If the body is not perfectly conducting, but is instead composed of a nonmagnetic lossy dielectric, the logical extension of the physical optics approximation is to use the reflection coefficients for a dielectric half space. For the incident plane wave,

$$\mathbf{E}^i = \mathbf{E}_0 \, e^{ik_0 \hat{\mathbf{k}}_i \cdot \hat{\mathbf{r}}} \; , \quad \mathbf{H}^i = \mathbf{H}_0 \, e^{ik_0 \hat{\mathbf{k}}_i \cdot \hat{\mathbf{r}}} \tag{3.15}$$

with

$$\mathbf{H}_0 = \eta_0^{-1} \hat{\mathbf{k}}_i \times \mathbf{E}_0 \; , \tag{3.16}$$

the plane of incidence is defined by the unit vectors $\hat{\mathbf{k}}$ and $\hat{\mathbf{n}}$. Let $\hat{\mathbf{t}}$ be a unit vector perpendicular to this plane. Then, if

$$\mathbf{E}^i = \{A(\hat{\mathbf{k}}_i \times \hat{\mathbf{t}}) + B\hat{\mathbf{t}}\} e^{ik_0 \hat{\mathbf{k}}_i \cdot \mathbf{r}} \; , \tag{3.17}$$

implying

$$\mathbf{H}^i = \eta_0^{-1} \{B(\hat{\mathbf{k}}_i \times \hat{\mathbf{t}}) - A\hat{\mathbf{t}}\} e^{ik_0 \hat{\mathbf{k}}_i \cdot \mathbf{r}} \; , \tag{3.18}$$

the reflected plane wave is

$$\mathbf{E}^r = \{AR_H(\hat{\mathbf{k}}_r \times \hat{\mathbf{t}}) + BR_E\hat{\mathbf{t}}\} e^{ik_0 \hat{\mathbf{k}}_r \cdot \mathbf{r}} \; , \tag{3.19}$$

implying

$$\mathbf{H}^r = \eta_0^{-1} \{BR_E(\hat{\mathbf{k}}_r \times \hat{\mathbf{t}}) - AR_H\hat{\mathbf{t}}\} e^{ik_0 \hat{\mathbf{k}}_r \cdot \mathbf{r}} \; , \tag{3.20}$$

where

$$\hat{\mathbf{k}}_r = \hat{\mathbf{k}}_i - 2(\hat{\mathbf{k}}_i \cdot \hat{\mathbf{n}})\hat{\mathbf{n}} \tag{3.21}$$

and $R_E$, $R_H$ are the Fresnel reflection coefficients (Stratton, 1941, pp. 492–494):

$$R_E = \frac{\cos\phi - \sqrt{\epsilon_r - \sin^2\phi}}{\cos\phi + \sqrt{\epsilon_r - \sin^2\phi}}$$

$$R_H = \frac{\epsilon_r \cos\phi - \sqrt{\epsilon_r - \sin^2\phi}}{\epsilon_r \cos\phi + \sqrt{\epsilon_r - \sin^2\phi}} \tag{3.22}$$

with

$$\cos\phi = -\hat{\mathbf{n}}\cdot\hat{\mathbf{k}}_i \ . \tag{3.23}$$

Clearly,

$$A = -\eta_0 \mathbf{H}_0\cdot\hat{\mathbf{t}} \ , \quad B = \mathbf{E}_0\cdot\hat{\mathbf{t}} \ , \tag{3.24}$$

and in the lit region the physical optics approximations to the surface currents are

$$\mathbf{J}_e = \hat{\mathbf{n}} \times (\mathbf{H}^i + \mathbf{H}^r) \ , \quad \mathbf{J}_m = -\hat{\mathbf{n}} \times (\mathbf{E}^i + \mathbf{E}^r) \ , \tag{3.25}$$

where

$$\mathbf{J}_e = -\left\{\eta_0^{-1}(1 - R_E)(\mathbf{E}^i\cdot\hat{\mathbf{t}})(\hat{\mathbf{n}}\cdot\hat{\mathbf{k}}_i)\hat{\mathbf{t}} - (1 + R_H)(\mathbf{H}^i\cdot\hat{\mathbf{t}})\hat{\mathbf{n}} \times \hat{\mathbf{t}}\right\} \tag{3.26}$$

$$\mathbf{J}_m = -\left\{\eta_0(1 - R_H)(\mathbf{H}^i\cdot\hat{\mathbf{t}})(\hat{\mathbf{n}}\cdot\hat{\mathbf{k}}_i)\hat{\mathbf{t}} + (1 + R_E)(\mathbf{E}^i\cdot\hat{\mathbf{t}})\hat{\mathbf{n}} \times \hat{\mathbf{t}}\right\} \ . \tag{3.27}$$

On the assumption that the loss is sufficient to make the body impenetrable, we then have

$$\mathbf{J}_e = \mathbf{J}_m = 0 \tag{3.28}$$

on the geometrically shadowed portion of the surface, and this completes the specification of the surface currents.

The expressions given in (3.26) and (3.27) are equivalent to, but simpler than, those given by Beckmann (1968, p. 85), and their symmetry is noteworthy. They provide the extension of the physical optics approximation to a dielectric surface, and in the particular case of perfect conductivity for which $R_H = -R_E = 1$, $\mathbf{J}_e = 2\hat{\mathbf{n}} \times \mathbf{H}^i$ and $\mathbf{J}_m = 0$ in accordance with (3.13) and (3.12) respectively.

For sufficiently large $|\epsilon_r|$ one method for simulating the material properties of a surface is to impose the impedance boundary condition (Senior, 1981):

$$\hat{\mathbf{n}} \times (\hat{\mathbf{n}} \times \mathbf{E}) = -\eta \hat{\mathbf{n}} \times \mathbf{H} \ , \tag{3.29}$$

where $\eta = (\mu_0/\epsilon)^{1/2}$. The corresponding surface currents are such that

$$\mathbf{J}_e = \frac{1}{\eta} \hat{\mathbf{n}} \times \mathbf{J}_m \ , \quad \mathbf{J}_m = -\eta \hat{\mathbf{n}} \times \mathbf{J}_e \ , \tag{3.30}$$

satisfying duality and showing that the electric current now dominates. When applied to a dielectric half-space, the boundary condition (3.29) implies the reflection coefficients:

$$R'_E = -\frac{\eta_0 - \eta \cos\phi}{\eta_0 + \eta \cos\phi} \ , \quad R'_H = \frac{\eta_0 \cos\phi - \eta}{\eta_0 \cos\phi + \eta} \ , \tag{3.31}$$

which are valid approximations to (3.22) if $|\eta| \ll \eta_0$. In view of (3.30), the boundary condition somewhat simplifies the physical optics approximation (Maffett, 1989, pp. 300–306), and because

$$\eta_0(1+R'_E) = -\eta(1-R'_E)(\hat{\mathbf{n}}\cdot\hat{\mathbf{k}}_i) \ , \quad \eta_0(1-R'_H)(\hat{\mathbf{n}}\cdot\hat{\mathbf{k}}_i) = -\eta(1+R'_H) \ , \tag{3.32}$$

we can verify that the resulting approximations to (3.26) and (3.27) satisfy (3.30). Thus, the two currents are no longer independent.

In the case of a thin slab or plate, the currents (3.26) and (3.27) are those present on the illuminated face, but because the field may now penetrate through the plate, the reflection coefficients required are for an infinite slab of the same thickness $t$ (Willis *et al.*, 1988). If $t$ is sufficiently small, the slab can be approximated as an infinitesimally thin resistive sheet supporting only a total electric current (Senior *et al.*, 1987):

$$\mathbf{J}_e = \Gamma_{E,H} \mathbf{J}_e^{p.c.} \tag{3.33}$$

where $\mathbf{J}_e^{p.c.}$ is the current for a perfectly conducting surface and $\Gamma_E(\Gamma_H)$ is the reflection coefficient for $\mathbf{E}^i(\mathbf{H}^i)$ perpendicular to the plane of incidence:

$$\begin{aligned} \Gamma_E &= \left(1 + \tfrac{2R}{\eta_0} \cos\phi\right)^{-1} \\ \Gamma_H &= \left(1 + \tfrac{2R}{\eta_0} \sec\phi\right)^{-1} , \end{aligned} \tag{3.34}$$

with

$$R = \frac{i\eta_0}{k_0 t(\epsilon_r - 1)} \tag{3.35}$$

and $\cos\phi$ as given in (3.23). We will use this method in conjunction with the physical optics approximation to $\mathbf{J}_e^{p.c.}$ to compute the scattering from a plate.

### 3-2.2 Flat Plate

A thin flat plate is illuminated by the plane wave (3.15), and if the plate is perfectly conducting, the physical optics approximation gives

$$\mathbf{\Pi}_e(\mathbf{r}) = \frac{i\eta_0}{2\pi k_0}\hat{\mathbf{n}}' \times \mathbf{H}_0 \int_s e^{ik_0\hat{\mathbf{k}}_i\cdot\mathbf{r}'}\frac{e^{ik_0|\mathbf{r}-\mathbf{r}'|}}{|\mathbf{r}-\mathbf{r}'|}dS' \tag{3.36}$$

where $\hat{\mathbf{n}}'$ is the normal to the illuminated face and the phase origin is at the center of the plate. For observation in the direction $\hat{\mathbf{k}}_s$ in the far zone

$$\mathbf{\Pi}_e(\mathbf{r}) = \frac{e^{ik_0 r}}{r}\hat{\mathbf{n}}' \times H_0\frac{i\eta_0}{2\pi k_0}\int_s e^{ik_0(\hat{\mathbf{k}}_i-\hat{\mathbf{k}}_s)\cdot\mathbf{r}'}dS' \ , \tag{3.37}$$

and hence

$$\mathbf{S} = \left[(\hat{\mathbf{n}}'\cdot\mathbf{E}_0)\left\{(\hat{\mathbf{k}}_i\cdot\hat{\mathbf{k}}_s)\hat{\mathbf{k}}_s - \hat{\mathbf{k}}_i\right\} - (\hat{\mathbf{n}}'\cdot\hat{\mathbf{k}}_i)\left\{(\hat{\mathbf{k}}_s\cdot\mathbf{E}_0)\hat{\mathbf{k}}_s - \mathbf{E}_0\right\}\right]K \ , \tag{3.38}$$

where

$$K = -\frac{i}{\lambda_0}\int_s e^{ik_0(\hat{\mathbf{k}}_i-\hat{\mathbf{k}}_s)\cdot\mathbf{r}'}dS' \ . \tag{3.39}$$

In the case of a circular disk of radius $a$,

$$K = -i\frac{a}{|(\hat{\mathbf{k}}_i-\hat{\mathbf{k}}_s)\cdot\hat{\rho}'|}J_1(|(\hat{\mathbf{k}}_i-\hat{\mathbf{k}}_s)\cdot\hat{\rho}'|k_0 a) \tag{3.40}$$

where $\hat{\rho}'$ is a unit vector in the plane of the disk and $J_1$ is the Bessel function of the first order. For a rectangular plate with $-a/2 < y' < a/2$ and $-b/2 < z' < b/2$, where $y'$ and $z'$ are Cartesian coordinates in its plane, we have

$$K = -\frac{iab}{\lambda_0}\frac{\sin U}{U}\frac{\sin V}{V} \tag{3.41}$$

with

$$U = \frac{1}{2}k_0 a(\hat{\mathbf{k}}_i-\hat{\mathbf{k}}_s)\cdot\hat{\mathbf{y}}' \ , \quad V = \frac{1}{2}k_0 b(\hat{\mathbf{k}}_i-\hat{\mathbf{k}}_s)\cdot\hat{\mathbf{z}}' \ , \tag{3.42}$$

and for other shapes, the integral expression (3.39) can be evaluated numerically.

As is evident from (3.39), if $(\hat{\mathbf{k}}_i - \hat{\mathbf{k}}_s)\cdot\mathbf{r}' = 0$,

$$K = -\frac{iA}{\lambda_0} \tag{3.43}$$

where $A$ is the plate area.

Some particular cases of the above results for a rectangular plate are as follows: in the forward scattering direction ($\hat{\mathbf{k}}_s = \hat{\mathbf{k}}_i$),

$$\mathbf{S} = -\frac{iab}{\lambda_0}(\hat{\mathbf{n}}'\cdot\hat{\mathbf{k}}_i)\mathbf{E}_0 \; ; \tag{3.44}$$

in the backscattering direction ($\hat{\mathbf{k}}_s = -\hat{\mathbf{k}}_i$),

$$\mathbf{S} = -\frac{iab}{\lambda_0}(\hat{\mathbf{n}}'\cdot\hat{\mathbf{k}}_i)\mathbf{E}_0 \frac{\sin(k_0 a\hat{\mathbf{k}}_i\cdot\hat{\mathbf{y}}')}{k_0 a\hat{\mathbf{k}}_i\cdot\hat{\mathbf{y}}'}\frac{\sin(k_0 b\hat{\mathbf{k}}_i\cdot\hat{\mathbf{z}}')}{k_0 b\hat{\mathbf{k}}_i\cdot\hat{\mathbf{z}}'} \; ; \tag{3.45}$$

in the specular direction ($\hat{\mathbf{k}}_s\cdot\hat{\mathbf{r}}' = \hat{\mathbf{k}}_i\cdot\hat{\mathbf{r}}'$ but $\hat{\mathbf{k}}_s\cdot\hat{\mathbf{n}}' = -\hat{\mathbf{k}}_i\cdot\hat{\mathbf{n}}'$),

$$\mathbf{S} = -\frac{iab}{\lambda_0}(\hat{\mathbf{n}}'\cdot\hat{\mathbf{k}}_i)\left\{\mathbf{E}_0 - 2(\hat{\mathbf{n}}'\cdot\mathbf{E}_0)\hat{\mathbf{n}}'\right\} \; ; \tag{3.46}$$

and for normal incidence backscatter,

$$\mathbf{S} = \frac{iab}{\lambda_0}\mathbf{E}_0 \; . \tag{3.47}$$

Note the dependence on $A/\lambda_0$.

If the plate is not perfectly conducting, the resistive sheet simulation is employed, and to resolve $\hat{\mathbf{n}}' \times \mathbf{H}_0$ into components perpendicular and parallel to the plane of incidence is then necessary. Let

$$\hat{\mathbf{t}}_1 = \frac{\hat{\mathbf{n}}' \times \hat{\mathbf{k}}_i}{|\hat{\mathbf{n}}' \times \hat{\mathbf{k}}_i|} \, , \quad \hat{\mathbf{t}}_2 = \frac{\hat{\mathbf{k}}_i \times (\hat{\mathbf{n}}' \times \hat{\mathbf{k}}_i)}{|\hat{\mathbf{n}}' \times \hat{\mathbf{k}}_i|} \, , \tag{3.48}$$

in terms of which

$$\hat{\mathbf{n}}' \times \mathbf{H}_0 = c_1\hat{\mathbf{t}}_1 + c_2\hat{\mathbf{t}}_2 \, , \tag{3.49}$$

where

$$c_1 = -\frac{(\hat{\mathbf{n}}'\cdot\mathbf{H}_0)(\hat{\mathbf{n}}'\cdot\hat{\mathbf{k}}_i)}{|\hat{\mathbf{n}}' \times \hat{\mathbf{k}}_i|} \, , \quad c_2 = -\frac{\eta_0^{-1}(\hat{\mathbf{n}}'\cdot\mathbf{E}_0)(\hat{\mathbf{n}}'\cdot\hat{\mathbf{k}}_i)}{|\hat{\mathbf{n}}' \times \hat{\mathbf{k}}_i|} \; . \tag{3.50}$$

The physical optics approximation is

$$\hat{\mathbf{n}}' \times \mathbf{H} = 2(c_1\Gamma_H\hat{\mathbf{t}}_1 + c_2\Gamma_E\hat{\mathbf{t}}_2)\, e^{ik_0\hat{\mathbf{k}}_i\cdot\mathbf{r}'} , \tag{3.51}$$

where $\Gamma_E$ and $\Gamma_H$ are the reflection coefficients given in (3.34), and in the far zone

$$\begin{aligned}\mathbf{\Pi}_e(\mathbf{r}) = \frac{e^{ik_0r}}{r}&\left\{(\hat{\mathbf{n}}'\cdot\mathbf{E}_0)\Gamma_E\hat{\mathbf{k}}_i\times(\hat{\mathbf{n}}'\times\hat{\mathbf{k}}_i) + (\hat{\mathbf{n}}'\cdot\mathbf{H}_0)\Gamma_H(\hat{\mathbf{n}}'\times\hat{\mathbf{k}}_i)\right\}\\ &\cdot\frac{K(\hat{\mathbf{n}}'\cdot\hat{\mathbf{k}}_i)}{k_0^2|\hat{\mathbf{n}}'\times\hat{\mathbf{k}}_i|^2} .\end{aligned} \tag{3.52}$$

The scattering amplitude is therefore

$$\begin{aligned}\mathbf{S} \;=\; &\left\{(\hat{\mathbf{n}}'\cdot\mathbf{E}_0)\Gamma_E\hat{\mathbf{k}}_s\times\hat{\mathbf{k}}_s\times\left[(\hat{\mathbf{n}}'\cdot\hat{\mathbf{k}}_i)\hat{\mathbf{k}}_i - \hat{\mathbf{n}}'\right]\right.\\ &\left.+(\hat{\mathbf{n}}'\cdot\mathbf{H}_0)\eta_0\Gamma_H\left[(\hat{\mathbf{k}}_s\cdot\hat{\mathbf{k}}_i)(\hat{\mathbf{n}}'\times\hat{\mathbf{k}}_s) + (\hat{\mathbf{n}}'\cdot\hat{\mathbf{k}}_s)(\hat{\mathbf{k}}_s\times\hat{\mathbf{k}}_i)\right]\right\}\frac{\hat{\mathbf{n}}'\cdot\hat{\mathbf{k}}_i}{|\hat{\mathbf{n}}'\times\hat{\mathbf{k}}_i|^2}K,\end{aligned} \tag{3.53}$$

where $K$ is given in (3.39), and the elements of the scattering matrix are

$$\begin{aligned}S_{vv} = &\left\{(\hat{\mathbf{n}}'\cdot\hat{\mathbf{v}}_i)\left[(\hat{\mathbf{n}}'\cdot\hat{\mathbf{v}}_s) - (\hat{\mathbf{n}}'\cdot\hat{\mathbf{k}}_i)(\hat{\mathbf{k}}_i\cdot\hat{\mathbf{v}}_s)\right]\Gamma_E\right.\\ &\left.+(\hat{\mathbf{n}}'\cdot\hat{\mathbf{h}}_i)\left[(\hat{\mathbf{k}}_i\cdot\hat{\mathbf{k}}_s)(\hat{\mathbf{n}}'\cdot\hat{\mathbf{h}}_s) - (\hat{\mathbf{k}}_i\cdot\hat{\mathbf{h}}_s)(\hat{\mathbf{n}}'\cdot\hat{\mathbf{k}}_s)\right]\Gamma_H\right\}\frac{\hat{\mathbf{n}}'\cdot\hat{\mathbf{k}}_i}{|\hat{\mathbf{n}}'\times\hat{\mathbf{k}}_i|^2}K\end{aligned} \tag{3.54}$$

$$\begin{aligned}S_{vh} = &\left\{(\hat{\mathbf{n}}'\cdot\hat{\mathbf{h}}_i)\left[(\hat{\mathbf{n}}'\cdot\hat{\mathbf{v}}_s) - (\hat{\mathbf{n}}'\cdot\hat{\mathbf{k}}_i)(\hat{\mathbf{k}}_i\cdot\hat{\mathbf{v}}_s)\right]\Gamma_E\right.\\ &\left.-(\hat{\mathbf{n}}'\cdot\hat{\mathbf{v}}_i)\left[(\hat{\mathbf{k}}_i\cdot\hat{\mathbf{k}}_s)(\hat{\mathbf{n}}'\cdot\hat{\mathbf{h}}_s) - (\hat{\mathbf{k}}_i\cdot\hat{\mathbf{h}}_s)(\hat{\mathbf{n}}'\cdot\hat{\mathbf{k}}_s)\right]\Gamma_H\right\}\frac{\hat{\mathbf{n}}'\cdot\hat{\mathbf{k}}_i}{|\hat{\mathbf{n}}'\times\hat{\mathbf{k}}_i|^2}K\end{aligned} \tag{3.55}$$

$$\begin{aligned}S_{hv} = &\left\{(\hat{\mathbf{n}}'\cdot\hat{\mathbf{v}}_i)\left[(\hat{\mathbf{n}}'\cdot\hat{\mathbf{h}}_s) - (\hat{\mathbf{n}}'\cdot\hat{\mathbf{k}}_i)(\hat{\mathbf{k}}_i\cdot\hat{\mathbf{h}}_s)\right]\Gamma_E\right.\\ &\left.-(\hat{\mathbf{n}}'\cdot\hat{\mathbf{h}}_i)\left[(\hat{\mathbf{k}}_i\hat{\mathbf{k}}_s)(\hat{\mathbf{n}}'\cdot\hat{\mathbf{v}}_s) - (\hat{\mathbf{k}}_i\cdot\hat{\mathbf{v}}_s)(\hat{\mathbf{n}}'\cdot\hat{\mathbf{k}}_s)\right]\Gamma_H\right\}\frac{\hat{\mathbf{n}}'\cdot\hat{\mathbf{k}}_i}{|\hat{\mathbf{n}}'\times\hat{\mathbf{k}}_i|^2}K\end{aligned} \tag{3.56}$$

$$\begin{aligned}S_{hh} = &\left\{(\hat{\mathbf{n}}'\cdot\hat{\mathbf{h}}_i)\left[(\hat{\mathbf{n}}'\cdot\hat{\mathbf{h}}_s) - (\hat{\mathbf{n}}'\cdot\hat{\mathbf{k}}_i)(\hat{\mathbf{k}}_i\cdot\hat{\mathbf{h}}_s)\right]\Gamma_E\right.\\ &\left.+(\hat{\mathbf{n}}'\cdot\hat{\mathbf{v}}_i)\left[(\hat{\mathbf{k}}_i\cdot\hat{\mathbf{k}}_s)(\hat{\mathbf{n}}'\cdot\hat{\mathbf{v}}_s) - (\hat{\mathbf{k}}_i\cdot\hat{\mathbf{v}}_s)(\hat{\mathbf{n}}'\cdot\hat{\mathbf{k}}_s)\right]\Gamma_H\right\}\frac{\hat{\mathbf{n}}'\cdot\hat{\mathbf{k}}_i}{|\hat{\mathbf{n}}'\times\hat{\mathbf{k}}_i|^2}K .\end{aligned} \tag{3.57}$$

The corresponding scattering cross sections are

$$\sigma_{ij} = 4\pi |S_{ij}|^2 \tag{3.58}$$

with $i, j = v, h$.

In Figs. 3.2 through 3.6, the physical optics expression for the (normalized) backscattering cross section of a rectangular plate having $a = 1.33\lambda_0$ and $b = 2\lambda_0$ is compared with measured or computed data. The latter were obtained by using a moment method solution for an infinite strip of width $a$, modified for a finite plate by assuming that the current is independent of $z'$. In Fig. 3.2, the perfectly conducting plate is oriented with its shorter side parallel to the plane of incidence. We observe that physical optics is in good agreement with the numerical solution over the main lobe of the pattern. Most of the energy is concentrated here, and the discrepancies at wider angles are attributable to traveling wave and edge effects that are not accurately simulated by physical optics. A comparison with measured data is shown in Fig. 3.3 for the same plate with its larger principal ($z'$) axis tilted back 8°. The angle with the plane of incidence is therefore 82°, and again there is good agreement over the main lobe.

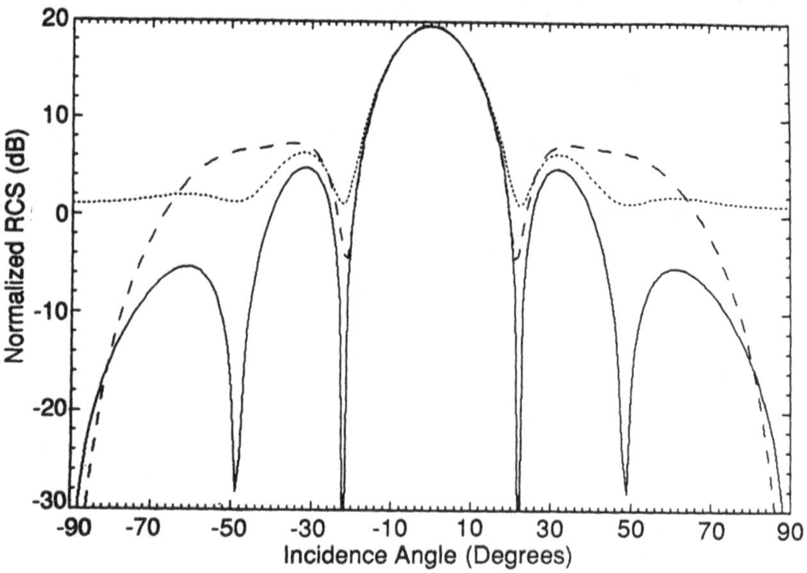


**Figure 3.2** Normalized backscattering cross section $\sigma/\lambda_0^2$ of a rectangular perfectly conducting plate with $a = 1.33\lambda_0$ and $b = 2\lambda_0$: (—) physical optics, (- - - ) moment method E-polarization, (– –) moment method H-polarization.

For a lossy plate, the physical optics expression is polarization-dependent, and because traveling waves and edge effects are less significant, the accuracy is better. In Figs. 3.4 and 3.5, $\sigma/\lambda_0^2$, calculated by using physical optics for a resistive plate having $R = 0.163 + i0.442$, is compared with moment method data for E- and H-polarizations, respectively. The agreement now extends to wider angles. For the tilted plate, a similar comparison with measured data for E-polarization is given in Fig. 3.6, and for completeness we have included here the cross-polarized cross section computed by using physical optics. Note that, because physical optics is a high frequency technique, its accuracy generally improves as the frequency or plate dimensions increase.

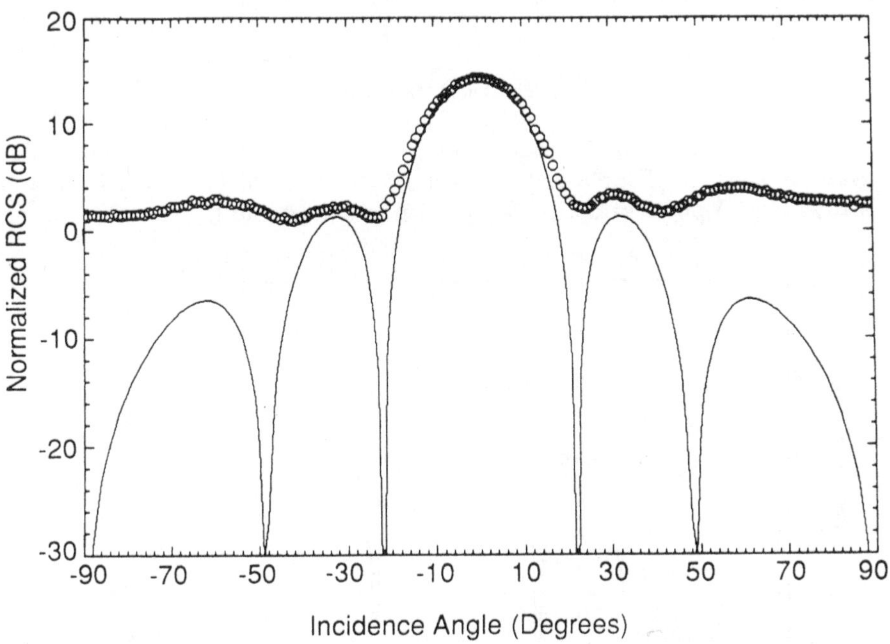


**Figure 3.3** Normalized backscattering cross section $\sigma/\lambda_0^2$ of a rectangular perfectly conducting plate for E-polarization with $a = 1.33\lambda_0$ and $b = 2\lambda_0$ and tilt angle $\beta = 8°$: (—) physical optics, (o o o) measurements.

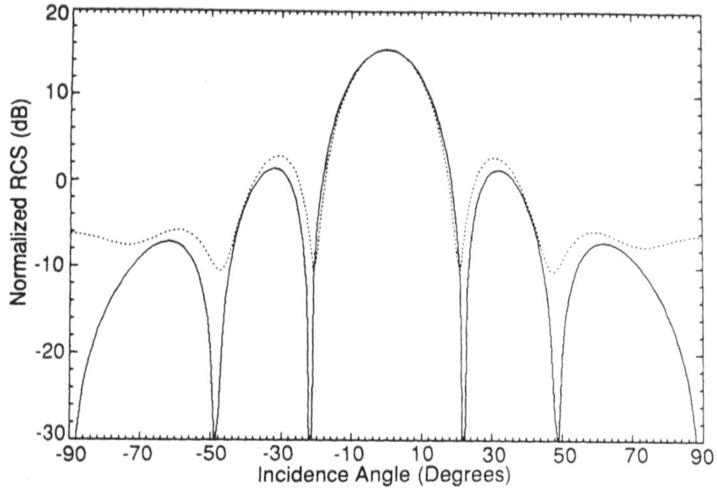


**Figure 3.4** Normalized backscattering cross section $\sigma/\lambda_0^2$ of a rectangular resistive sheet for E-polarization with $R = 0.163 + i0.442, a = 1.33\lambda_0$ and $b = 2\lambda_0$: (—) physical optics, (- - - - -) moment method.

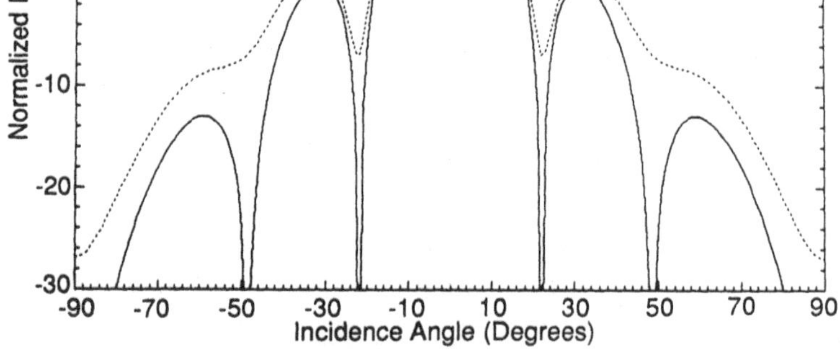


**Figure 3.5** Normalized backscattering cross section $\sigma/\lambda_0^2$ of a rectangular resistive sheet for H-polarization with $R = 0.163 + i0.442, a = 1.33\lambda_0$ and $b = 2\lambda_0$: (—) physical optics, (- - - - -) moment method.

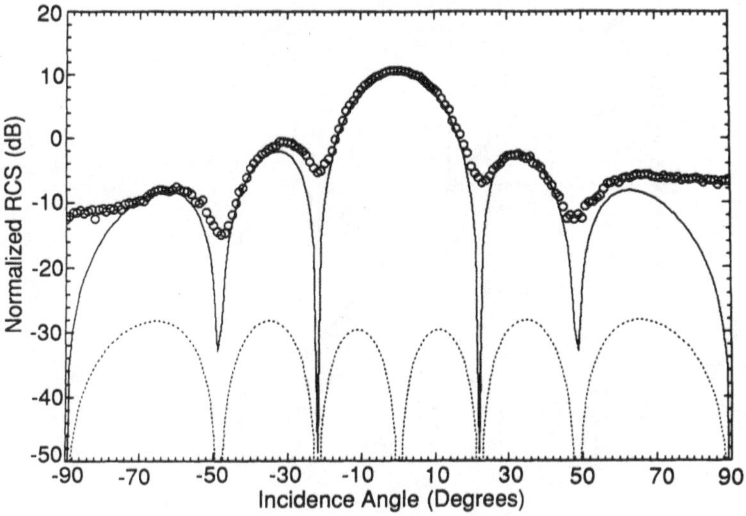


**Figure 3.6** Normalized backscattering cross section $\sigma/\lambda_0^2$ of a rectangular resistive sheet for E-polarization with $R = 0.163 + i0.442, a = 1.33\lambda_0, b = 2\lambda_0$ and tilt angle $\beta = 8°$: (—) like-polarized physical optics, (- - - - -) cross-polarized physical optics, (o o o) measurement.

### 3-2.3 Curved Plates

Most leaves are not planar, and to examine the effect that curvature has on the scattering is of interest. To model this situation, we imagine a rectangular plate given a constant radius of curvature $\rho$ in a principal plane (Sarabandi *et al.*, 1988). As a result of the bending, the plate conforms to a portion of the surface of a right circular cylinder of radius $\rho_1$ as shown in Fig. 3.7.

If the flat plate has width $a$ in the $y'$ direction and length $b$ in the $z'$ direction, $b$ is also the length of the curved section and $a = 2\rho_1\phi_1$. On the surface:

$$\mathbf{r}' = \rho_1 \left[(\cos\phi' - 1)\hat{\mathbf{x}}' + \sin\phi'\,\hat{\mathbf{y}}'\right] + z'\,\hat{\mathbf{z}}' \,, \tag{3.59}$$

where we have chosen the origin at the front of the cylinder. For perfect conductivity, the physical optics approximation to $\mathbf{\Pi}_e(\mathbf{r})$ in the far zone is

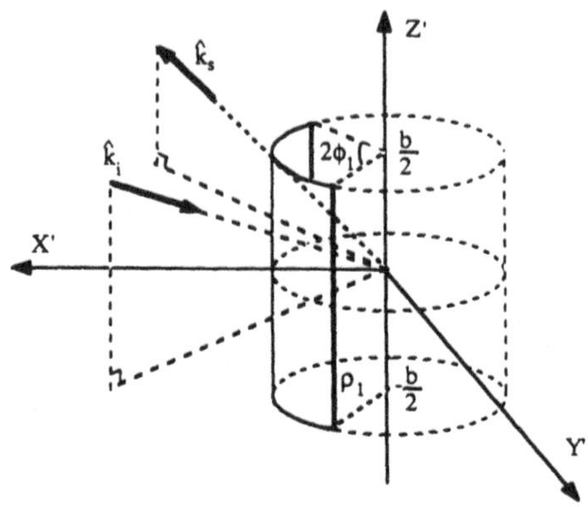

**Figure 3.7** Geometry of a curved plate and the coordinate system.

$$\begin{aligned}\mathbf{\Pi}_e(\mathbf{e}) &= \frac{e^{ik_0 r}}{r} e^{ik_0\rho_1(\hat{\mathbf{k}}_s-\hat{\mathbf{k}}_i)\cdot\hat{\mathbf{x}}'} \frac{i\eta_0\rho_1}{2\pi k_0} \int_{-\frac{b}{2}}^{\frac{b}{2}} \int_{-\phi_1}^{\phi_1} \hat{\mathbf{n}}'\times\mathbf{H}_0 e^{-ik_0 B\rho_1\cos(\phi'-\tilde{\phi})} \\ &\cdot e^{-ik_0(\hat{\mathbf{k}}_s-\hat{\mathbf{k}}_i)\cdot\hat{\mathbf{z}}' z'} d\phi' dz' \end{aligned} \tag{3.60}$$

where

$$\cos\tilde{\phi} = \frac{1}{B}(\hat{\mathbf{k}}_s - \hat{\mathbf{k}}_i)\cdot\hat{\mathbf{x}}' \ , \quad \sin\tilde{\phi} = \frac{1}{B}(\hat{\mathbf{k}}_s - \hat{\mathbf{k}}_i)\cdot\hat{\mathbf{y}}' \tag{3.61}$$

and

$$B = \left\{\left[(\hat{\mathbf{k}}_s - \hat{\mathbf{k}}_i)\cdot\hat{\mathbf{x}}'\right]^2 + \left[(\hat{\mathbf{k}}_s - \hat{\mathbf{k}}_i)\cdot\hat{\mathbf{y}}'\right]^2\right\}^{1/2} . \tag{3.62}$$

To ensure that the outer surface of the plate is completely illuminated, the following is necessary:

$$\hat{\mathbf{k}}_i\cdot\hat{\mathbf{x}}' < -\sin\phi_1 \ . \tag{3.63}$$

The $z'$ integration can be carried out immediately and yields $b\sin V/V$, where $V$ is given in (3.42), and the $\phi'$ integral can be evaluated by using the *stationary phase* (SP) approximation. The SP point is $\phi' = \tilde{\phi}$, and on the assumption that $k_0 B\phi_1 \gg 1$ with $-\phi_1 < \tilde{\phi} < \phi_1$ (so that the SP point lies within the range of integration), the nonexponential portion of the integrand is assigned its value at the SP point. The expression for the scattering amplitude is then

$$\mathbf{S} = e^{ik_0\rho_1\{(\hat{\mathbf{k}}_s-\hat{\mathbf{k}}_i)\cdot\hat{\mathbf{x}}'-B\}}\Big\{(\hat{\mathbf{n}}'\cdot\mathbf{E}_0)\Big[(\hat{\mathbf{k}}_i\cdot\hat{\mathbf{k}}_s)\hat{\mathbf{k}}_s-\hat{\mathbf{k}}_i\Big] -(\hat{\mathbf{n}}'\cdot\hat{\mathbf{k}}_i)\Big[(\hat{\mathbf{k}}_s\cdot\mathbf{E}_0)\hat{\mathbf{k}}_s-\mathbf{E}_0\Big]\Big\}K\,, \tag{3.64}$$

(see (3.38)) with

$$K=-ib\sqrt{\frac{\rho_1}{\pi\lambda_0 B}}\left\{\mathcal{F}\left[\sqrt{\frac{k_0B\rho_1}{2}}(\phi_1+\tilde{\phi})\right]+\mathcal{F}\left[\sqrt{\frac{k_0B\rho_1}{2}}(\phi_1-\tilde{\phi})\right]\right\}\frac{\sin V}{V} \tag{3.65}$$

corresponding to the substitution:

$$a\frac{\sin U}{U}\rightarrow\sqrt{\frac{\rho_1\lambda_0}{\pi B}}\left\{\mathcal{F}\left[\sqrt{\frac{k_0B\rho_1}{2}}(\phi_1+\tilde{\phi})\right]+\mathcal{F}\left[\sqrt{\frac{k_0B\rho_1}{2}}(\phi_1-\tilde{\phi})\right]\right\} \tag{3.66}$$

in the expression (3.41) for a flat plate.

In (3.65),

$$\mathcal{F}(\tau)=\int_0^\tau e^{iu^2}\,du \tag{3.67}$$

is the finite range Fresnel integral (Jahnke and Emde, 1945, p. 36). For $|\tau|\ll 1$

$$\mathcal{F}(\tau)=\tau+O(\tau^3)\,, \tag{3.68}$$

whereas for $|\tau|\gg 1$,

$$\mathcal{F}(\tau)\sim\frac{1}{2}\sqrt{\pi}\,e^{i(\pi/4)}\,. \tag{3.69}$$

The bracketed terms in (3.64) are evaluated at the SP point and, as indicated by the expression (3.61) for $\cos\tilde{\phi}$, the point is located where the bisector of the angle between $\hat{\mathbf{k}}_s$ and $-\hat{\mathbf{k}}_i$ intersects the surface. This is the "specular" point of geometrical (ray) optics, and manifests itself at visible wavelengths as a bright streak parallel to the $z'$ axis on the surface. Close to the forward scattering direction $\hat{\mathbf{k}}_s=\hat{\mathbf{k}}_i$, $k_0B\rho_1$ is no longer large, and a method of evaluation appropriate to this range is discussed in Section 3-2.4.

For a curved resistive plate the analysis is similar, and the result is (see (3.53)):

$$\mathbf{S} = e^{ik_0\rho_1\{(\hat{\mathbf{k}}_s - \hat{\mathbf{k}}_i)\cdot\hat{\mathbf{x}}' - B\}} \left\{(\hat{\mathbf{n}}'\cdot\mathbf{E}_0)\Gamma_E \hat{\mathbf{k}}_s \times \hat{\mathbf{k}}_s \times \left[(\hat{\mathbf{n}}'\cdot\hat{\mathbf{k}}_i)\hat{\mathbf{k}}_i - \hat{\mathbf{n}}'\right]\right.$$

$$\left. + (\hat{\mathbf{n}}'\cdot\hat{\mathbf{h}}_0)\eta_0\Gamma_H \left[(\hat{\mathbf{k}}_s\cdot\hat{\mathbf{k}}_i)(\hat{\mathbf{n}}' \times \hat{\mathbf{k}}_s) + (\hat{\mathbf{n}}'\cdot\hat{\mathbf{k}}_s)(\hat{\mathbf{k}}_s \times \hat{\mathbf{k}}_i)\right]\right\} \frac{\hat{\mathbf{n}}'\cdot\hat{\mathbf{k}}_i}{|\hat{\mathbf{n}}' \times \hat{\mathbf{k}}_i|^2} K \,, \tag{3.70}$$

where the terms in brackets are evaluated at the SP point and $K$ is as shown in (3.65).

It is instructive to examine the special case of backscattering for symmetric normal incidence on the plate. Then $\hat{\mathbf{k}}_s = -\hat{\mathbf{k}}_i = \hat{\mathbf{x}}'$, implying $B = 2$ and $V = 0$, and

$$\mathbf{S} = \Gamma_E \mathbf{E}_0 \frac{ib}{\pi} \sqrt{k_0\rho_1}\, \mathcal{F}\left(\sqrt{k_0\rho_1}\,\phi_1\right) \,. \tag{3.71}$$

Since $\phi_1 = a/(2\rho_1)$, the argument of the Fresnel integral is $a/2\sqrt{k_0/\rho_1}$, and if $\rho_1 \gg k_0(a^2/4)$, the approximation (3.68) gives

$$\mathbf{S} = \frac{iab}{\lambda_0} \Gamma_E \mathbf{E}_0 \tag{3.72}$$

in agreement with the result for a flat plate. To this order, the curvature has no effect. If $\rho_1 \ll k_0(a^2/4)$, however, the surface now encompasses many Fresnel zones, and

$$\mathbf{S} = -\frac{a}{2}\sqrt{\frac{k_0\rho_1}{\pi}}\, \Gamma_E \mathbf{E}_0 e^{-i(\pi/4)} \,. \tag{3.73}$$

To add to the plate, and thereby complete more of the cylinder, would not affect the scattering, and (3.73) is, in fact, the backscattering amplitude for a resistive cylindrical shell of length $b$ and radius $\rho_1$ at normal incidence. Note that the curved plate, and hence the cylindrical shell, must be so lossy that no field penetrates to the far side, and a more general result for a right circular cylinder is given in Section 3-2.5.

For intermediate values of $\rho_1$ (and for all $\rho_1$ if the incidence or scattering direction is not symmetric), the Fresnel integral must be retained, and when the scattering amplitude (3.71) is normalized to its flat plate value (3.72), denoted by the superscript $fp$,

$$\frac{S}{S^{fp}} = \frac{1}{\gamma_1} \mathcal{F}(\gamma_1) \,, \tag{3.74}$$

independent of the resistivity, where

$$\gamma_1 = \frac{a}{2}\sqrt{\frac{k_0}{\rho_1}} \; . \tag{3.75}$$

In Fig. 3.8, $\gamma_1^{-2}\{\mathcal{F}(\gamma_1)\}^2$ is plotted as a function of $\gamma_1$ and compared with the cross section reduction measured using a coleus leaf. The agreement is excellent. The results of a numerical evaluation of the $\phi'$ integral in (3.60) are also included, and confirm the accuracy of the SP approximation.

To examine the effect of curvature in two dimensions, imagine giving the cylindrical plate a constant radius of curvature $\rho_2$ in the plane $y' = 0$. If, for simplicity, we confine attention to backscattering for symmetric normal incidence, the solution can be obtained from (3.65) by making the substitution:

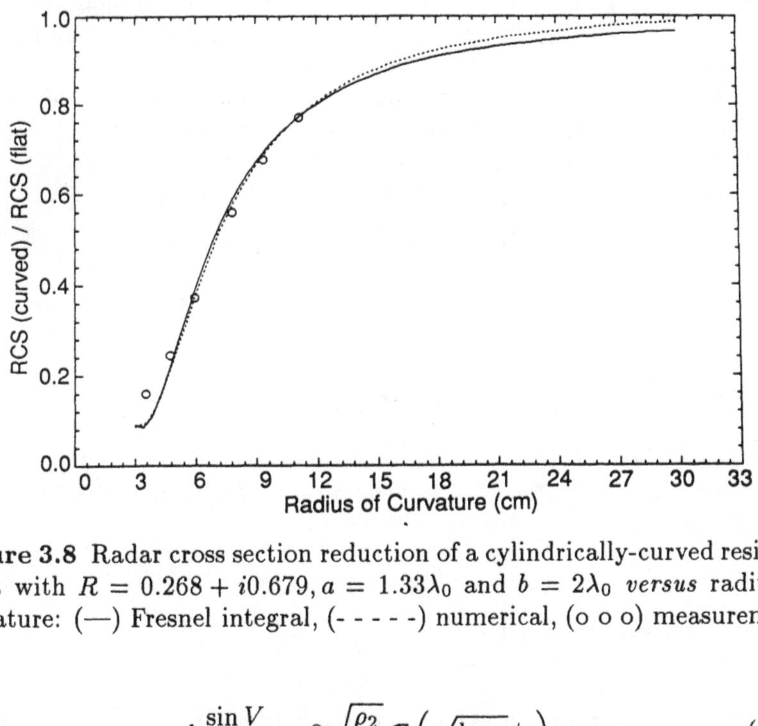


**Figure 3.8** Radar cross section reduction of a cylindrically-curved resistive sheet with $R = 0.268 + i0.679, a = 1.33\lambda_0$ and $b = 2\lambda_0$ *versus* radius of curvature: (—) Fresnel integral, (- - - - -) numerical, (o o o) measurement.

$$b\,\frac{\sin V}{V} \rightarrow 2\sqrt{\frac{\rho_2}{k_0}}\,\mathcal{F}\left(\sqrt{k_0\rho_2}\phi_2\right) \; . \tag{3.76}$$

Compare (3.66), where $\phi_2 = b/(2\rho_2)$. Hence,

$$\mathbf{S} = \frac{2i}{\pi}\sqrt{\rho_1\rho_2}\,\Gamma_E\mathbf{E}_0\mathcal{F}\left(\sqrt{k_0\rho_1}\phi_1\right)\mathcal{F}\left(\sqrt{k_0\rho_2}\phi_2\right) \; . \tag{3.77}$$

If $\rho_2 \gg \phi_2^2/k_0 = k_0(b^2/4)$, the approximation (3.68) applied to the second Fresnel integral reduces (3.77) to (3.71), showing that the curvature in the plane $y' = 0$ has no effect. If $\rho_2 \ll k_0(b^2/4)$, however, (3.68) gives

$$\mathbf{S} = -\sqrt{\frac{\rho_1\rho_2}{\pi}}\Gamma_E \mathbf{E}_0 e^{-i(\pi/4)} \mathcal{F}\left(\sqrt{k_0\rho_1}\phi_1\right) , \tag{3.78}$$

and, if also $\rho_1 \ll k_0(a^2/4)$,

$$\mathbf{S} = -\frac{1}{2}\sqrt{\rho_1\rho_2}\,\Gamma_E \mathbf{E}_0 , \tag{3.79}$$

which is the backscattering amplitude for a resistive spheroidal cap having principal radii of curvature $\rho_1$ and $\rho_2$ at the specular point. The above is also the result for a complete spheroidal shell, provided that the material loss is sufficiently high that no field penetrates to the far side and, in the special case $\rho_1 = \rho_2$, the spheroid becomes a sphere. A more general solution for a dielectric sphere is given in Section 3-2.6.

For intermediate values of $\rho_1$ and $\rho_2$ such that neither of the approximations (3.68) and (3.69) is applicable, the Fresnel integrals in (3.77) must be retained, and then

$$\frac{S}{S^{fp}} = \frac{1}{\gamma_1}\mathcal{F}(\gamma_1)\frac{1}{\lambda_2}\mathcal{F}(\lambda_2) , \tag{3.80}$$

where $S^{fp}$ is the backscattering amplitude (3.72) for a flat plate, and

$$\gamma_2 = \frac{b}{2}\sqrt{\frac{k_0}{\rho_2}} . \tag{3.81}$$

Sarabandi *et al.* (1988) present a comparison of (3.80) with measured values for a coleus leaf, and also generalize (3.77) to backscattering in a principal plane.

### 3-2.4 Convex Bodies

The previous results for a cylinder and sphere were based on the idea of a shell having a conductivity so large that the local reflection process can be simulated by using a resistive sheet supporting only an electric current. Instead of approaching the closed body problem in this manner, we now consider a solid body from the outset, and develop the general physical optics approximation to the scattered field.

For the plane wave (3.15) incident on a dielectric half-space, the reflected wave can be attributed to the surface electric and magnetic currents $\mathbf{J}_e$ and $\mathbf{J}_m$. If $\hat{\mathbf{n}}'$ is the unit vector outward normal to the interface and

$$\hat{\mathbf{t}}' = \frac{\hat{\mathbf{n}}' \times \hat{\mathbf{k}}_i}{|\hat{\mathbf{n}}' \times \hat{\mathbf{k}}_i|} , \tag{3.82}$$

the currents are (see (3.26) and (3.27)):

$$\begin{aligned}\mathbf{J}_e &= -\left\{\eta_0^{-1}(1-R_E)(\mathbf{E}_0\cdot\hat{\mathbf{t}}')(\hat{\mathbf{n}}'\cdot\hat{\mathbf{k}}_i)\hat{\mathbf{t}}' - (1+R_H)(\mathbf{H}_0\cdot\hat{\mathbf{t}}')\hat{\mathbf{n}}'\times\hat{\mathbf{t}}'\right\} e^{ik_0\hat{\mathbf{k}}_i\cdot\mathbf{r}} \\ \mathbf{J}_m &= -\left\{\eta_0(1-R_H)(\mathbf{H}_0\cdot\hat{\mathbf{t}}')(\hat{\mathbf{n}}'\cdot\hat{\mathbf{k}}_i)\hat{\mathbf{t}}' + (1+R_E)(\mathbf{E}_0\cdot\hat{\mathbf{t}}')\hat{\mathbf{n}}'\times\hat{\mathbf{t}}'\right\} e^{ik_0\hat{\mathbf{k}}_i\cdot\mathbf{r}} ,\end{aligned} \tag{3.83}$$

where $R_E$ and $R_H$ are the Fresnel reflection coefficients (3.22). On the assumption of a body with all of its dimensions being large compared with the wavelength, and loss sufficient to prevent any penetration through, the physical optics approximations to the electric and magnetic Hertz vectors in the far zone are

$$\begin{aligned}\mathbf{\Pi}_e(\mathbf{r}) &= \frac{e^{ik_0 r}}{r}\left(-\frac{i}{4\pi k_0}\right)\int_{\text{lit}} \{(1-R_E)(\mathbf{E}_0\cdot\hat{\mathbf{t}}')(\hat{\mathbf{n}}'\cdot\hat{\mathbf{k}}_i)\hat{\mathbf{t}}' - \eta_0(1+R_H) \\ &\quad \cdot (\mathbf{H}_0\cdot\hat{\mathbf{t}}')\hat{\mathbf{n}}'\times\hat{\mathbf{t}}'\Big\} e^{ik_0(\hat{\mathbf{k}}_i-\hat{\mathbf{k}}_s)\cdot\mathbf{r}'} dS' \\ \mathbf{\Pi}_m(\mathbf{r}) &= \frac{e^{ik_0 r}}{r}\left(-\frac{i}{4\pi k_0}\right)\int_{\text{lit}} \{(1-R_H)(\mathbf{H}_0\cdot\hat{\mathbf{t}}')(\hat{\mathbf{n}}'\cdot\hat{\mathbf{k}}_i)\hat{\mathbf{t}}' + \eta_0^{-1}(1+R_E) \\ &\quad \cdot (\mathbf{E}_0\cdot\hat{\mathbf{t}}')\hat{\mathbf{n}}'\times\hat{\mathbf{t}}'\Big\} e^{ik_0(\hat{\mathbf{k}}_i-\hat{\mathbf{k}}_s)\cdot\mathbf{r}'} dS' ,\end{aligned} \tag{3.84}$$

and the scattering amplitude is therefore

$$\begin{aligned}\mathbf{S} = \frac{ik_0}{4\pi}\int_{\text{lit}} &\{(1-R_E)(\mathbf{E}_0\cdot\hat{\mathbf{t}}')(\hat{\mathbf{n}}'\cdot\hat{\mathbf{k}}_i)\hat{\mathbf{k}}_s\times(\hat{\mathbf{k}}_s\times\hat{\mathbf{t}}') \\ &+(1+R_E)(\mathbf{E}_0\cdot\hat{\mathbf{t}}')\hat{\mathbf{k}}_s\times(\hat{\mathbf{n}}'\times\hat{\mathbf{t}}') + \eta_0(1-R_H)(\mathbf{H}_0\cdot\hat{\mathbf{t}}')(\hat{\mathbf{n}}'\cdot\hat{\mathbf{k}}_i)\hat{\mathbf{k}}_s\times\hat{\mathbf{t}}' \\ &-\eta_0(1+R_H)(\mathbf{H}_0\cdot\hat{\mathbf{t}}')\hat{\mathbf{k}}_s\times\hat{\mathbf{k}}_s\times(\hat{\mathbf{n}}'\times\hat{\mathbf{t}}')\Big\} e^{ik_0(\hat{\mathbf{k}}_i-\hat{\mathbf{k}}_s)\cdot\mathbf{r}'} dS' ,\end{aligned} \tag{3.85}$$

where the integration is confined to the illuminated portion of the surface, the outward unit vector normal of which is $\hat{\mathbf{n}}'$.

For any body with local reflection coefficients $R_E$ and $R_H$ that can be determined, the integral can be evaluated, and a convenient method to employ is the stationary phase approximation, which leads to an analytical expression for $\mathbf{S}$. If, for simplicity, we assume that the body is convex, there is a single SP point which is also the specular reflection point of geometrical optics. When the nonexponential portion of the integrand is evaluated at the SP point, the elements of the scattering matrix are found to be

$$S_{vv} = -\left\{R_H(\hat{\mathbf{n}}'\cdot\hat{\mathbf{v}}_i)(\hat{\mathbf{n}}'\cdot\hat{\mathbf{v}}_s) + R_E(\hat{\mathbf{n}}'\cdot\hat{\mathbf{h}}_i)(\hat{\mathbf{n}}'\cdot\hat{\mathbf{h}}_s)\right\}\frac{\hat{\mathbf{n}}'\cdot\hat{\mathbf{k}}_i}{|\hat{\mathbf{n}}'\times\hat{\mathbf{k}}_i|^2}Q \tag{3.86}$$

$$S_{vh} = -\left\{R_H(\hat{\mathbf{n}}'\cdot\hat{\mathbf{h}}_i)(\hat{\mathbf{n}}'\cdot\hat{\mathbf{v}}_s) - R_E(\hat{\mathbf{n}}'\cdot\hat{\mathbf{v}}_i)(\hat{\mathbf{n}}'\cdot\hat{\mathbf{h}}_s)\right\}\frac{\hat{\mathbf{n}}'\cdot\hat{\mathbf{k}}_i}{|\hat{\mathbf{n}}'\times\hat{\mathbf{k}}_i|^2}Q \tag{3.87}$$

$$S_{hv} = -\left\{R_H(\hat{\mathbf{n}}'\cdot\hat{\mathbf{v}}_i)(\hat{\mathbf{n}}'\cdot\hat{\mathbf{h}}_s) - R_E(\hat{\mathbf{n}}'\cdot\hat{\mathbf{h}}_i)(\hat{\mathbf{n}}'\cdot\hat{\mathbf{v}}_s)\right\}\frac{\hat{\mathbf{n}}'\cdot\hat{\mathbf{k}}_i}{|\hat{\mathbf{n}}'\times\hat{\mathbf{k}}_i|^2}Q \tag{3.88}$$

$$S_{hh} = -\left\{R_H(\hat{\mathbf{n}}'\cdot\hat{\mathbf{h}}_i)(\hat{\mathbf{n}}'\cdot\hat{\mathbf{h}}_s) + R_E(\hat{\mathbf{n}}'\cdot\hat{\mathbf{v}}_i)(\hat{\mathbf{n}}'\cdot\hat{\mathbf{v}}_s)\right\}\frac{\hat{\mathbf{n}}'\cdot\hat{\mathbf{k}}_i}{|\hat{\mathbf{n}}'\times\hat{\mathbf{k}}_i|^2}Q\,, \tag{3.89}$$

where we have used (3.82) and (3.21) with $\hat{\mathbf{k}}_r$ replaced by $\hat{\mathbf{k}}_s$ to simplify the expressions, and

$$Q = -\frac{ik_0}{2\pi}\int_{\text{lit}} e^{ik_0(\hat{\mathbf{k}}_i-\hat{\mathbf{k}}_s)\cdot\mathbf{r}'}\,dS'\,. \tag{3.90}$$

The boundary of the lit region is the curve $\hat{\mathbf{n}}'\cdot\hat{\mathbf{k}}_i = 0$ on the surface, and as $\hat{\mathbf{k}}_s$ approaches $\hat{\mathbf{k}}_i$, the specular point approaches the boundary. More importantly, the parameter which is assumed large in the SP approximation tends to zero as $\hat{\mathbf{k}}_s \to \hat{\mathbf{k}}_i$ and the method fails, but an alternative evaluation is now possible. For $\hat{\mathbf{k}}_s = \hat{\mathbf{k}}_i$, $R_E = R_H = -1$ and the terms in braces in the integrand of (3.85) reduce to $-2(\hat{\mathbf{n}}'\cdot\hat{\mathbf{k}}_i)\mathbf{E}_0$, giving

$$\mathbf{S} = -\frac{i}{\lambda_0}\mathbf{E}_0\int_{\text{lit}}\hat{\mathbf{n}}'\cdot\hat{\mathbf{k}}_i\,dS' = \frac{iA_p}{\lambda_0}\mathbf{E}_0\,, \tag{3.91}$$

where $A_p$ is the planar area enclosed by the terminator (i.e., the boundary of the illuminated and shadowed regions on the surface). There is no depolarization and we observe that (3.91) is identical to the result for a flat plate (see (3.44)), having a projected area perpendicular to $\hat{\mathbf{k}}_i$ that is $A_p$.

The above results are valid for any impenetrable convex body of large dimensions with local reflection coefficients $R_E$ and $R_H$ that are known, and can be generalized to a nonconvex body, which may possess more than one SP point. We shall now examine the special cases of a circular cylinder and sphere for which the integral expression for $Q$ can be evaluated analytically.

## 3-2.5 Circular Cylinder

If the body is a right circular cylinder of radius $\rho_1$ and length $b$ illuminated by the plane wave (3.15), to introduce the cylindrical polar coordinates $\rho'$, $\phi'$, $z'$ having the $z'$ axis coincident with the axis of the cylinder is convenient. On the surface $\rho' = \rho_1$,

$$\begin{aligned} \hat{\mathbf{n}}' &= \cos\phi'\,\hat{\mathbf{x}}' + \sin\phi'\,\hat{\mathbf{y}}' \\ \mathbf{r}' &= \rho_1(\cos\phi'\,\hat{\mathbf{x}}' + \sin\phi'\,\hat{\mathbf{y}}') + z'\,\hat{\mathbf{z}}'\ . \end{aligned} \tag{3.92}$$

Hence, if $\phi'$ is measured from the plane defined by $\hat{\mathbf{k}}_i$ and $\hat{\mathbf{z}}$, the illuminated region is $|\phi'| < \pi/2$ with $-b/2 < z' < b/2$. Then,

$$Q = -\frac{ik_0\rho_1}{2\pi}\int_{-\frac{b}{2}}^{\frac{b}{2}}\int_{-\frac{\pi}{2}}^{\frac{\pi}{2}} \mathrm{e}^{-ik_0B\rho_1\cos(\phi'-\tilde{\phi})}\mathrm{e}^{-ik_0(\hat{\mathbf{k}}_s-\hat{\mathbf{k}}_i)\cdot\hat{\mathbf{z}}'z'}\mathrm{d}\phi'\mathrm{d}z' \tag{3.93}$$

where $\tilde{\phi}$ and $B$ are given in (3.61) and (3.62), respectively, with $\hat{\mathbf{k}}_i\cdot\hat{\mathbf{y}}' = 0$. The $z'$ integration can be done immediately, giving $b(\sin V)/V$, where $V$ is defined in (3.42), and provided that $k_0B\rho_1 \gg 1$, the $\phi'$ integral can be evaluated using the SP method. The SP point is $\phi' = \tilde{\phi}$, giving

$$\begin{aligned} Q = -\tfrac{ib}{\pi}\sqrt{\tfrac{k_0\rho_1}{2B}}\mathrm{e}^{-ik_0B\rho_1}\ &\left\{\mathcal{F}\left[\sqrt{\tfrac{k_0B\rho_1}{2}}\left(\tfrac{\pi}{2}+\tilde{\phi}\right)\right]\right. \\ &\left.+\mathcal{F}\left[\sqrt{\tfrac{k_0B\rho_1}{2}}\left(\tfrac{\pi}{2}-\tilde{\phi}\right)\right]\right\}\frac{\sin V}{V}\ , \end{aligned} \tag{3.94}$$

and at aspects for which $\sqrt{(k_0B\rho_1)/2}(\pi/2 \pm \tilde{\phi}) \gg 1$, the expression can be simplified by using (3.69):

$$Q = b\sqrt{\frac{k_0\rho_1}{2B\pi}}\mathrm{e}^{-i(\pi/4)-ik_0B\rho_1}\frac{\sin V}{V}\ . \tag{3.95}$$

The above results fail in the forward scattering direction for which $B = 0$, but in directions close to forward, an alternative approximation is possible. As $\hat{\mathbf{n}}'\cdot\hat{\mathbf{k}}_i = \hat{\mathbf{k}}_i\cdot\hat{\mathbf{x}}'\cos\phi'$,

$$\begin{aligned}
\mathbf{S} &\approx -\frac{ib\rho_1}{\lambda_0}\mathbf{E}_0\frac{\sin V}{V}\int_{-\frac{\pi}{2}}^{-\frac{\pi}{2}} \hat{\mathbf{n}}'\cdot\hat{\mathbf{k}}_i e^{-ik_0\rho_1\left\{(\hat{\mathbf{k}}_s-\hat{\mathbf{k}}_i)\cdot\hat{\mathbf{x}}'\cos\phi'+\hat{\mathbf{k}}_s\cdot\hat{\mathbf{y}}'\sin\phi'\right\}}d\phi' \\
&\approx -\frac{ib\rho_1}{\lambda_0}(\hat{\mathbf{k}}_i\cdot\hat{\mathbf{x}}')\frac{\sin V}{V}\mathbf{E}_0\int_{-\frac{\pi}{2}}^{-\frac{\pi}{2}} \cos\phi' e^{-ik_0\rho_1(\hat{\mathbf{k}}_s\cdot\hat{\mathbf{y}}')\sin\phi'}d\phi' \\
&= -\frac{2ib\rho_1}{\lambda_0}(\hat{\mathbf{k}}_i\cdot\hat{\mathbf{x}}')\frac{\sin V}{V}\frac{\sin W}{W}\mathbf{E}_0 ,
\end{aligned} \tag{3.96}$$

with

$$W = k_0\rho_1(\hat{\mathbf{k}}_s\cdot\hat{\mathbf{y}}') . \tag{3.97}$$

When $\hat{\mathbf{k}}_s = \hat{\mathbf{k}}_i$ implying $\hat{\mathbf{k}}_s\cdot\hat{\mathbf{y}}' = 0$ and, hence, $W = V = 0$, (3.96) reduces to (3.91).

Because of the importance of the solution for a circular cylinder, as is desirable, we express the results in a global coordinate system $x, y, z$, in terms of which

$$\hat{\mathbf{k}}_i = \sin\theta_i\cos\phi_i\,\hat{\mathbf{x}} + \sin\theta_i\,\hat{\mathbf{y}} + \cos\theta_i\,\hat{\mathbf{z}} \tag{3.98}$$

and

$$\mathbf{E}^i = (E_v^i\hat{\mathbf{v}}_i + E_h^i\hat{\mathbf{h}}_i)\,e^{ik_0\hat{\mathbf{k}}_i\cdot\mathbf{r}} \tag{3.99}$$

with

$$\begin{aligned}
\hat{\mathbf{v}}_i &= \cos\theta_i\cos\phi_i\,\hat{\mathbf{x}} + \cos\theta_i\sin\phi_i\,\hat{\mathbf{y}} - \sin\theta_i\,\hat{\mathbf{z}} \\
\hat{\mathbf{h}}_i &= -\sin\phi_i\,\hat{\mathbf{x}} + \cos\phi_i\,\hat{\mathbf{y}} ,
\end{aligned} \tag{3.100}$$

so that

$$\hat{\mathbf{h}}_i = \hat{\mathbf{k}}_i \times \hat{\mathbf{v}}_i , \quad \hat{\mathbf{v}}_i = -\hat{\mathbf{k}}_i \times \hat{\mathbf{h}}_i . \tag{3.101}$$

For the scattered electric field, the same definitions apply with subscript $i$ replaced by $s$.

In terms of the global coordinates, the $z'$ axis of the cylinder is

$$\hat{\mathbf{z}}' = \sin\theta_c\cos\phi_c\,\hat{\mathbf{x}} + \sin\theta_c\sin\phi_c\,\hat{\mathbf{y}} + \cos\theta_c\,\hat{\mathbf{z}} , \tag{3.102}$$

and, hence,

$$\hat{\mathbf{k}}_i = (\hat{\mathbf{k}}_i\cdot\hat{\mathbf{x}}')\hat{\mathbf{x}}' + (\hat{\mathbf{k}}_i\cdot\hat{\mathbf{z}}')\hat{\mathbf{z}}' \tag{3.103}$$

with

$$\hat{\mathbf{k}}_i \cdot \hat{\mathbf{x}}' = -C$$
$$\hat{\mathbf{k}}_i \cdot \hat{\mathbf{z}}' = \cos\theta_c \cos\theta_i + \sin\theta_c \sin\theta_i \cos(\phi_i - \phi_c) \;, \tag{3.104}$$

where

$$C = \left\{1 - \left[\cos\theta_c \cos\theta_i + \sin\theta_c \sin\theta_i \cos(\phi_i - \phi_c)\right]^2\right\}^{1/2} \;. \tag{3.105}$$

In Section 3-4, $\hat{\mathbf{k}}_i \cdot \hat{\mathbf{x}}'$ and $\hat{\mathbf{k}}_i \cdot \hat{\mathbf{z}}'$ are written as $-\sin\beta$ and $\cos\beta$, respectively. Also,

$$\begin{aligned} \hat{\mathbf{v}}_i \cdot \hat{\mathbf{z}}' &= -\cos\theta_c \sin\theta_i + \sin\theta_c \cos\theta_i \cos(\phi_i - \phi_c) \\ \hat{\mathbf{h}}_i \cdot \hat{\mathbf{z}}' &= -\sin\theta_c \sin(\phi_i - \phi_c) \;, \end{aligned} \tag{3.106}$$

and for the scattered field:

$$\begin{aligned} \hat{\mathbf{k}}_s \cdot \hat{\mathbf{x}}' = \tfrac{1}{C} \{ &\sin^2\theta_c \sin\theta_i \sin\theta_s \cos(\phi_i - \phi_c)\cos(\phi_s - \phi_c) + \sin\theta_c \cos\theta_c \\ &\cdot [\sin\theta_i \cos\theta_s \cos(\phi_i - \phi_c) + \cos\theta_i \sin\theta_s \cos(\phi_s - \phi_c)] \\ &- \sin^2\theta_c \cos\theta_i \cos\theta_s - \sin\theta_i \sin\theta_s \cos(\phi_i - \phi_s)\} \end{aligned}$$

$$\begin{aligned} \hat{\mathbf{k}}_s \cdot \hat{\mathbf{y}}' = \tfrac{1}{C} \{ &\cos\theta_c \sin\theta_i \sin\theta_s \sin(\phi_i - \phi_s) + \sin\theta_c \cos\theta_i \sin\theta_s \sin(\phi_s - \phi_c) \\ &- \sin\theta_c \sin\theta_i \cos\theta_s \sin(\phi_i - \phi_c)\} \end{aligned}$$

$$\hat{\mathbf{k}}_s \cdot \hat{\mathbf{z}}' = \cos\theta_c \cos\theta_s + \sin\theta_c \sin\theta_s \cos(\phi_s - \phi_c) \tag{3.107}$$

with

$$\begin{aligned} \hat{\mathbf{v}}_s \cdot \hat{\mathbf{z}}' &= -\cos\theta_c \sin\theta_s + \sin\theta_c \cos\theta_s \cos(\phi_s - \phi_c) \\ \hat{\mathbf{h}}_s \cdot \hat{\mathbf{z}}' &= -\sin\theta_c \sin(\phi_s - \phi_c) \;, \end{aligned} \tag{3.108}$$

and these are required in Section 3-4.

In (3.86) to (3.89) all quantities with the possible exception of $Q$ are evaluated at the stationary phase point such that

$$\hat{\mathbf{k}}_s = \hat{\mathbf{k}}_i - 2(\hat{\mathbf{k}}_i \cdot \hat{\mathbf{n}}')\hat{\mathbf{n}}' \;. \tag{3.109}$$

At the SP point,

$$\hat{\mathbf{n}}' = \cos\tilde{\phi}\,\hat{\mathbf{x}}' + \sin\tilde{\phi}\,\hat{\mathbf{y}}' \,, \tag{3.110}$$

where $\tilde{\phi}$ is defined in (3.61) with $\hat{\mathbf{k}}_i \cdot \hat{\mathbf{y}}' = 0$. Hence,

$$\hat{\mathbf{n}}' = \frac{1}{B}\left\{(\hat{\mathbf{k}}_s - \hat{\mathbf{k}}_i)\cdot\hat{\mathbf{x}}'\hat{\mathbf{x}}' + (\hat{\mathbf{k}}_s \cdot \hat{\mathbf{y}}')\hat{\mathbf{y}}'\right\} \,, \tag{3.111}$$

and we can show that in scattering directions other than forward:

$$B = -2\hat{\mathbf{n}}' \cdot \hat{\mathbf{k}}_i \,. \tag{3.112}$$

Then,

$$\begin{aligned} \hat{\mathbf{n}}' \cdot \hat{\mathbf{v}}_i &= \frac{1}{BC}\left\{\left[(\hat{\mathbf{k}}_s - \hat{\mathbf{k}}_i)\cdot\hat{\mathbf{x}}'\right](\hat{\mathbf{k}}_i \cdot \hat{\mathbf{z}}')(\hat{\mathbf{v}}_i \cdot \hat{\mathbf{z}}') - (\hat{\mathbf{k}}_s \cdot \hat{\mathbf{y}}')(\hat{\mathbf{h}}_i \cdot \hat{\mathbf{z}}')\right\} \\ \hat{\mathbf{n}}' \cdot \hat{\mathbf{h}}_i &= \frac{1}{BC}\left\{\left[(\hat{\mathbf{k}}_s - \hat{\mathbf{k}}_i)\cdot\hat{\mathbf{x}}'\right](\hat{\mathbf{k}}_i \cdot \hat{\mathbf{z}}')(\hat{\mathbf{h}}_i \cdot \hat{\mathbf{z}}') + (\hat{\mathbf{k}}_s \cdot \hat{\mathbf{y}}')(\hat{\mathbf{v}}_i \cdot \hat{\mathbf{z}}')\right\} \end{aligned} \tag{3.113}$$

which can be evaluated by using (3.104) to (3.107), and

$$\begin{aligned} \hat{\mathbf{n}}' \cdot \hat{\mathbf{v}}_s &= \frac{1}{B}\left\{\cos\theta_i \sin\theta_s - \sin\theta_i \cos\theta_s \cos(\phi_i - \phi_s)\right\} \\ \hat{\mathbf{n}}' \cdot \hat{\mathbf{h}}_s &= -\frac{1}{B}\sin\theta_i \sin(\phi_i - \phi_s) \,. \end{aligned} \tag{3.114}$$

Principal plane bistatic scattering patterns for a homogeneous cylinder having $\rho_1 = 2\lambda_0$ and $\epsilon_r = 10 + i5$, computed using the physical optics approximations, are shown in Figs. 3.9 and 3.10 for $E$ (or TM) and $H$ (or TE) polarizations respectively. The quantity plotted is the normalized scattering cross section $\sigma/\sigma_0$, where $\sigma_0 = k_0 ab^2$ is the physical optics value for the backscattering cross section of a perfectly conducting cylinder, and the ratio is therefore independent of the length $b$ of the cylinder. The solid curves were computed by using (3.86) and (3.89) with $Q$ given by (3.95), and we note that the angular dependence (including the Brewster angle effect for $H$-polarization) is attributable to the reflection coefficients $R_E$ and $R_H$. The (common) dashed curves are based on (3.96). Comparison with the exact solution (see Section 3-4.1) shows that the best approximation is obtained by using the near-forward scattering expression (3.96) for the main lobe and (perhaps) the first one or two sidelobes, and (3.86), (3.89), and (3.95) at other angles.

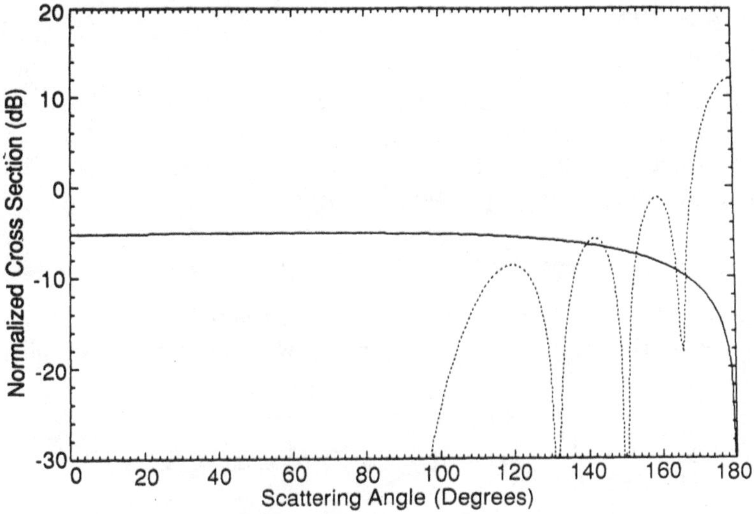


**Figure 3.9** Normalized bistatic scattering cross section $\sigma/k_0ab^2$ of a finite cylinder with radius $\rho_1 = 2\lambda_0$ and $\epsilon_r = 10 + i5$ for E-polarization using physical optics: (—) stationary phase expression, (- - - - -) forward scattering expression.

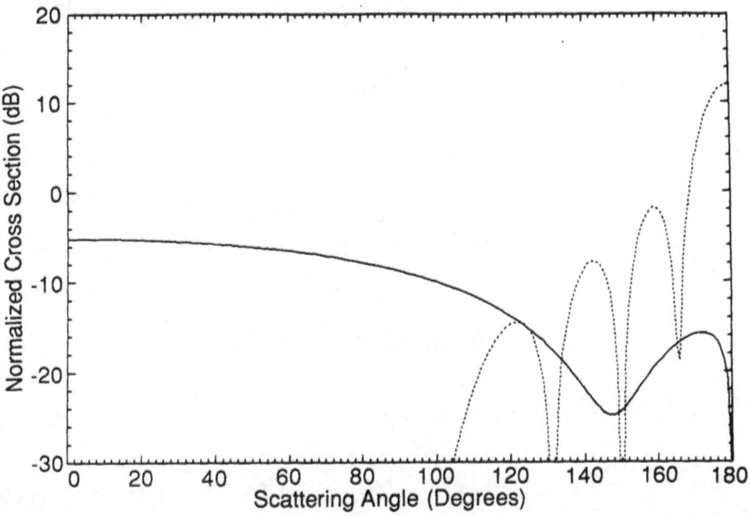


**Figure 3.10** Normalized bistatic scattering cross section $\sigma/k_0ab^2$ of a finite cylinder with radius $\rho_1 = 2\lambda_0$ and $\epsilon_r = 10 + i5$ for H- polarization using physical optics: (—) stationary phase expression, (- - - - -) forward scattering expression.

### 3-2.6 Sphere

The solution for a sphere of a radius $a$ is easy to obtain. Without loss of generality we can choose the local coordinate system $x'$, $y'$, $z'$ such that $\hat{\mathbf{k}}_i = -\hat{\mathbf{z}}'$ with $\hat{\mathbf{k}}_s \cdot \hat{\mathbf{y}}' = 0$, and then introduce the spherical polar coordinates $r'$, $\theta'$, $\phi'$ in terms of which a position vector is

$$\mathbf{r}' = r'(\sin\theta' \cos\phi' \, \hat{\mathbf{x}}' + \sin\theta' \sin\phi' \, \hat{\mathbf{y}}' + \cos\theta' \, \hat{\mathbf{z}}') \ . \tag{3.115}$$

On the surface $r' = a$, $\hat{\mathbf{n}}' = \hat{\mathbf{r}}'$ and the lit region is $0 \leq \phi' < 2\pi$, $0 \leq \theta' < \pi/2$. Conveniently, we can now work directly with the far field amplitude, and from (3.85)

$$\begin{aligned} \mathbf{S} = \ & \frac{ik_0 a^2}{4\pi} \int_0^{2\pi} \int_0^{\pi/2} \Big\{ (1 - R_E)(\mathbf{E}_0 \cdot \hat{\mathbf{t}}')(\hat{\mathbf{n}}' \cdot \hat{\mathbf{k}}_i) \hat{\mathbf{k}}_s \times (\hat{\mathbf{k}}_s \times \hat{\mathbf{t}}') \\ & + (1 + R_E)(\mathbf{E}_0 \cdot \hat{\mathbf{t}}') \hat{\mathbf{k}}_s \times (\hat{\mathbf{n}}' \times \hat{\mathbf{t}}') + \eta_0 (1 - R_H)(\mathbf{H}_0 \cdot \hat{\mathbf{t}}')(\hat{\mathbf{n}}' \cdot \hat{\mathbf{k}}_i) \hat{\mathbf{k}}_s \times \hat{\mathbf{t}}' \\ & - \eta_0 (1 + R_H)(\hat{\mathbf{h}}_0 \cdot \hat{\mathbf{t}}') \hat{\mathbf{k}}_s \times \hat{\mathbf{k}}_s \times (\hat{\mathbf{n}}' \times \hat{\mathbf{t}}') \Big\} \\ & \mathrm{e}^{-ik_0 a \{\cos\theta'(1 + \cos\theta'_s) + \sin\theta'_s \sin\theta' \cos\phi'\}} \sin\theta' \mathrm{d}\theta' \mathrm{d}\phi' \ , \end{aligned} \tag{3.116}$$

where we have chosen the scattering direction as

$$\hat{\mathbf{k}}_s = \sin\theta' \, \hat{\mathbf{x}}' + \cos\theta'_s \, \hat{\mathbf{z}}' \ . \tag{3.117}$$

The SP points for the $\phi'$ and $\theta'$ integrations are 0 and $\theta'_s/2$, respectively, and because $\hat{\mathbf{t}}' = \hat{\mathbf{y}}'$ at the (double) SP point, the bracketted terms in the integrand of (3.116) become

$$-2 \Big\{ R_E (\mathbf{E}_0 \cdot \hat{\mathbf{y}}') \hat{\mathbf{y}}' - \eta_0 R_H (\mathbf{H}_0 \cdot \hat{\mathbf{y}}') \hat{\mathbf{k}}_s \times \hat{\mathbf{y}}' \Big\} \cos\theta' \ . \tag{3.118}$$

Hence, if $k_0 a \ \sin\theta'_s \gg 1$,

$$\mathbf{S} = \frac{a}{2} \Big\{ R_E (\mathbf{E}_0 \cdot \hat{\mathbf{y}}') \hat{\mathbf{y}}' - \eta_0 R_H (\mathbf{H}_0 \cdot \hat{\mathbf{y}}') \hat{\mathbf{k}}_s \times \hat{\mathbf{y}}' \Big\} \mathrm{e}^{-2ik_0 a \cos(\theta'_s/2)} \ . \tag{3.119}$$

In the forward scattering direction $\theta'_s = \pi$, the quantity assumed large in the SP approximation is actually zero and, though it is not evident from (3.119), the formula is no longer valid. To derive an alternative expression which holds in directions close to forward, we first observe that if $\hat{\mathbf{k}}_s = \hat{\mathbf{k}}_i$,

$R_E = R_H = -1$ and (3.118) reduces to $2\mathbf{E}_0 \cos\theta'$. Near forward, the second term in the exponent in (3.116) dominates, and because

$$\int_0^{2\pi} e^{-ik_0 a \sin\theta_s' \sin\theta' \cos\phi'} d\phi' = 2\pi J_0(k_0 a \sin\theta_s' \sin\theta') \ , \tag{3.120}$$

$$\int_0^{2\pi} J_0(k_0 a \sin\theta_s' \sin\theta') \sin\theta' \cos\theta' d\theta' = \frac{J_1(k_0 a \sin\theta_s')}{k_0 a \sin\theta_s'} \ , \tag{3.121}$$

we have

$$\mathbf{S} = ik_0 a^2 \mathbf{E}_0 \frac{J_1(k_0 a \sin\theta_s')}{k_0 a \sin\theta_s'} \tag{3.122}$$

for $\pi - \theta_s'$ small, where $J_1$ is the Bessel function of the first order. We remark that (3.122) is identical to the near-forward scattering pattern of a circular disk of radius $a$ at normal incidence.

# 3-3 LOW FREQUENCY SCATTERING

The antithesis of high frequency scattering is low frequency scattering when the wavelength is large compared with *all* dimensions of the body (i.e., $k_0 D \ll 1$, where $D$ is a typical dimension). The condition can be achieved either by decreasing the frequency for a body of fixed dimensions so that in the limit the frequency is zero, or by imagining the body's size to decrease at a fixed frequency. The latter justifies using the material properties of the body appropriate to the actual frequency, and is the situation of interest here, but for brevity we will still speak about expansion in powers of the small quantity $k_0$.

The first study of the low frequency regime was made by Rayleigh (1897), who showed that when a small perfectly conducting body is illuminated by an electromagnetic wave, the far zone scattered field can be attributed to radiating electric and magnetic dipoles with moments that can be expressed in terms of electrostatic and magnetostatic potentials. In his honor, this is often referred to as *Rayleigh scattering,* and constitutes the leading term in an expansion in powers of $k_0$. The determination of the subsequent terms can also be reduced to problems in potential theory, and methods for doing so were proposed by Stevenson (1953a,b) and refined by Kleinman (1967a, 1978), but for most practical purposes the first term is sufficient.

### 3-3.1 Formulation

Let $S$ be a smooth, finite, closed surface containing a volume $V$, and let $\hat{\mathbf{n}}$ be the unit vector normal to $S$ directed into the surrounding free space region. If the surface is illuminated by the incident plane wave (3.15), the scattering field can be expressed in terms of the Hertz vectors (3.2) with

$$\mathbf{J}_e(\mathbf{r}') = \hat{\mathbf{n}}' \times \mathbf{H} \ , \quad \mathbf{J}_m(\mathbf{r}') = -\hat{\mathbf{n}}' \times \mathbf{E} \ , \tag{3.123}$$

and in the far zone:

$$\begin{aligned} \mathbf{E}^s(\mathbf{r}) = \frac{e^{ik_0 r}}{r}\left(\frac{ik_0}{4\pi}\right)\Big\{\hat{\mathbf{k}}_s \times \int_s \hat{\mathbf{n}}' \times \mathbf{E}(\mathbf{r}')e^{-ik_0\hat{\mathbf{k}}_s\cdot\mathbf{r}'}dS' \\ -\eta_0\hat{\mathbf{k}}_s \times \hat{\mathbf{k}}_s \times \int_s \hat{\mathbf{n}}' \times \mathbf{H}(\mathbf{r}')e^{-ik_0\hat{\mathbf{k}}_s\cdot\mathbf{r}'}dS'\Big\} \ . \end{aligned} \tag{3.124}$$

The scattering amplitude therefore is

$$\begin{aligned} \mathbf{S} \ &= \ \frac{k_0^2}{4\pi}\Big\{\hat{\mathbf{k}}_s \times \hat{\mathbf{k}}_s \times \int_s \mathbf{r}'\left[\eta_0\hat{\mathbf{k}}_s\cdot\hat{\mathbf{n}}' \times \mathbf{H}(\mathbf{r}') - \hat{\mathbf{n}}'\cdot\mathbf{E}(\mathbf{r})\right]e^{-ik_0\hat{\mathbf{k}}_s\cdot\mathbf{r}'}dS' \\ &-\hat{\mathbf{k}}_s \times \int_s \mathbf{r}'\left[\hat{\mathbf{k}}_s\cdot\hat{\mathbf{n}}' \times \mathbf{E}(\mathbf{r}') + \eta_0\hat{\mathbf{n}}'\cdot\mathbf{H}(\mathbf{r}')\right]e^{-ik_0\hat{\mathbf{k}}_s\cdot\mathbf{r}'}dS'\Big\} \ , \end{aligned} \tag{3.125}$$

where we have used the vector identities in Kleinman (1967b). The fields appearing in the integrands can be either total or scattered.

Because the dimensions are electrically small, in the immediate vicinity of $S$

$$\mathbf{E}(\mathbf{r}') = \sum_{n=0}^{\infty}(ik_0)^n\mathbf{E}^n(\mathbf{r}') \ , \quad \mathbf{H}(\mathbf{r}') = \sum_{n=0}^{\infty}(ik_0)^n\mathbf{H}^n(\mathbf{r}') \tag{3.126}$$

with similar expansions for the scattered and incident fields. To the lowest (zero) order:

$$\mathbf{E}^i = \mathbf{E}_0 = \hat{\mathbf{a}}E_0 \ , \quad \mathbf{H}^i = \mathbf{H}_0 = \hat{\mathbf{b}}H_0 \ , \tag{3.127}$$

where $\hat{\mathbf{a}}$ and $\hat{\mathbf{b}}$ are orthogonal unit vectors such that

$$\hat{\mathbf{b}} = \hat{\mathbf{k}}_i \times \hat{\mathbf{a}} \ , \quad \hat{\mathbf{a}}\cdot\hat{\mathbf{k}}_i = \hat{\mathbf{b}}\cdot\hat{\mathbf{k}}_i = 0 \ , \tag{3.128}$$

describing the orientation of the incident electric and magnetic fields. Moreover, for the scattered field:

$$\nabla \times \mathbf{E}_0^s = 0 = \nabla \times \mathbf{H}_0^s \,, \tag{3.129}$$

and thus

$$\mathbf{E}_0^s = -E_0 \nabla \Phi^s \,, \ \mathbf{H}_0^s = -H_0 \nabla \Psi^s \,, \tag{3.130}$$

where $\Phi^s$ and $\Psi^s$ are electrostatic and magnetostatic potentials, satisfying

$$\nabla^2 \Phi^s = 0 = \nabla^2 \Psi^s \tag{3.131}$$

in the exterior region. For the incident field, the corresponding potentials are

$$\Phi^i = -\hat{\mathbf{a}} \cdot \mathbf{r} + c \,, \quad \Psi^i = -\hat{\mathbf{b}} \cdot \mathbf{r} + d \,, \tag{3.132}$$

where $c$ and $d$ are arbitrary constants.

To the lowest order in $k_0$, the scattering amplitude is

$$\begin{aligned} \mathbf{S} \;=\; & \frac{k_0^2}{4\pi} \left\{ \hat{\mathbf{k}}_s \times \hat{\mathbf{k}}_s \times \int_s \mathbf{r}' \left[ -\eta_0 \hat{\mathbf{k}}_s \cdot \hat{\mathbf{n}}' \times \nabla'(H_0 \Psi^s) + E_0 \frac{\partial \Phi^s}{\partial n'} \right] \mathrm{d}S' \right. \\ & \left. + \hat{\mathbf{k}}_s \times \int_s \mathbf{r}' \left[ \hat{\mathbf{k}}_s \cdot \hat{\mathbf{n}}' \times \nabla'(E_0 \Phi^s) + \eta_0 H_0 \frac{\partial \Psi^s}{\partial n'} \right] \mathrm{d}S' \right\} \end{aligned} \tag{3.133}$$

and because (Jones, 1964; p. 531):

$$\int_s \mathbf{r}' \hat{\mathbf{k}}_s \cdot \hat{\mathbf{n}}' \times \nabla' \Psi^s dS' = -\hat{\mathbf{k}}_s \times \int_s \hat{\mathbf{n}}' \Psi^s \mathrm{d}S' \,, \tag{3.134}$$

with a similar result for the potential $\Phi^s$:

$$\mathbf{S} = -\frac{k_0^2}{4\pi} \left\{ \frac{1}{\epsilon_0} \hat{\mathbf{k}}_s \times \hat{\mathbf{k}}_s \times \mathbf{p} + \eta_0 \hat{\mathbf{k}}_s \times \mathbf{m} \right\} \,, \tag{3.135}$$

where $\mathbf{p}$ and $\mathbf{m}$ are electric and magnetic dipoles with moments that are

$$\begin{aligned} \mathbf{p} &= \epsilon_0 E_0 \int_s \left\{ \hat{\mathbf{n}}' \Phi^s - \mathbf{r}' \frac{\partial \Phi^s}{\partial n'} \right\} \mathrm{d}S' \\ \mathbf{m} &= H_0 \int_s \left\{ \hat{\mathbf{n}}' \Psi^s - \mathbf{r}' \frac{\partial \Psi^s}{\partial n'} \right\} \mathrm{d}S' \,, \end{aligned} \tag{3.136}$$

respectively. As is evident, the moments have dimensions (length cubed), and when presenting data, we conveniently normalize $\mathbf{p}$ and $\mathbf{m}$ with respect to the volume $V$ of the body.

In view of (3.132), the dependence of the dipole moments on $\hat{\mathbf{a}}$ and $\hat{\mathbf{b}}$ can be made explicit by writing

$$\Phi^s = \sum_{i=1}^{3} (\hat{\mathbf{a}}\cdot\hat{\mathbf{x}}_i)\Phi_i^s \ , \quad \Psi^s = \sum_{i=1}^{3} (\hat{\mathbf{b}}\cdot\hat{\mathbf{x}}_i)\Psi_i^s \ , \tag{3.137}$$

where $x_i, i = 1, 2, 3$, are Cartesian coordinates. The boundary values of $\Phi_i^s$ and $\Psi_i^s$, and hence the potentials themselves, are now independent of $\hat{\mathbf{a}}$ and $\hat{\mathbf{b}}$. We can therefore write [1]

$$\mathbf{p} = \epsilon_0 \mathcal{P}\cdot\hat{\mathbf{a}}E_0 \ , \tag{3.138}$$

where $\mathcal{P}$ is the electric polarizability tensor, the elements of which

$$P_{ij} = \int_s \left(\hat{\mathbf{n}}'\cdot\hat{\mathbf{x}}_i\Phi_j^s - x_i'\frac{\partial \Phi_j^s}{\partial n'}\right) \mathrm{d}S' \tag{3.139}$$

are functions only of the geometry and relative permittivity of the body. The elements are real if the permittivity is real. Similarly, the magnetic dipole moment can be written as

$$\mathbf{m} = -\mathcal{M}\cdot\hat{\mathbf{b}}H_0 \tag{3.140}$$

where $\mathcal{M}$ is the magnetic polarizability tensor with elements

$$M_{ij} = -\int_s \left(\hat{\mathbf{n}}'\cdot\hat{\mathbf{x}}_i\Psi_j^s - x_i'\frac{\partial \Psi_j^s}{\partial n'}\right) \mathrm{d}S' \ . \tag{3.141}$$

These are functions only of the geometry and relative permeability of the body, and are real if the permeability is real.

The computation of the scattered field is now reduced to the determination of the polarizability tensors. From (3.135), (3.138), and (3.140),

$$\begin{aligned} \mathbf{S}\cdot\hat{\mathbf{a}} &= \tfrac{k_0^2}{4\pi}\left\{\hat{\mathbf{a}}\cdot\mathcal{P}\cdot\hat{\mathbf{a}} - (\hat{\mathbf{k}}_s \times \hat{\mathbf{a}})\cdot\mathcal{M}\cdot\hat{\mathbf{b}}\right\} E_0 \\ \mathbf{S}\cdot\hat{\mathbf{b}} &= \tfrac{k_0^2}{4\pi}\left\{\hat{\mathbf{b}}\cdot\mathcal{P}\cdot\hat{\mathbf{a}} - (\hat{\mathbf{k}}_s \times \hat{\mathbf{b}})\cdot\mathcal{M}\cdot\hat{\mathbf{b}}\right\} E_0 \ , \end{aligned} \tag{3.142}$$

showing that the scattering cross sections (including $\sigma_T$) are $O(k_0^4)$. In contrast, the extinction cross section is

$$\sigma_{\mathrm{EXT}} = k_0 \,\mathrm{Im}\left\{\hat{\mathbf{a}}\cdot\mathcal{P}\cdot\hat{\mathbf{a}} - \hat{\mathbf{b}}\cdot\mathcal{M}\cdot\hat{\mathbf{b}}\right\} \ , \tag{3.143}$$

and absorption therefore dominates the scattering at low frequencies.

[1] To avoid confusion between tensors and vectors, the former are represented by upper-case script letters.

### 3-3.2 Homogeneous Dielectric Bodies

If the homogeneous dielectric is nonmagnetic, as we shall assume, $\mu = \mu_0$ and the body is invisible to a static magnetic field. Hence, $\Psi^s = 0$ implying $\mathbf{m} = 0$, and the scattered field can be attributed to an electric dipole alone.

The determination of the potential $\Phi_j^s$ is an example of a transition problem in electrostatics. If $\Phi_j^{int}$ is the corresponding interior potential, the conditions satisfied by $\Phi_j^s$ and $\Phi_j^{int}$ at the surface $S$ are

$$\begin{aligned} \Phi_j^s &= \Phi_j^{int} + x_j - c_j \\ \frac{\partial \Phi_j^s}{\partial n} &= \epsilon_r \frac{\partial \Phi_j^{int}}{\partial n} + \hat{\mathbf{n}} \cdot \hat{\mathbf{x}}_j \ , \end{aligned} \tag{3.144}$$

and these satisfy the zero induced charge condition

$$\int_s \frac{\partial \Phi_j^s}{\partial n'} = 0 \tag{3.145}$$

with the constant $c_j$ set equal to zero.

From (3.139), we may appear to require a knowledge of both $\Phi_j^s$ and $\partial \Phi_j^s / \partial n'$ on $S$ to calculate $P_{ij}$, but this is not in fact so. As shown by Kleinman and Senior (1986), an alternative expression for $P_{ij}$ is

$$P_{ij} = (1 - \epsilon_r) \int_s \hat{\mathbf{n}}' \cdot \hat{\mathbf{x}}_i \Phi_j^t \, \mathrm{d}S' \tag{3.146}$$

where $\Phi_j^t = \Phi_j^s - x_j$ is the total potential outside $S$, and yet another form is

$$P_{ij} = \frac{1 - \epsilon_r}{\epsilon_r} \int_s x_i' \frac{\partial \Phi_j^t}{\partial n'} \, \mathrm{d}S' \ . \tag{3.147}$$

Moreover, for real $\epsilon_r > 0$,

$$\epsilon_r - 1 \geq \frac{P_{ii}}{V} \geq \frac{\epsilon_r - 1}{\epsilon_r} \ , \tag{3.148}$$

and the upper and lower bounds are achieved by thin prolate and oblate spheroids, respectively. Obviously, if $\epsilon_r$ is close to unity, these bound $P_{ii}$ extremely closely, and whatever geometry dependence then exists must be minimal.

Because $P_{ij} = P_{ji}$, $\mathcal{P}$ is a symmetric tensor which is real if $\epsilon_r$ is real, and has at most six independent elements. If $S$ is symmetric about (for example) the $x_1 = 0$ plane, the only nonzero elements are $P_{11}$ and $P_{ij}$ for

$i, j = 2, 3$, and of these no more than four are independent. If, in addition, the body is symmetric about a second perpendicular plane, which we take to be the plane $x_2 = 0$, then $P_{23} = P_{32} = 0$. The tensor is now diagonal and has at most three nonzero elements. In general, all three are distinct, but in the special case of rotational symmetry about the $x_3$ axis, $P_{22} = P_{11}$.

A body for which analytical expressions for the tensor elements are available is the ellipsoid. If the ellipsoid is defined as

$$\frac{x_1^2}{a_1^2} + \frac{x_2^2}{a_2^2} + \frac{x_3^2}{a_3^2} = 1 \tag{3.149}$$

so that the principal axes coincide with the coordinate axes $x_i$, then the tensor is diagonal, and

$$\frac{P_{ii}(\epsilon_r)}{V} = \frac{\epsilon_r - 1}{L_i\epsilon_r + 1 - L_i}, \tag{3.150}$$

where

$$L_1 = \frac{1}{2}a_1a_2a_3 \int_0^\infty (u + a_1^2)^{-3/2}(u + a_2^2)^{-1/2}(u + a_3^2)^{-1/2}\, du \tag{3.151}$$

with $L_2$ and $L_3$ given by cyclical changes of the suffices. In addition,

$$L_1 + L_2 + L_3 = 1 \tag{3.152}$$

and some computed values of the $L_i$ are quoted by van de Hulst (1957, p. 72). We can show that

$$\left(\frac{1}{P_{11}} + \frac{1}{P_{22}} + \frac{1}{P_{33}}\right) V = \frac{\epsilon_r + 2}{\epsilon_r - 1}. \tag{3.153}$$

A special case of an ellipsoid is a spheroid, prolate or oblate. If the $x_3$ axis is chosen as the axis of symmetry, $L_2 = L_1$ so that $P_{22} = P_{11}$, and the tensor elements can be written as ratios of Legendre functions of the first and second kinds. On inserting the expressions for these functions, the results for a prolate spheroid are found to be

$$\begin{aligned} P_{11}(\epsilon_r) &= -2\left\{\tfrac{1}{2}\zeta(\zeta^2 - 1)\ln\tfrac{\zeta+1}{\zeta-1} - \zeta^2 - \frac{2}{\epsilon_r - 1}\right\}^{-1} V \\ P_{33}(\epsilon_r) &= \left\{\tfrac{1}{2}\zeta(\zeta^2 - 1)\ln\tfrac{\zeta+1}{\zeta-1} - \zeta^2 - \frac{\epsilon_r}{\epsilon_r - 1}\right\}^{-1} V \end{aligned} \tag{3.154}$$

where $\zeta(1 \leq \zeta \leq \infty)$ is the radial spheroidal variable, $2d$ is the interfocal distance, and

$$V = \frac{4}{3}\pi d^3 \zeta(\zeta^2 - 1) \tag{3.155}$$

is the volume. In terms of $\zeta$, the length-to-width ratio of the body is

$$\frac{l}{w} = \zeta(\zeta^2 - 1)^{-1/2}\,, \tag{3.156}$$

ranging from infinity for a long thin spheroid or "rod" ($\zeta = 1$) down to unity for a sphere ($\zeta = \infty$). The analogous results for an oblate spheroid can be obtained by replacing $d$ by $-id$ and $\zeta$ by $i\zeta$, where now $0 \leq \zeta \leq \infty$. The length-to-width ratio is then

$$\frac{l}{w} = \zeta(\zeta^2 + 1)^{-1/2} \tag{3.157}$$

and varies from unity for a sphere ($\zeta = \infty$) down to zero for a "disk" ($\zeta = 0$).

Figures 3.11 and 3.12 show $P_{11}(\epsilon_r)$ and $P_{33}(\epsilon_r)$ as functions of $l/w$, $0.1 \leq l/w \leq 10$, computed using (3.154) for a variety of real $\epsilon_r \geq 0$. For any given $\epsilon_r > 1$, $P_{11}/V$ and $P_{33}/V$ are decreasing and increasing functions respectively of $l/w$. The identity

$$P_{33}(\tau) = -\frac{1}{2}P_{11}\left(\frac{2}{\tau} - 1\right) \tag{3.158}$$

holding for any spheroid allows us to deduce the values of $P_{11}$ and $P_{33}$ for $\epsilon_r < 1$ from those of $P_{33}$ and $P_{11}$ for the corresponding $\epsilon_r > 1$. The simple analytical expressions for $P_{11}/V$ and $P_{33}/V$ for a disk, a sphere and a rod are listed in Table 3.1.

**Table 3.1** Polarizability Tensor Elements

| Shape | $l/w$ | $P_{11}/V$ | $P_{33}/V$ |
|---|---|---|---|
| disk | 0 | $\epsilon_r - 1$ | $(\epsilon_r - 1)/\epsilon_r$ |
| sphere | 1 | $3(\epsilon_r - 1)/(\epsilon_r + 2)$ | $3(\epsilon_r - 1)/(\epsilon_r + 2)$ |
| rod | $\infty$ | $2(\epsilon_r - 1)/(\epsilon_r + 1)$ | $\epsilon_r - 1$ |

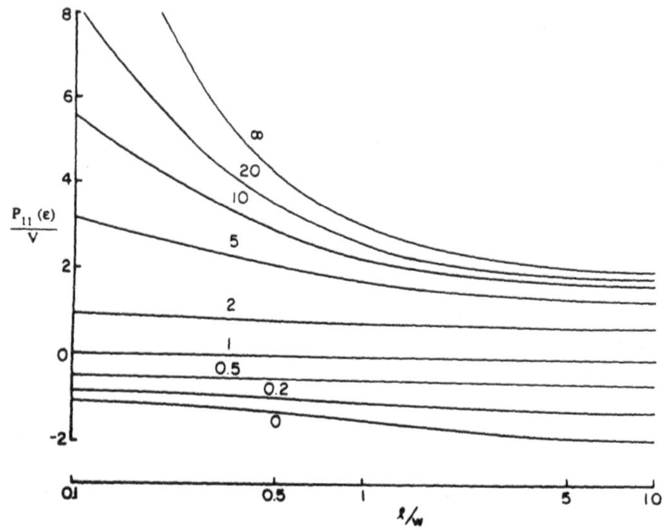


**Figure 3.11** $P_{11}(\epsilon_r)/V$ for a spheroid as a function of the length-to-width ratio $l/w$ for different $\epsilon_r$.

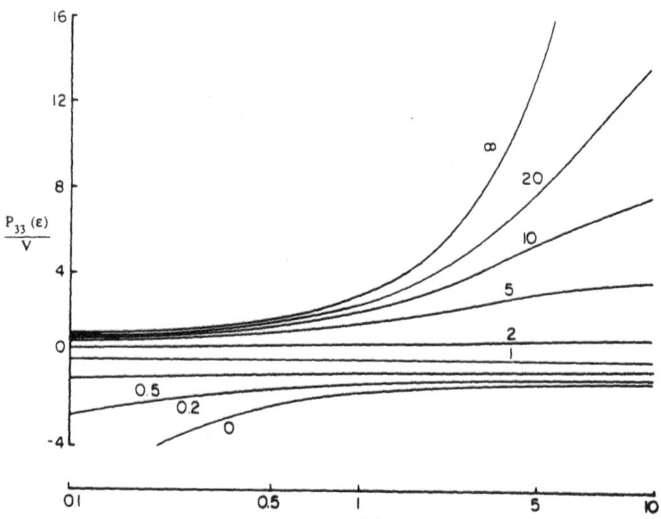


**Figure 3.12** $P_{33}(\epsilon_r)/V$ for a spheroid as a function of the length-to-width ratio $l/w$ for different $\epsilon_r$.

For other homogeneous dielectric bodies, computer programs have been written to solve the integral equations for $\Phi_j^s$ by the moment method, and hence compute the tensor elements. One such program, designated DIELCOM (Senior and Willis, 1982), is applicable to any body of revolution. Concave (re-entrant) shapes can be treated as well as multiply-connected ones (e.g., a torus). Bodies having two or more disjoint portions can also be considered, provided their permittivities are the same and they have a common axis of symmetry. The program can therefore be used to examine the effect of body shape on scattering and absorption, and a similar program is available (Herrick and Senior, 1977) for rectangular parallelepipeds.

Data for a variety of shapes are presented by Senior and Weil (1982), who discuss the validity of modeling Rayleigh scatterers by spheroids. They show that for a convex symmetric body having $\epsilon_r' > 0$ a spheroid of the same length-to-width ratio can be used to approximate the real and imaginary parts of $P_{11}$ and $P_{33}$ with an accuracy of a few percent, and the manner in which the absorption depends on body shape is described by Senior (1980). If the body is not symmetric, a spheroid still gives results which are accurate enough for most practical purposes, but, for concave or multiply-connected shapes that do not even superficially resemble a spheroid, a spheroid having the same $l/w$ is generally inadequate. The situation is very different if $\epsilon_r' < 0$. Both the scattering and absorption can now show extreme dependence on the body's shape, and no such modeling is possible. Many crystalline materials have permittivities with real parts that are negative at frequencies in the infrared and optical ranges, and some of the phenomena which then occur are discussed by Weil *et al.* (1985).

For any body of revolution about the $x_3$ axis, $\mathcal{P} = P_{11}\mathcal{I}+(P_{33}-P_{11})\hat{\mathbf{x}}_3\hat{\mathbf{x}}_3$ where $\mathcal{I}$ is the identity tensor, and the elements of the scattering matrix are

$$\begin{aligned}
S_{vv} &= \frac{k_0^2}{4\pi}\Big\{P_{11}\hat{\mathbf{v}}_i\cdot\hat{\mathbf{v}}_s + (P_{33}-P_{11})(\hat{\mathbf{v}}_i\cdot\hat{\mathbf{x}}_3)(\hat{\mathbf{v}}_s\cdot\hat{\mathbf{x}}_3)\Big\} \\
S_{vh} &= \frac{k_0^2}{4\pi}\Big\{P_{11}\hat{\mathbf{h}}_i\cdot\hat{\mathbf{v}}_s + (P_{33}-P_{11})(\hat{\mathbf{h}}_i\cdot\hat{\mathbf{x}}_3)(\hat{\mathbf{v}}_s\cdot\hat{\mathbf{x}}_3)\Big\} \\
S_{hv} &= \frac{k_0^2}{4\pi}\Big\{P_{11}\hat{\mathbf{v}}_i\cdot\hat{\mathbf{h}}_s + (P_{33}-P_{11})(\hat{\mathbf{v}}_i\cdot\hat{\mathbf{x}}_3)(\hat{\mathbf{h}}_s\cdot\hat{\mathbf{x}}_3)\Big\} \\
S_{hh} &= \frac{k_0^2}{4\pi}\Big\{P_{11}\hat{\mathbf{h}}_i\cdot\hat{\mathbf{h}}_s + (P_{33}-P_{11})(\hat{\mathbf{h}}_i\cdot\hat{\mathbf{x}}_3)(\hat{\mathbf{h}}_s\cdot\hat{\mathbf{x}}_3)\Big\}.
\end{aligned} \tag{3.159}$$

### 3-3.3 Perfectly Conducting Bodies and Plates

At the surface of a perfect conductor,

$$\hat{\mathbf{n}} \times \mathbf{E} = 0 \; , \quad \hat{\mathbf{n}} \cdot \mathbf{H} = 0 \; , \tag{3.160}$$

where $(\mathbf{E}, \mathbf{H})$ is the total external field, and in the static limit when $\mathbf{E}$ and $\mathbf{H}$ are no longer coupled, the two conditions are independent. The boundary conditions for the scattered field potentials on $S$ are therefore

$$\Phi_j^s = x_j - c_j \quad \text{and} \quad \frac{\partial \Psi_j^s}{\partial n} = \hat{\mathbf{n}} \cdot \hat{\mathbf{x}}_j \; , \tag{3.161}$$

and the latter implies that $\mu = 0$ in the static limit for a perfect conductor. As a consequence, $\mathbf{m} \neq 0$.

The results previously obtained for a homogeneous dielectric body can be used to derive the tensor elements in the present case. Thus,

$$P_{ij} = P_{ij}(\infty) \; , \tag{3.162}$$

and from a comparison of (3.139) and (3.141),

$$M_{ij} = -P_{ij}(0) \; . \tag{3.163}$$

From Table 3.1, for a thin oblate spheroid (or "disk"), we thus have

$$\frac{P_{11}}{V} = \infty \; , \quad \frac{P_{33}}{V} = 1 \; , \quad \frac{M_{11}}{V} = 1 \; , \quad \frac{M_{33}}{V} = \infty \; ;$$

for a sphere:

$$\frac{P_{11}}{V} = \frac{P_{33}}{V} = 3 \; , \quad \frac{M_{11}}{V} = \frac{M_{33}}{V} = \frac{3}{2} \; ;$$

and for a thin prolate spheroid (or "rod"),

$$\frac{P_{11}}{V} = 2 \; , \quad \frac{P_{33}}{V} = \infty \; , \quad \frac{M_{11}}{V} = 2 \; , \quad \frac{M_{33}}{V} = 1 \; ;$$

where the infinities are attributable to the vanishing volume. For spheroids with other length-to-width ratios, the tensor elements can be found from the curves corresponding to $\epsilon_r = 0$ and $\infty$ in Figs. 3.11 and 3.12.

More generally, for any body of revolution about the $x_3$ axis,

$$M_{33} = \frac{1}{2} P_{11} \; , \tag{3.164}$$

and of the six nonzero tensor elements $P_{ii}$ and $M_{ii}$, $i = 1,2,3$, only $P_{11}$, $P_{33}$ and $M_{11}$ are independent. Programs are available to solve the integral equations for $\Phi_1^s$, $\Phi_3^s$, and $\Psi_1^s$, by the moment method, and to compute the elements. Data for a variety of shapes are presented by Kleinman and Senior (1986). In addition, there are a few geometries such as a hemisphere and a hemispherical bowl for which analytical expressions are known, and these are also cited. For convex bodies, at least, the normalized elements are very similar to those for a spheroid with the same length-to-width ratio.

In the case of an infinitesimally thin plate lying in the plane $x_3 = 0$,

$$P_{ij} = -\int_s x_i' \frac{\partial \Phi_j^s}{\partial n'} \Big|_-^+ \mathrm{d}S' \tag{3.165}$$

where $|\pm$ denotes the discontinuity across the plate. Hence, $P_{3j} = 0$, and because the tensor is symmetric, it has at most three independent elements. Also, $\Psi_i^s(\mathbf{r}) = 0$ unless $i = 3$, and the magnetic polarizability has only the single nonzero element:

$$M_{33} = -\int_S \Psi_3^s \Big|_-^+ \mathrm{d}S' \,. \tag{3.166}$$

One of the few plates for which analytical expressions are available is the elliptical disk. If the equation of the perimeter is

$$\frac{x_1^2}{a^2} + \frac{x_2^2}{b^2} = 1 \tag{3.167}$$

and $e = (1 - b^2/a^2)^{1/2}$ is the ellipticity, we have

$$\begin{aligned} P_{11} &= \tfrac{4}{3}\pi a^3 e^2 \left\{K(e^2) - E(e^2)\right\}^{-1} \\ P_{22} &= \tfrac{4}{3}\pi a^3 e^2 \left\{(1-e^2)^{-1}E(e^2) - K(e^2)\right\}^{-1} \\ M_{33} &= \tfrac{4}{3}\pi a^3 (1-e^2)\{E(e^2)\}^{-1} \end{aligned} \tag{3.168}$$

with $P_{ij} = 0$, $i \neq j$, where $K(e^2)$ and $E(e^2)$ are the complete elliptic integrals of the first and second kinds, respectively, as defined by Abramowitz and Stegun (1964, pp. 608–609). We note that

$$\frac{1}{M_{33}} = \frac{1}{P_{11}} + \frac{1}{P_{22}} \,. \tag{3.169}$$

In the particular case of zero ellipticity when the disk is circular,

$$P_{11} = P_{22} = \frac{16}{3}a^3 \,, \quad M_{33} = \frac{8}{3}a^3 \,, \tag{3.170}$$

where $a$ is the radius of the disk. In place of the volume $V$ used to normalize the tensor elements for a solid body, it is now customary to use $A^{3/2}$, where $A$ is the area. For plates other than those listed above, the accuracy with which the normalized elements can be estimated using the values for an elliptical disk is discussed by Kleinman and Senior (1986).

# 3-4 EXACT SOLUTIONS

The Rayleigh contribution is only the leading term in the low frequency expansion of the scattered field in powers of $k_0$, and although we can increase the frequency coverage by adding more terms, the radius of convergence of the series is close to the *beginning* of the resonance region (Darling and Senior, 1965). We can attempt to penetrate the region from above by using high frequency techniques which are more sophisticated than physical optics, and include edge contributions, nonlocal (e.g., surface wave) effects, and the interactions between them, but even in cases where this is feasible, the accuracy achieved is difficult to quantify. One way to do so is by application to a simple body for which the exact solution is known, and we now consider two such bodies: a circular cylinder of length $b \gg \lambda_0$ and a sphere. In the former case, the solution is exact only as regards the transverse resonances, and assumes that the currents borne by the curved surface are the same as if the cylinder were infinite in length.

## 3-4.1 Circular Cylinder

An infinite right circular cylinder of radius $\rho_1$ is composed of a homogeneous nonmagnetic dielectric and is oriented with its axis coincident with the $z'$ axis of a local Cartesian coordinate system $x'$, $y'$, $z'$. The cylinder is illuminated by a plane wave of arbitrary polarization, the propagation vector $\hat{\mathbf{k}}_i$ of which lies in the $x'$-$z'$ plane with

$$\hat{\mathbf{k}}_i \cdot \hat{\mathbf{x}}' = -\sin\beta\ , \quad \hat{\mathbf{k}}_i \cdot \hat{\mathbf{z}}' = \cos\beta \tag{3.171}$$

(see (3.104)), and we choose

$$\begin{aligned} \mathbf{E}^i \cdot \hat{\mathbf{z}}' &= e_z\, \mathrm{e}^{ik_0(-x'\sin\beta + z'\cos\beta)} \\ \mathbf{H}^i \cdot \hat{\mathbf{z}}' &= \eta_0^{-1} h_z\, \mathrm{e}^{ik_0(-x'\sin\beta + z'\cos\beta)}\ , \end{aligned} \tag{3.172}$$

from which the other field components can be found. By expanding the incident and scattered fields in terms of Bessel functions, we can show that at the surface $\rho = \rho_1$ the tangential components of the total field are

$$
\begin{aligned}
E_z &= \sum_{m=-\infty}^{+\infty} (-i)^m \left\{ e_z J_m(x_0) + A_m H_m^{(1)}(x_0) \right\} e^{im\phi' + ik_0 z' \cos\beta} \\
H_z &= \eta_0^{-1} \sum_{m=-\infty}^{+\infty} (-i)^m \left\{ h_z J_m(x_0) + B_m H_m^{(1)}(x_0) \right\} e^{im\phi' + ik_0 z' \cos\beta}
\end{aligned}
\tag{3.173}
$$

$$
\begin{aligned}
E_\phi &= \frac{\eta_0^{-1}}{ik_0 \sin^2\beta} \sum_{m=-\infty}^{\infty} (-i)^m \Bigg\{ k_0 \sin\beta \left[ h_z J_m'(x_0) + B_m H_m^{(1)\,\prime}(x_0) \right] \\
&\quad - \frac{im\cos\beta}{\rho_1} \left[ e_z J_m(x_0) + A_m H_m^{(1)}(x_0) \right] \Bigg\} e^{im\phi' + ik_0 z' \cos\beta} \\
H_\phi &= \frac{\eta_0^{-1}}{ik_0 \sin^2\beta} \sum_{m=-\infty}^{\infty} (-i)^m \Bigg\{ -k_0 \sin\beta \left[ e_z J_m'(x_0) + A_m H_m^{(1)\,\prime}(x_0) \right] \\
&\quad - \frac{im\cos\beta}{\rho_1} \left[ h_z J_m(x_0) + B_m H_m^{(1)}(x_0) \right] \Bigg\} e^{im\phi' + ik_0 z' \cos\beta} ,
\end{aligned}
\tag{3.174}
$$

where $J_m$ and $H_m^{(1)}$ respectively are the Bessel and Hankel functions of order $m$, a prime attached to a function denotes the derivative, and $x_0 = k_0\rho_1 \sin\beta$. When the boundary conditions at $\rho = \rho_1$ are imposed, we obtain (Ruck *et al.*, 1970, pp. 273–274):

$$
\begin{aligned}
A_m &= C_m^{TM} e_z + i\overline{C}_m h_z \\
B_m &= C_m^{TE} h_z - i\overline{C}_m e_z ,
\end{aligned}
\tag{3.175}
$$

where

$$C_m^{TM} = -\frac{V_m P_m - q_m^2 J_m(x_0) H_m^{(1)}(x_0) J_m^2(x_1)}{P_m N_m - \left[q_m H_m^{(1)}(x_0) J_m(x_1)\right]^2}$$

$$C_m^{TE} = -\frac{M_m N_m - q_m^2 J_m(x_0) H_m^{(1)}(x_0) J_m^2(x_1)}{P_m N_m - \left[q_m H_m^{(1)}(x_0) J_m(x_1)\right]^2} \tag{3.176}$$

$$\overline{C}_m = i\frac{2}{\pi x_0 \sin\beta} \frac{q_m J_m^2(x_1)}{P_m N_m - \left[q_m H_m^{(1)}(x_0) J_m(x_1)\right]^2}$$

with

$$x_1 = k_0 \rho_1 \sqrt{\epsilon_r - \cos^2\beta}$$

$$q_m = m k_0 \rho_1 \cos\beta \left(\frac{1}{x_1^2} - \frac{1}{x_0^2}\right)$$

$$V_m = k_0 \rho_1 \left\{\frac{\epsilon_r}{x_1} J_m(x_0) J_m'(x_1) - \frac{1}{x_0} J_m'(x_0) J_m(x_1)\right\}$$

$$P_m = k_0 \rho_1 \left\{\frac{1}{x_1} H_m^{(1)}(x_0) J_m'(x_1) - \frac{1}{x_0} H_m^{(1)\prime}(x_0) J_m(x_1)\right\} \tag{3.177}$$

$$N_m = k_0 \rho_1 \left\{\frac{\epsilon_r}{x_1} H_m^{(1)}(x_0) J_m'(x_1) - \frac{1}{x_0} H_m^{(1)\prime}(x_0) J_m(x_1)\right\}$$

$$M_m = k_0 \rho_1 \left\{\frac{1}{x_1} J_m(x_0) J_m'(x_1) - \frac{1}{x_0} J_m'(x_0) J_m(x_1)\right\} .$$

The surface currents are therefore

$$\begin{aligned} \mathbf{J}_e = \; & \eta_0^{-1}(\sin\phi' \, \hat{\mathbf{x}}' - \cos\phi' \, \hat{\mathbf{y}}') \\ & \cdot \sum_{m=-\infty}^{+\infty} (-i)^m \{h_z J_m(x_0) + B_m H_m^{(1)}(x_0)\} e^{im\phi' + ik_0 z' \cos\beta} \\ & - \frac{\eta_0^{-1}}{ik_0 \sin^2\beta} \hat{\mathbf{z}}' \sum_{m=-\infty}^{+\infty} (-i)^m \left\{ k_0 \sin\beta \left[e_z J_m'(x_0) + A_m H_m^{(1)\prime}(x_0)\right] \right. \\ & \left. + \frac{im\cos\beta}{\rho_1} \left[h_z J_m(x_0) + B_m H_m^{(1)}(x_0)\right] \right\} e^{im\phi' + ik_0 z' \cos\beta} , \end{aligned} \tag{3.178}$$

$$\begin{aligned}\mathbf{J}_m = & -(\sin\phi'\,\hat{\mathbf{x}}' - \cos\phi'\,\hat{\mathbf{y}}') \\ & \cdot \sum_{m=-\infty}^{+\infty} (-i)^m \{e_z J_m(x_0) + A_m H_m^{(1)}(x_0)\} e^{im\phi' + ik_0 z' \cos\beta} \\ & - \frac{\eta_0^{-1}}{ik_0 \sin^2\beta} \hat{\mathbf{z}}' \sum_{m=-\infty}^{+\infty} (-i)^m \left\{ k_0 \sin\beta \left[ h_z J_m'(x_0) + B_m H_m^{(1)\prime}(x_0) \right] \right. \\ & \left. - \frac{im\cos\beta}{\rho_1} \left[ e_z J_m(x_0) + A_m H_m^{(1)\prime}(x_0) \right] \right\} e^{im\phi' + ik_0 z' \cos\beta} .\end{aligned} \tag{3.179}$$

For a cylinder of finite length $b$, $-b/2 < z' < b/2$, we assume that the currents on the curved sides are the same as those on the infinite cylinder. If the effect of the end caps is ignored, the Hertz vectors describing the scattered field are

$$\mathbf{\Pi}_e(\mathbf{r}) = \frac{e^{ik_0 r}}{r} \left( \frac{i\eta_0}{4\pi k_0} \right) \int_{-\frac{b}{2}}^{\frac{b}{2}} \int_0^{2\pi} \mathbf{J}_e \, e^{-ik_0 \rho_1 B \cos(\phi' - \tilde{\phi})} e^{ik_0 (\hat{\mathbf{k}}_i - \hat{\mathbf{k}}_s)\cdot\hat{\mathbf{z}}' z'} \rho_1 \, d\phi' \, dz'$$

$$\mathbf{\Pi}_m(\mathbf{r}) = \frac{e^{ik_0 r}}{r} \left( \frac{i\eta_0^{-1}}{4\pi k_0} \right) \int_{-\frac{b}{2}}^{\frac{b}{2}} \int_0^{2\pi} \mathbf{J}_m \, e^{-ik_0 \rho_1 B \cos(\phi' - \tilde{\phi})} e^{ik_0 (\hat{\mathbf{k}}_i - \hat{\mathbf{k}}_s)\cdot\hat{\mathbf{z}}' z'} \rho_1 \, d\phi' \, dz' \tag{3.180}$$

where $\tilde{\phi}$ and $B$ are given in (3.61) and (3.62) respectively with $\hat{\mathbf{k}}_i \cdot \hat{\mathbf{y}}' = 0$. The integrations in (3.180) can be carried out exactly. The $z'$ integration produces $b(\sin V)/V$, where $V$ is defined in (3.42), and

$$\begin{aligned}\int_0^{2\pi} e^{-ik_0\rho_1 B\cos(\phi'-\tilde{\phi})} e^{im\phi'} \, d\phi' &= 2\pi(-i)^m J_m(y_0) e^{im\tilde{\phi}} \\ \int_0^{2\pi} \cos\phi' e^{-ik_0\rho_1 B\cos(\phi'-\tilde{\phi})} e^{im\phi'} \, d\phi' &= 2\pi(-i)^m \left\{ i\cos\tilde{\phi} J_m'(y_0) \right. \\ & \left. + \sin\tilde{\phi} \frac{m}{y_0} J_m(y_0) \right\} e^{im\tilde{\phi}} \\ \int_0^{2\pi} \sin\phi' e^{-ik_0\rho_1 B\cos(\phi'-\tilde{\phi})} e^{im\phi'} \, d\phi' &= 2\pi(-i)^m \left\{ i\sin\tilde{\phi} J_m'(y_0) \right. \\ & \left. - \cos\tilde{\phi} \frac{m}{y_0} J_m(y_0) \right\} e^{im\tilde{\phi}}\end{aligned} \tag{3.181}$$

where $y_0 = k_0\rho_1 B$. Hence, the scattering amplitude is

$$\mathbf{S} = \frac{k_0 \rho_1 b}{2\sin\beta} \left\{ (D^e e_z + i\overline{D}^e h_z) \hat{\mathbf{k}}_s \times (\hat{\mathbf{k}}_s \times \hat{\mathbf{z}}') + (D^h h_z - i\overline{D}^h e_z) \hat{\mathbf{k}}_s \times \hat{\mathbf{z}}' \right\} \frac{\sin V}{V} \tag{3.182}$$

with

$$D^e = \sum_{m=-\infty}^{+\infty} (-1)^m \left\{ J'_m(x_0)J_m(y_0) - \frac{\sin\beta}{B} J_m(x_0)J'_m(y_0) \right.$$
$$+ C_m^{TM} \left[ H_m^{(1)\prime}(x_0)J_m(y_0) - \frac{\sin\beta}{B} H_m^{(1)}(x_0)J'_m(y_0) \right]$$
$$\left. + \frac{m\cos\beta}{x_0}\left(1 - \frac{\hat{\mathbf{k}}_s\cdot\hat{\mathbf{z}}'}{\cos\beta}\frac{x_0\sin\beta}{y_0 B}\right) \overline{C}_m H_m^{(1)}(x_0)J_m(y_0) \right\} e^{im\tilde{\phi}}$$

$$D^h = \sum_{m=-\infty}^{+\infty} (-1)^m \left\{ J'_m(x_0)J_m(y_0) - \frac{\sin\beta}{B} J_m(x_0)J'_m(y_0) \right.$$
$$+ C_m^{TE} \left[ H_m^{(1)\prime}(x_0)J_m(y_0) - \frac{\sin\beta}{B} H_m^{(1)}(x_0)J'_m(y_0) \right]$$
$$\left. + \frac{m\cos\beta}{x_0}\left(1 - \frac{\hat{\mathbf{k}}_s\cdot\hat{\mathbf{z}}'}{\cos\beta}\frac{x_0\sin\beta}{y_0 B}\right) \overline{C}_m H_m^{(1)}(x_0)J_m(y_0) \right\} e^{im\tilde{\phi}}$$

$$\overline{D}^e = \sum_{m=-\infty}^{+\infty} (-1)^m \left\{ \overline{C}_m \left[ H_m^{(1)\prime}(x_0)J_m(y_0) - \frac{\sin\beta}{B} H_m^{(1)}(x_0)J'_m(y_0) \right] \right.$$
$$\left. + \frac{m\cos\beta}{x_0}\left(1 - \frac{\hat{\mathbf{k}}_s\cdot\hat{\mathbf{z}}'}{\cos\beta}\frac{x_0\sin\beta}{y_0 B}\right) \left[ J_m(x_0) + C_m^{TE} H_m^{(1)}(x_0) \right] J_m(y_0) \right\} e^{im\tilde{\phi}}$$

$$\overline{D}^h = \sum_{m=-\infty}^{+\infty} (-1)^m \left\{ \overline{C}_m \left[ H_m^{(1)\prime}(x_0)J_m(y_0) - \frac{\sin\beta}{B} H_m^{(1)}(x_0)J'_m(y_0) \right] \right.$$
$$\left. + \frac{m\cos\beta}{x_0}\left(1 - \frac{\hat{\mathbf{k}}_s\cdot\hat{\mathbf{z}}'}{\cos\beta}\frac{x_0\sin\beta}{y_0 B}\right) \left[ J_m(x_0) + C_m^{TM} H_m^{(1)}(x_0) \right] J_m(y_0) \right\} e^{im\tilde{\phi}}, \tag{3.183}$$

and the elements of the scattering matrix are

$$S_{vv} = -\left\{ \left[ D^e(\hat{\mathbf{v}}_i\cdot\hat{\mathbf{z}}') + i\overline{D}^e(\hat{\mathbf{h}}_i\cdot\hat{\mathbf{z}}') \right] (\hat{\mathbf{v}}_s\cdot\hat{\mathbf{z}}') + \left[ D_h(\hat{\mathbf{h}}_i\cdot\hat{\mathbf{z}}') \right.\right.$$
$$\left.\left. - i\overline{D}^h(\hat{\mathbf{v}}_i\cdot\hat{\mathbf{z}}') \right] (\hat{\mathbf{h}}_s\cdot\hat{\mathbf{z}}') \right\} \frac{k_0\rho_1 b}{2\sin\beta}\frac{\sin V}{V} \tag{3.184}$$

$$S_{vh} = -\left\{ \left[ D^e(\hat{\mathbf{h}}_i\cdot\hat{\mathbf{z}}') - i\overline{D}^e(\hat{\mathbf{v}}_i\cdot\hat{\mathbf{z}}') \right] (\hat{\mathbf{v}}_s\cdot\hat{\mathbf{z}}') - \left[ D_h(\hat{\mathbf{v}}_i\cdot\hat{\mathbf{z}}') \right.\right.$$
$$\left.\left. + i\overline{D}^h(\hat{\mathbf{h}}_i\cdot\hat{\mathbf{z}}') \right] (\hat{\mathbf{h}}_s\cdot\hat{\mathbf{z}}') \right\} \frac{k_0\rho_1 b}{2\sin\beta}\frac{\sin V}{V} \tag{3.185}$$

$$S_{hv} = -\Big\{\Big[D^e(\hat{\mathbf{v}}_i\cdot\hat{\mathbf{z}}') + i\overline{D}^e(\hat{\mathbf{h}}_i\cdot\hat{\mathbf{z}}')\Big](\hat{\mathbf{h}}_s\cdot\hat{\mathbf{z}}') - \Big[D^h(\hat{\mathbf{h}}_i\cdot\hat{\mathbf{z}}') - i\overline{D}^h(\hat{\mathbf{v}}_i\cdot\hat{\mathbf{z}}')\Big](\hat{\mathbf{v}}_s\cdot\hat{\mathbf{z}}')\Big\}\frac{k_0\rho_1 b}{2\sin\beta}\frac{\sin V}{V} \tag{3.186}$$

$$S_{hh} = -\Big\{\Big[D^e(\hat{\mathbf{h}}_i\cdot\hat{\mathbf{z}}') - i\overline{D}^e(\hat{\mathbf{v}}_i\cdot\hat{\mathbf{z}}')\Big](\hat{\mathbf{h}}_s\cdot\hat{\mathbf{z}}') + \Big[D^h(\hat{\mathbf{v}}_i\cdot\hat{\mathbf{z}}') + i\overline{D}^h(\hat{\mathbf{h}}_i\cdot\hat{\mathbf{z}}')\Big](\hat{\mathbf{v}}_s\cdot\hat{\mathbf{z}}')\Big\}\frac{k_0\rho_1 b}{2\sin\beta}\frac{\sin V}{V}\,. \tag{3.187}$$

These can be expressed in terms of the global coordinate system defined in Section 3-2.5 by using (3.106) and (3.108).

The above results differ from, and are potentially more accurate than, those given by Ruck *et al.* (1970). In effect, these authors determine the far field of the infinite cylinder and then account for the finite length $b$ by multiplying by a factor proportional to $b(\sin V)/V$. Because the field of the infinite cylinder exists only on the forward scattering cone having $\hat{\mathbf{k}}_s\cdot\hat{\mathbf{z}}' = \hat{\mathbf{k}}_i\cdot\hat{\mathbf{z}}' = \cos\beta$, the polarization characteristics are those appropriate to these aspects alone. On the cone $B = \sin\beta$, so that $y_0 = x_0$. Equation (3.182) then reduces to

$$\mathbf{S} = -\frac{ib}{\pi\sin^2\beta}\frac{\sin V}{V}\sum_{m=-\infty}^{\infty}(-1)^m\Big\{A_m\hat{\mathbf{k}}_s\times(\hat{\mathbf{k}}_s\times\hat{\mathbf{z}}') + B_m\hat{\mathbf{k}}_s\times\hat{\mathbf{z}}'\Big\}e^{im\phi}\,, \tag{3.188}$$

that is

$$\mathbf{S} = -\frac{ib}{\pi\sin^2\beta}\frac{\sin V}{V}\sum_{m=-\infty}^{\infty}(-1)^m\Big\{-\cos\beta A_m\hat{\rho} + B_m\hat{\phi} + \sin\beta A_m\hat{\mathbf{z}}\Big\}e^{im\phi}\,, \tag{3.189}$$

and the sum is proportional to the far zone scattered electric field in the two-dimensional case.

The improved accuracy that (3.182) provides, however, is limited. Because the effect of the cylinder caps is ignored and the termination of the curved sides is not correctly modeled by simply truncating the current, the accuracy of (3.182) and, hence, the scattering matrix elements, deteriorates with increasing $\hat{\mathbf{k}}_i\cdot\hat{\mathbf{z}}'$ or $\hat{\mathbf{k}}_s\cdot\hat{\mathbf{z}}'$, but for aspects that are not too oblique, the transverse resonances of the cylinder should be accurately simulated.

From a physical point of view, the scattering from a dielectric cylinder of radius comparable to or greater than the wavelength can be attributed to two types of mechanism: ray reflections from the surface and from multiple internal bounces, for example, the glory ray; and surface waves which propagate around the cylinder, shedding energy as they go. The latter are the main source of the field in the geometric shadow of perfectly conducting and very lossy cylinders. For a lossless cylinder, the internally-reflected rays produce a backscattered field which is an extremely complicated function of frequency and is almost noise-like, even for the largest values of $k_0\rho_1$. If the cylinder is lossy, however, these resonances diminish as $k_0\rho_1$ or the loss tangent increases. This is illustrated in Figs. 3.13 and 3.14 where the normalized backscattering cross section $\sigma/(k_0\rho_1 b^2)$ is plotted as a function of $k_0\rho_1$ for $E$ (or $TM$) or $H$ (or $TE$) polarizations, respectively, for $\epsilon_r = 10 + i\epsilon_r''$ with $\epsilon_r'' = 1, 2$, and 5. In Figs. 3.15 and 3.16 the results for the most lossy cylinder are compared with the physical optics prediction, and the agreement is excellent for $k_0\rho_1 \geq 3$. For this same material, the bistatic scattering patterns for a cylinder of radius $\rho_1 = 2\lambda_0$ are compared in Figs. 3.17 and 3.18. The physical optics results were shown previously in Figs. 3.9 and 3.10, and provide a good approximation to the exact data at most angles of scatter.

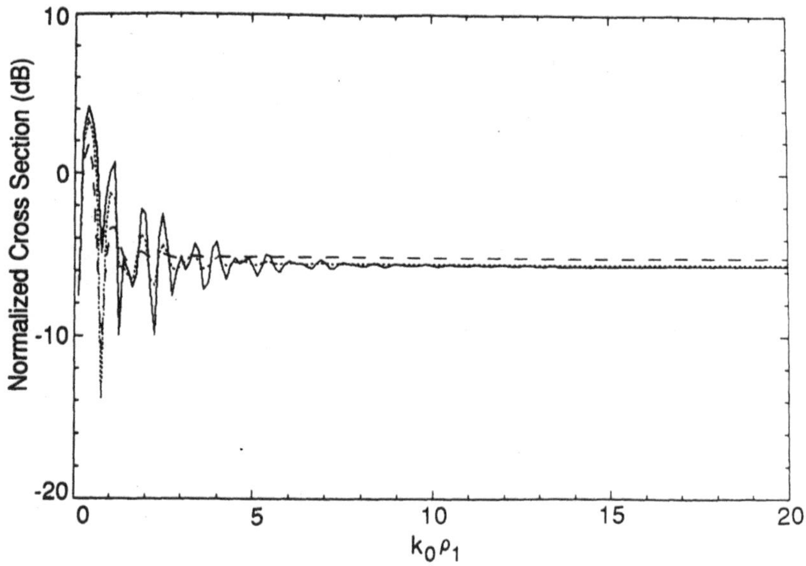


**Figure 3.13** Normalized backscattering cross section $\sigma/(k_0\rho_1 b^2)$ of a circular cylinder *versus* $k_0\rho_1$ for E-polarization and different loss tangents: (—) $\epsilon_r = 10 + i1$, (- - - - -) $\epsilon_r = 10 + i2$, (– –) $\epsilon_r = 10 + i5$.

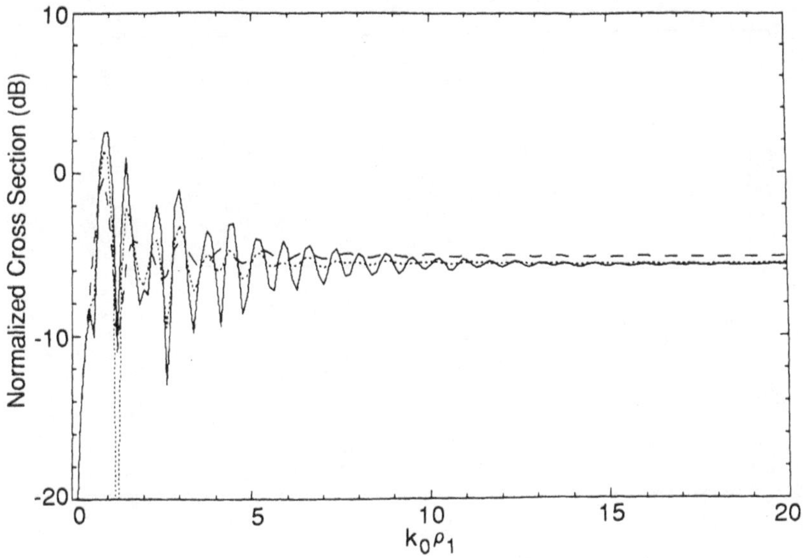


**Figure 3.14** Normalized backscattering cross section $\sigma/(k_0\rho_1 b^2)$ of a circular cylinder *versus* $k_0\rho_1$ for H-polarization and different loss tangents: (—) $\epsilon_r = 10 + i1$, (- - - - -) $\epsilon_r = 10 + i2$, (– –) $\epsilon_r = 10 + i5$.

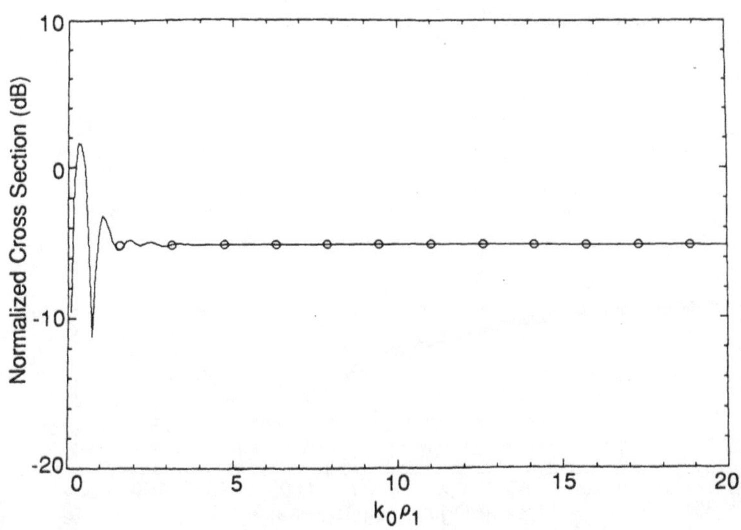


**Figure 3.15** Normalized backscattering cross section $\sigma/(k_0\rho_1 b^2)$ of a circular cylinder with $\epsilon_r = 10 + i5$ *versus* $k_0\rho_1$ for E-polarization: (—) exact solution, (o o o) physical optics.

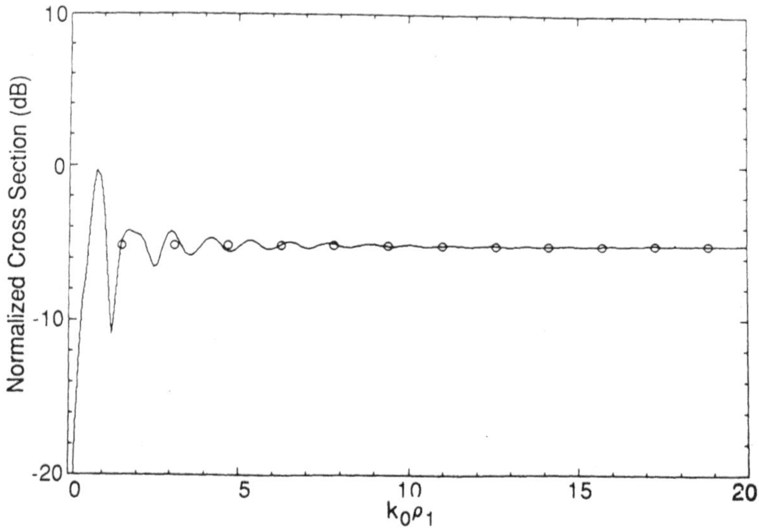


**Figure 3.16** Normalized backscattering cross section $\sigma/(k_0\rho_1 b^2)$ of a circular cylinder with $\epsilon_r = 10 + i5$ *versus* $k_0\rho_1$ for H-polarization: (—) exact solution, (o o o) physical optics.

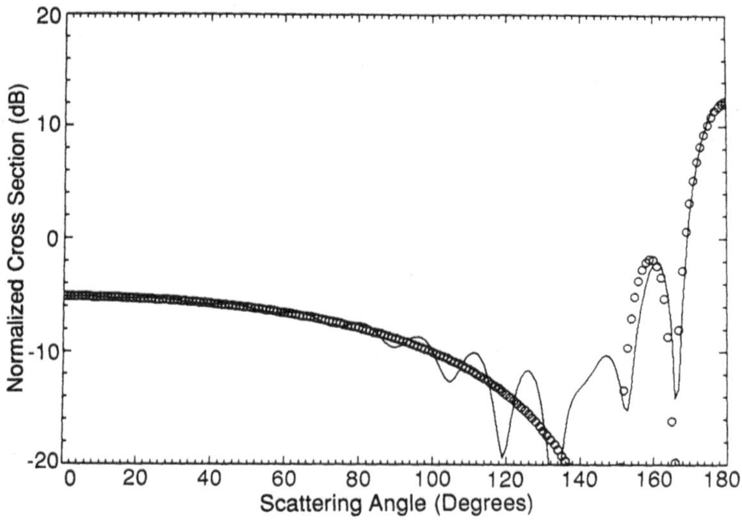


**Figure 3.17** Normalized bistatic scattering cross section $\sigma/(k_0\rho_1 b^2)$ of a circular cylinder with $\rho_1 = 2\lambda_0$ and $\epsilon_r = 10 + i5$ for E-polarization: (—) exact solution, (o o o) physical optics.

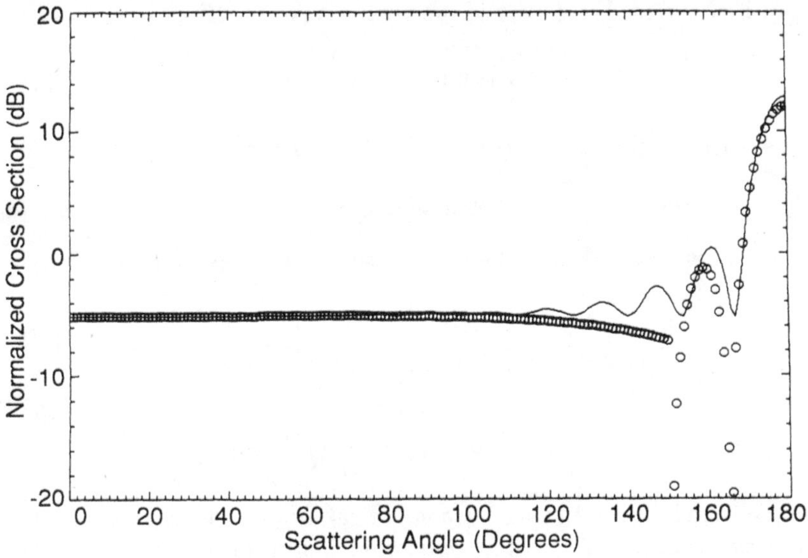


**Figure 3.18** Normalized bistatic scattering cross section $\sigma/(k_0\rho_1 b^2)$ of a circular cylinder with $\rho_1 = 2\lambda_0$ and $\epsilon_r = 10 + i5$ for H-polarization:(—) exact solution, (o o o) physical optics.

## 3-4.2 Sphere

The sphere is the only finite shape for which obtaining a truly exact analytic solution is possible. For either a homogeneous or spherically stratified sphere, the solution is expressed as an infinite sum of spherical Bessel and Legendre functions, and this is often called the *Mie series* in honor of G. Mie (1908), who obtained the solution for a perfectly conducting sphere.

A homogeneous nonmagnetic dielectric sphere of radius $a$, with center at the origin of the local coordinate system $x'$, $y'$, $z'$, is illuminated by a plane wave of arbitrary polarization. Without loss of generality, we can assume that the wave is incident in the $-\hat{\mathbf{z}}'$ direction with

$$\mathbf{E}^i \cdot \hat{\mathbf{x}}' = e_x \, \mathrm{e}^{-ik_0 z'} \, , \quad \mathbf{H}^i \cdot \hat{\mathbf{x}}' = \eta_0^{-1} h_x \, \mathrm{e}^{-ik_0 z'} \, . \tag{3.190}$$

The scattered field is a solution of the vector wave equation:

$$\nabla \times \nabla \times \mathbf{F} - k^2 \mathbf{F} = 0 \, , \tag{3.191}$$

and can be expressed in terms of the vector wave functions:

$$\mathbf{M} = \nabla \times (\mathbf{r}\Psi) , \quad \mathbf{N} = \frac{1}{k}\nabla \times \mathbf{M} , \tag{3.192}$$

where $\Psi$ satisfies the scalar wave equation (Stratton, 1941, pp. 392–395):

$$\nabla^2\Psi + k^2\Psi = 0 . \tag{3.193}$$

At a point inside or outside the sphere, the scalar wave function required is

$$\begin{aligned} \Psi_{\substack{e\\o}mn} &= j_n(kr')P_n^m(\cos\theta') \begin{matrix}\cos\\ \sin\end{matrix} m\phi' \\ \Psi_{\substack{e\\o}mn} &= h_n^{(1)}(kr')P_n^m(\cos\theta') \begin{matrix}\cos\\ \sin\end{matrix} m\phi' , \end{aligned} \tag{3.194}$$

respectively, where $j_n$ and $h_n^{(1)}$ are spherical Bessel and Hankel functions of the first kind, and $P_n^m$ is the associated Legendre function of degree $n$ and order $m$. The even (odd) azimuthal dependence is indicated by the subscript $e(o)$.

Following the procedure described by Goodrich *et al.* (1961), the fields are expanded in terms of the vector wave functions $\mathbf{M}_{\substack{e\\o}1n}$ and $\mathbf{N}_{\substack{e\\o}1n}$ corresponding to $\Psi_{\substack{e\\o}1n}$, and for the scattered field outside the sphere:

$$\begin{aligned} \mathbf{E}^s = & \sum_{n=1}^{\infty}(-i)^n \frac{2n+1}{n(n+1)} \{e_x(A_n\mathbf{M}_{o1n} + B_n\mathbf{N}_{e1n}) \\ & -ih_x(A_n\mathbf{M}_{e1n} + B_n\mathbf{N}_{o1n})\} \\ \mathbf{H}^s = & -\eta_0^{-1}\sum_{n=1}^{\infty}(-i)^n \frac{2n+1}{n(n+1)} \{h_x(A_n\mathbf{N}_{e1n} + B_n\mathbf{M}_{o1n}) \\ & +ie_x(A_n\mathbf{N}_{o1n} + B_n\mathbf{M}_{e1n})\} . \end{aligned} \tag{3.195}$$

Application of the boundary conditions at $r = a$ then gives

$$\begin{aligned} A_n &= -\frac{j_n(x_1)[x_0 j_n(x_0)]' - j_n(x_0)[x_1 j_n(x_1)]'}{j_n(x_1)[x_0 h_n^{(1)}(x_0)]' - h_n^{(1)}(x_0)[x_1 j_n(x_1)]'} \\ B_n &= \frac{\epsilon_r j_n(x_1)[x_0 j_n(x_0)]' - j_n(x_0)[x_1 j_n(x_1)]'}{\epsilon_r j_n(x_1)[x_0 h_n^{(1)}(x_0)]' - h_n^{(1)}(x_0)[x_1 j_n(x_1)]'} , \end{aligned} \tag{3.196}$$

where

$$x_0 = k_0 a \ , \quad x_1 = \sqrt{\epsilon_r} k_0 a \ , \tag{3.197}$$

and the prime denotes differentiation with respect to either $x_0$ or $x_1$ as appropriate.

In the direction:

$$\hat{\mathbf{k}}_s = \sin\theta_s' \cos\phi_s' \, \hat{\mathbf{x}}' + \sin\theta_s' \sin\phi_s' \, \hat{\mathbf{y}}' + \cos\theta_s' \, \hat{\mathbf{z}}' \tag{3.198}$$

in the far zone, the scattering amplitude is

$$\begin{aligned} \mathbf{S} = \frac{1}{k_0} \Big\{ & [e_x S_1(\theta_s') \cos\phi_s' + i h_x S_1'(\theta_s') \sin\phi_s'] \hat{\boldsymbol{\theta}}_s' \\ & - [e_x S_2(\theta_s') \sin\phi_s' - i h_x S_2'(\theta_s') \cos\phi_s'] \hat{\boldsymbol{\phi}}_s' \Big\} \ , \end{aligned} \tag{3.199}$$

where

$$\begin{aligned} S_1(\theta_s') &= -i \sum_{n=1}^{\infty} (-1)^n \frac{2n+1}{n(n+1)} \left\{ A_n \frac{P_n^1(\cos\theta_s')}{\sin\theta_s'} + i B_n \frac{\partial}{\partial\theta_s'} P_n^1(\cos\theta_s') \right\} \\ S_1'(\theta_s') &= -i \sum_{n=1}^{\infty} (-1)^n \frac{2n+1}{n(n+1)} \left\{ A_n \frac{P_n^1(\cos\theta_s')}{\sin\theta_s'} - i B_n \frac{\partial}{\partial\theta_s'} P_n^1(\cos\theta_s') \right\} \\ S_2(\theta_s') &= -i \sum_{n=1}^{\infty} (-1)^n \frac{2n+1}{n(n+1)} \left\{ A_n \frac{\partial}{\partial\theta_s'} P_n^1(\cos\theta_s') + i B_n \frac{P_n^1(\cos\theta_s')}{\sin\theta_s'} \right\} \\ S_2'(\theta_s') &= -i \sum_{n=1}^{\infty} (-1)^n \frac{2n+1}{n(n+1)} \left\{ A_n \frac{\partial}{\partial\theta_s'} P_n^1(\cos\theta_s') - i B_n \frac{P_n^1(\cos\theta_s')}{\sin\theta_s'} \right\} \ . \end{aligned} \tag{3.200}$$

For backscattering, $\theta_s' = 0$ and

$$\begin{aligned} S_1(0) = S_2(0) &= -i \sum_{n=1}^{\infty} (-1)^n (n + \frac{1}{2})(A_n + i B_n) \\ S_1'(0) = S_2'(0) &= -i \sum_{n=1}^{\infty} (-1)^n (n + \frac{1}{2})(A_n - i B_n) \ , \end{aligned} \tag{3.201}$$

giving

$$\mathbf{S} = \frac{1}{k_0} \{ e_x S_1(0) \, \hat{\mathbf{x}} + i h_x S_1'(0) \, \hat{\mathbf{y}} \} \ , \tag{3.202}$$

whereas in the forward scattering direction, $\theta_s' = \pi$ and

$$S_1(\pi) = -S_2(\pi) = i\sum_{n=1}^{\infty}(n+\frac{1}{2})(A_n - iB_n)$$

$$S_1'(\pi) = -S_2'(\pi) = i\sum_{n=1}^{\infty}(n+\frac{1}{2})(A_n + iB_n)\ , \tag{3.203}$$

giving

$$\mathbf{S} = -\frac{1}{k_0}\{e_x S_1(\pi)\,\hat{\mathbf{x}} + ih_x S_1'(\pi)\,\hat{\mathbf{y}}\}\ . \tag{3.204}$$

Not surprisingly, the scattering mechanisms for a dielectric sphere are similar to those for a dielectric cylinder, and qualitatively, at least, the scattering behavior is similar. In Fig. 3.19 the normalized backscattering cross section is plotted as a function of $k_0a$ for three spheres having $\epsilon_r = 10 + i\epsilon_r''$ with $\epsilon_r'' = 1, 2$ and 5, and we observe that the oscillations decrease with increasing loss tangent, albeit not as rapidly as in the case of a cylinder. Because of this, $k_0a$ must be somewhat larger if physical optics is to provide the same accuracy of approximation (see Fig. 3.20). For $\epsilon_r = 10 + i5$ and $a = 2\lambda_0$, the normalized bistatic scattering cross sections are shown in Figs. 3.21 and 3.22 for $\mathbf{E}$ perpendicular and parallel to the plane of incidence, respectively, and we observe that, for most angles of scattering, physical optics provides an adequate approximation.

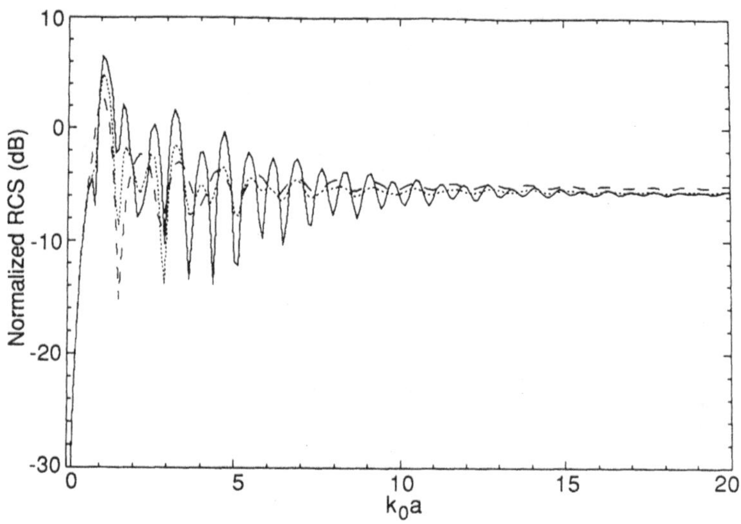


**Figure 3.19** Normalized backscattering cross section $\sigma/(\pi a^2)$ of a sphere *versus* $k_0a$ for different loss tangents: (—) $\epsilon_r = 10+i1$, (- - - - -) $\epsilon_r = 10+i2$, (- -) $\epsilon_r = 10 + i5$.

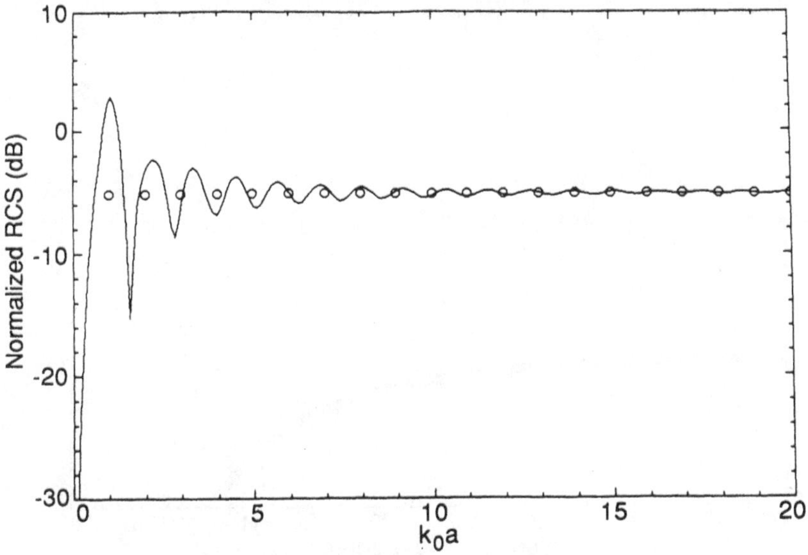


**Figure 3.20** Normalized backscattering cross section $\sigma/(\pi a^2)$ of a sphere with $\epsilon_r = 10 + i5$ *versus* $k_0 a$: (—) exact solution, (o o o) physical optics.

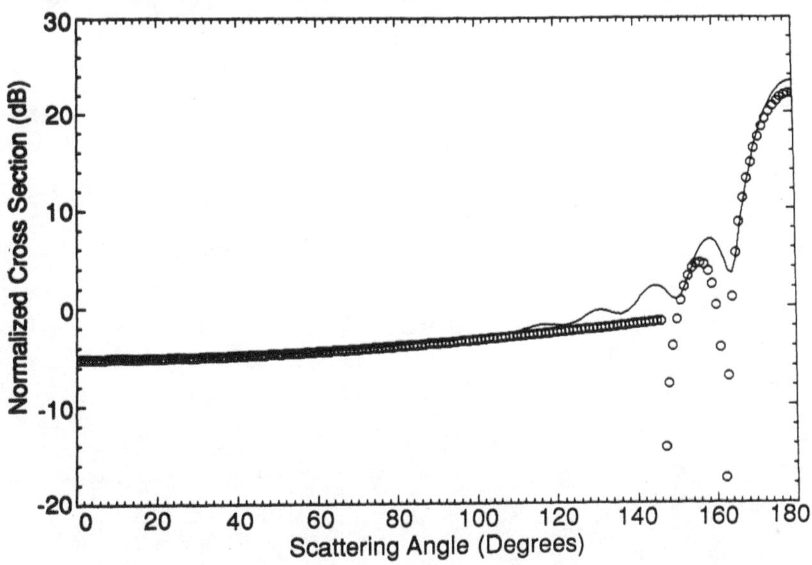


**Figure 3.21** Normalized bistatic scattering cross section $\sigma/(\pi a^2)$ of a sphere with a $a = 2\lambda_0$ and $\epsilon_r = 10 + i5$ for perpendicular polarization: (—) exact solution, (o o o) physical optics.

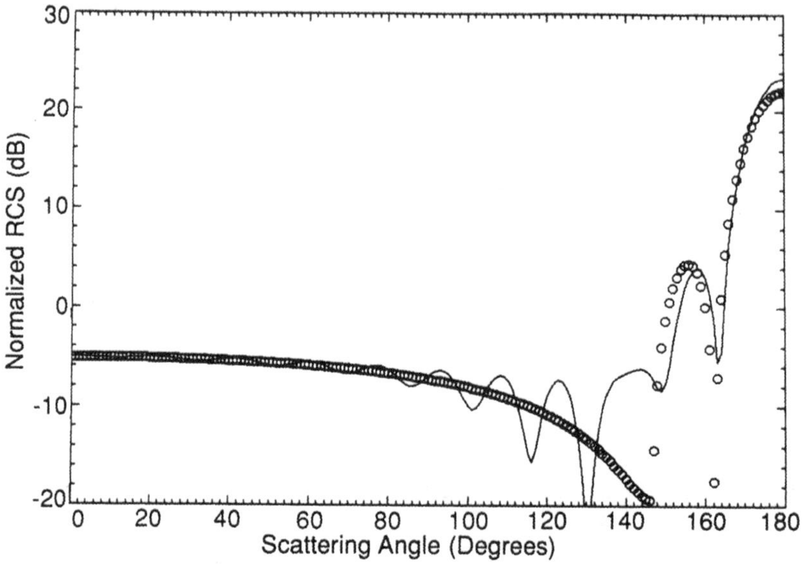


**Figure 3.22** Normalized bistatic scattering cross section $\sigma/(\pi a^2)$ of a sphere with $a = 2\lambda_0$ and $\epsilon_r = 10 + i5$ for parallel polarization: (—) exact solution, (o o o) physical optics.

# 3-5 SUMMARY TABLE

As a guide to some of the more significant formulas presented in this chapter, the following table lists some of the bodies discussed and indicates the equation numbers for the scattering matrix elements appropriate to different frequency ranges.

**Table 3.2** Scattering Matrix Elements Formulas for Different Targets

| Body | $D \ll \lambda_0$ | $D \gg \lambda_0$ | Exact |
|---|---|---|---|
| General convex body | — | (3.86)–(3.89) and (3.90) | — |
| Body of revolution | (3.59) | — | — |
| Ellipsoid | (3.150), (3.142) | — | — |
| Sphere | (3.15) with $L_i = \frac{1}{3}$ and (3.142) | (3.119), (3.122) | (3.119), (3.122), and a coordinate transformation |
| Finite length cylinder (long) | — | (3.94), (3.113), and (3.114) in (3.86)–(3.89); (3.96) in forward direction | (3.106), (3.107) (3.107) in (3.184)–(3.187) (semiexact) |
| Resistive flat plate | — | (3.54)–(3.57) | — |
| Resistive curved plate | — | (3.70) | — |
| Perfectly conducting plate | (3.168), (3.142) (elliptical) (3.170), (3.142) (circular) | (3.54)–(3.57) ($R = 0$) | — |

# References


Abramowitz, M., and I.E. Stegun (eds.), *Handbook of Mathematical Functions, Appl. Math. Sciences 55,* National Bureau of Standards, Washington, D.C., 1964.

Beckmann, P., *The Depolarization of Electromagnetic Waves,* The Golem Press, Boulder, CO, 1968.

Darling, D.A., and T.B.A. Senior, "Low-Frequency Expansions for Scattering by Separable and Nonseparable Bodies," *J. Acoust. Soc. Amer.,* Vol. 37, 1965, pp. 228–234.

Goodrich, R.F., B.A. Harrison, R.E. Kleinman, and T.B.A. Senior, "Diffraction and Scattering by Regular Bodies - I: The Sphere," University of Michigan Radiation Laboratory Report No. 3648-1-T, Ann Arbor, MI, December 1961.

Herrick, D.F., and T.B.A. Senior, "Low Frequency Scattering by Rectangular Dielectric Particles," *Appl. Phys.,* Vol. 13, 1977, pp. 175–183.

James, G.L., *Geometrical Theory of Diffraction for Electromagnetic Waves,* Peter Peregrinus, London, 1976.

Jahnke, E., and F. Emde, *Tables of Functions,* Dover Publications, New York, 1945.

Jones, D.S., *The Theory of Electromagnetism,* Macmillan, New York, 1964.

Kleinman, R.E., "Low Frequency Solutions of Electromagnetic Scattering Problems," in *Electromagnetic Wave Theory, Part II* (ed. J. Brown), Pergamon Press, New York, 1967a, pp. 891–905.

Kleinman, R.E., "Far Field Scattering at Low Frequencies," *Appl. Sci. Res.,* Vol. 18, 1967b, pp. 1–8.

Kleinman, R.E., "Low Frequency Electromagnetic Scattering," in *Electromagnetic Scattering* (ed. P.L.E. Uslenghi), Academic Press, New York, 1978, pp. 1–28.

Kleinman, R.E., and T.B.A. Senior, "Rayleigh Scattering," in *Low* and High Frequency Asymptotics" (eds. V.K. Varadan and V.V. Varadan), North-Holland, Amsterdam, 1986, pp. 1–70.

Knott, E.F., J.F. Shaeffer, and M.J. Tuley, *Radar Cross Section,* Artech House, Norwood, MA, 1985.

Maffett, A.L., *Topics for a Statistical Description of Radar Cross Section,* John Wiley and Sons, New York, 1989.

Mie, G., "Beiträge zur Optik trüber Medien, Speziell Kolloidaler Metallösungen," *Ann. d. Physik,* Vol. 25, 1908, pp. 377–442.

Poggio, A.J., and E.K. Miller, "Integral Equation Solutions of Three-Dimensional Scattering Problems," in *Computer Techniques for Electromagnetics* (ed. R. Mittra), Hemisphere, New York, 1987, pp. 159–264.

Rayleigh, Lord, "On the Incidence of Aerial and Electric Waves upon Small Obstacles in the Form of Ellipsoids or Elliptic Cylinders and on the Passage of Electric Waves through a Circular Aperture in a Conducting Screen," *Phil. Mag.,* Vol. 44, 1897, pp. 28–52.

Sarabandi, K., T.B.A. Senior, and F.T. Ulaby, "Effect of Curvature on the Backscattering from a Leaf," *J. Electromag. Waves Applics.,* Vol. 2, 1988, pp. 653–670.

Senior, T.B.A., "Effect of Particle Shape on Low Frequency Absorption," *Appl. Opt.,* Vol. 19, 1980, pp. 2483–2485.

Senior, T.B.A., "Approximate Boundary Conditions," *IEEE Trans. Antennas and Propagation,* Vol. AP-20, 1981, pp. 826–829.

Senior, T.B.A., and D.J. Ahlgren, "Rayleigh Scattering," *IEEE Trans. Antennas and Propagation,* Vol. AP-21, 1973, p. 134.

Senior, T.B.A., K. Sarabandi, and F.T. Ulaby, "Measuring and Modeling the Backscattering Cross Section of a Leaf," *Radio Sci.,* Vol. 22, 1987, pp. 1109–1116.

Senior, T.B.A., and H. Weil, "On the Validity of Modeling Rayleigh Scatterers by Spheroids," *Appl. Phys.,* Vol. B29, 1982, pp. 117–124.

Senior, T.B.A., and T.M. Willis, "Rayleigh Scattering by Dielectric Bodies," *IEEE Trans. Antennas and Propagation,* Vol. AP-30, 1982, p. 1271.

Stevenson, A.F., "Solution of Electromagnetic Scattering Problems as Power Series in the Ratio (Dimension of Scatterer/Wavelength)," *J. Appl. Phys.,* Vol. 24, 1953a, pp. 1134–1142.

Stevenson, A.F., "Electromagnetic Scattering by an Ellipsoid in the Third Approximation," *J. Appl. Phys.*, Vol. 24, 1953b, pp. 1143–1151.

Stratton, J.A., *Electromagnetic Theory*, McGraw-Hill, New York, 1941.

van de Hulst, H.C., *Light Scattering by Small Particles*, John Wiley and Sons, New York, 1957.

Weil, H., T.B.A. Senior, and T.M. Willis III, "Internal and Near Fields of Small Particles Irradiated in Spectral Absorption Bands," *J. Opt. Soc. Amer.*, Vol. A2, 1985, pp. 989–996.

Willis, T.M., H. Weil, and D.M. Le Vine, "Applicability of Physical Optics Thin Plate Scattering Formulas for Remote Sensing," *IEEE Trans. Geoscience and Remote Sensing*, Vol. 26, 1988, pp. 153–160.

# Chapter 4

# Scattering Models for Distributed Targets

**Yasuo Kuga, Michael W. Whitt, Kyle C. McDonald, and Fawwaz T. Ulaby**

Radar scattering from terrain involves complicated interactions because the scattering elements have complex geometries and are randomly distributed in space. When developing mathematical models for polarimetric radar scattering from terrain surfaces and volumes (such as vegetation) to deal with all possible terrain configurations and conditions is impossible, and indeed impractical. Hence, we usually focus our attention on the development of generic models that can help us understand the nature of wave propagation and scattering in random media, which, in turn, allows us to interpret radar observations and extract useful information about the observed scene.

In this chapter, we will present a select set of relatively simple models describing vector electromagnetic-wave scattering by random media. We begin by introducing polarimetric models for scattering by a rough surface, specifically those based on the classical Kirchhoff approximation and the small perturbation method. We then introduce the theory of vector radiative transfer and examine its use for modeling propagation in and scattering by media containing randomly distributed particles, such as tree canopies. This examination includes a presentation of two techniques for solving the radiative transfer equation, namely the iterative technique, which provides an approximate solution, and the numerical technique, which provides an exact solution. The iterative technique is simpler to use and provides solutions that are easier to interpret, but its validity is limited to weakly scattering media. In contrast, the numerical solution accounts for multiple scattering interactions in the random medium, but its formulation is more complex, and it is computationally more demanding. By way of example, in the final section of this chapter we present a first-order solution of the radiative transfer equation that has been developed to model radar scattering from continuous tree canopies.

# 4-1 ROUGH SURFACE SCATTERING MODELS

The scattering of electromagnetic waves by rough surfaces has been studied for many years, but no exact closed-form solutions have been obtained. Numerical techniques such as the method of moments can be used to compute the exact solution, but, in general, these techniques are computationally prohibitive and used only in evaluating the accuracy and range of validity of approximate models (Chen and Fung, 1988). Usually, when dealing with practical applications, approximate models are used because of their simplicity. Although valid only within a limited range of rough surface parameters, these models can still be used quite effectively in many situations. In the microwave region, the most common approximate methods for modeling rough surface scattering have been the *Kirchhoff approach* and the *small perturbation method* (SPM).

With the Kirchhoff approach, the scattered field is written in terms of the tangential field on the rough surface. The surface field is then approximated by the field that would be present if the rough surface was replaced by a planar surface tangential to the point of interest. This is known as the *tangent plane approximation* to the surface field, which is valid if the radius of curvature at every point on the surface is large relative to the wavelength.

Even with the tangent plane approximation, the scattered field is still in the form of an integral over the surface, which is difficult to solve analytically. As a result, additional assumptions are required to obtain an analytic solution. In the high frequency limit, the method of *stationary phase* can be used to obtain the *geometrical optics* (GO) solution for the scattered field. The GO solution is generally valid when both the average radius of curvature and the vertical roughness are large relative to the wavelength. Another commonly used approximation is to expand the integrand of the diffraction integral into terms about zero slope, keeping only the lower order terms. This approximation provides the ***physical optics*** (PO) solution for the scattered field, and the result is valid when the radius of curvature is large and the rms surface slope is small relative to the wavelength.

The small perturbation method assumes that variations in surface height are small relative to the wavelength and the surface slope is small. Under these assumptions, the tangential surface field can be written in series form and approximated by the lowest order terms. The scattered field can then be computed to the order of approximation for the surface field.

Although the Kirchhoff approach and the small perturbation method are the most common techniques for computing the scattering from rough surfaces, these are not the only methods to receive significant attention. Recently, there has been an interest in developing scattering models that are valid over a wider range of surface roughness and slope parameters (especially where the Kirchhoff and small perturbation methods fail). Two additional methods that have been considered for rough surface scattering are the phase perturbation method (Winebrenner, 1985; Broschat, 1989; Thorsos and Jackson, 1989) and the smoothing method (Watson and Keller, 1984).

In the next three sections, we will consider the classical solutions to the rough surface scattering problem (i.e., GO, PO, and SPM). Because these models depend upon the form of the ***surface correlation function*** $\rho(\xi)$, we will use one (and at times all) of the following three typical correlation functions whenever examples are required:

$$\rho_1(\xi) = \exp\left[-\xi^2/l^2\right] \tag{4.1}$$

$$\rho_2(\xi) = \exp\left[-\frac{\xi^2}{(l_s^4+\xi^2 l^2)^{0.5}}\right] \tag{4.2}$$

$$\rho_3(\xi) = \left[1+\frac{\xi^2}{l^2}\right]^{-1.5}. \tag{4.3}$$

The first correlation function is Gaussian, and the second approximates a Gaussian function when $\xi$ (the horizontal distance between two points on the random surface) is small, and approximates an exponential function when $\xi$ is large. The parameters $l_s$ and $l$ are the small scale and large scale ***correlation lengths*** for the surface, and $s$ is the ***standard deviation of surface height.*** The third correlation function is considered because it is probably more realistic (Ulaby *et al.*, 1986a, Appendix 12L), although computations using this function may be more involved. Table 4.1 lists the average radius of curvature and the rms slope for a rough surface with each of the three correlation functions.

**Table 4.1** Average Radius of Curvature, $R_c$, and rms Slope, $m$, for the Three Example Correlation Functions

| Correlation | $R_c$ | $m$ |
|---|---|---|
| $\rho_1(\xi)$ | $l^2/(2.76s)$ | $\sqrt{2}s/l$ |
| $\rho_2(\xi)$ | $l_s^2/(2.76s)$ | $\sqrt{2}s/l_s$ |
| $\rho_3(\xi)$ | $l^2/(5.35s)$ | $\sqrt{3}s/l$ |

Finally, we should state that the validity conditions quoted for the various models considered in this section are presented only as estimates of the regions where the models are applicable. This is especially true for the small perturbation model of Section 4-1.4. A more thorough treatment of the question of validity is available in the literature (Chen and Kung, 1988; Broschat, 1989; Thorsos and Jackson, 1989).

## 4-1.1 Kirchhoff Approach

Consider the rough surface geometry shown in Fig. 4.1 with the unit vectors $\hat{\mathbf{k}}_i$ and $\hat{\mathbf{k}}_s$ representing the incident and scattered directions, respectively. Using the forward scatter alignment (FSA) convention as defined in Chapter 2, we let $\hat{\mathbf{v}}_i$ and $\hat{\mathbf{h}}_i$ denote the unit polarization vectors for the vertical and horizontal components of the incident field, and we let $\hat{\mathbf{v}}_s$ and $\hat{\mathbf{h}}_s$ denote the unit polarization vectors for the scattered field. These unit vectors are given in terms of the angles $(\theta_i, \phi_i)$ and $(\theta_s, \phi_s)$ for the incident and scattered fields by the following relations:

$$\hat{\mathbf{k}}_f = \hat{\mathbf{x}}\sin\theta_f\cos\phi_f + \hat{\mathbf{y}}\sin\theta_f\sin\phi_f + \hat{\mathbf{z}}\cos\theta_f \tag{4.4}$$

$$\hat{\mathbf{h}}_f = -\hat{\mathbf{x}}\sin\phi_f + \hat{\mathbf{y}}\cos\phi_f \tag{4.5}$$

$$\hat{\mathbf{v}}_f = \hat{\mathbf{h}}_f \times \hat{\mathbf{k}}_f = \hat{\mathbf{x}}\cos\theta_f\cos\phi_f + \hat{\mathbf{y}}\cos\theta_f\sin\phi_f - \hat{\mathbf{z}}\sin\theta_f, \tag{4.6}$$

where the subscript $f \in \{i, s\}$ represents either the incident or the scattered wave. The rough surface has a height profile in the $x$-$y$ plane given by the function $h(\mathbf{r}_\perp)$, where the subscript $\perp$ represents the component perpendicular to the $z$-axis. The regions above and below the rough surface interface are denoted as region 1 and region 2, respectively.

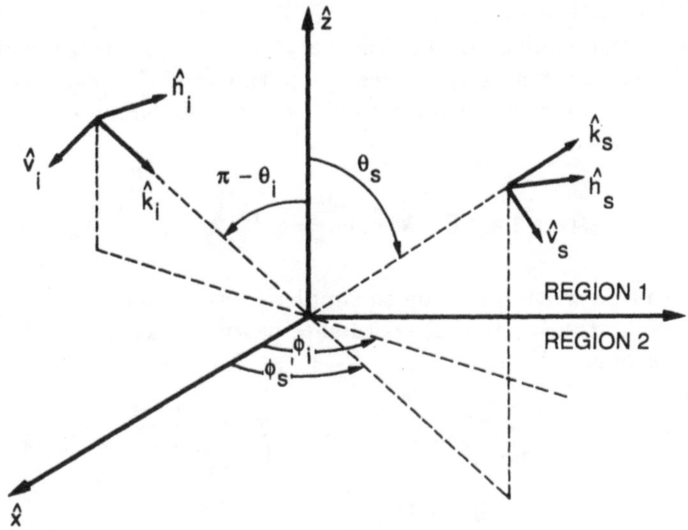


**Figure 4.1** Rough surface geometry.

For a plane wave incident upon the rough surface, the incident electric field is given by

$$\mathbf{E}^i(\mathbf{r}) \;=\; \hat{\mathbf{a}} E_0 e^{ik_1 \hat{\mathbf{n}}_i \cdot \mathbf{r}} \tag{4.7}$$

where $\hat{\mathbf{a}} \in \{\hat{\mathbf{h}}_i, \hat{\mathbf{v}}_i\}$ is the incident polarization, and $k_1 = \omega\sqrt{\mu_1 \epsilon_1}$ is the wavenumber in region 1. The scattered field in region 1 can be written in terms of the tangential electric and magnetic fields at the surface. In the form of a surface integral, the scattered field is (Tsang *et al.*, 1985)

$$\mathbf{E}^s(\mathbf{r}) = \int_{S'} \{i\omega\mu_1 \mathcal{G}(\mathbf{r},\mathbf{r}')\cdot[\hat{\mathbf{n}}_1 \times \mathbf{H}(\mathbf{r}')] + \nabla \times \mathcal{G}(\mathbf{r},\mathbf{r}')\cdot[\hat{\mathbf{n}}_1 \times \mathbf{E}(\mathbf{r}')]\} ds', \tag{4.8}$$

where $S'$ is the rough surface interface, and $\hat{\mathbf{n}}_1$ is the unit vector normal to the interface and pointing into region 1 as defined by

$$\hat{\mathbf{n}}_1 = \frac{-h_x(\mathbf{r}'_\perp)\hat{\mathbf{x}} - h_y(\mathbf{r}'_\perp)\hat{\mathbf{y}} + \hat{\mathbf{z}}}{\sqrt{1 + h_x^2(\mathbf{r}'_\perp) + h_y^2(\mathbf{r}'_\perp)}}. \tag{4.9}$$

The subscripts of $h_x(\mathbf{r}'_\perp)$ and $h_y(\mathbf{r}'_\perp)$ denote the partials with respect to the $x'$ and $y'$ variables. Because region 1 is a homogeneous space with wavenumber $k_1$, the dyadic Green's function is given by

$$\mathcal{G}(\mathbf{r},\mathbf{r}') = \left[\mathcal{I} + \frac{\nabla\nabla}{k_1^2}\right] \frac{e^{ik_1|\mathbf{r}-\mathbf{r}'|}}{4\pi|\mathbf{r}-\mathbf{r}'|}, \tag{4.10}$$

where $\mathbf{I}$ is the unit dyad.

An expression similar to (4.8) can be written for the transmitted field in region 2, but it is not needed when computing only the scattered field in region 1. For observation in the far field, the dyadic Green's function can be approximated by

$$\mathcal{G}(\mathbf{r},\mathbf{r}') \approx (\mathcal{I} - \hat{\mathbf{k}}_s\hat{\mathbf{k}}_s)\frac{e^{ik_1 r}}{4\pi r} e^{-ik_1 \hat{\mathbf{k}}_s \cdot \mathbf{r}'}, \tag{4.11}$$

and (4.8) can be written in a much simpler form. If the scattered field is to be written in the $(\hat{\mathbf{v}}_s, \hat{\mathbf{h}}_s, \hat{\mathbf{k}}_s)$ coordinate basis, $\mathcal{I} - \hat{\mathbf{k}}_s\hat{\mathbf{k}}_s = \hat{\mathbf{v}}_s\hat{\mathbf{v}}_s + \hat{\mathbf{h}}_s\hat{\mathbf{h}}_s$, and (4.8) becomes

$$\begin{aligned}\mathbf{E}^s(\mathbf{r}) \;=\; & ik_1\frac{e^{ik_1 r}}{4\pi r}(\hat{\mathbf{v}}_s\hat{\mathbf{v}}_s + \hat{\mathbf{h}}_s\hat{\mathbf{h}}_s)\cdot \int_{S'} \Big\{\hat{\mathbf{k}}_s \times [\hat{\mathbf{n}}_1 \times \mathbf{E}(\mathbf{r}')] \\ & + \eta_1 [\hat{\mathbf{n}}_1 \times \mathbf{H}(\mathbf{r}')]\Big\} e^{-ik_1 \hat{\mathbf{k}}_s \cdot \mathbf{r}'} ds'. \end{aligned} \tag{4.12}$$

The tangent plane approximation must now be used to compute the surface fields. It will be convenient to define a local coordinate system $(\hat{\mathbf{t}}, \hat{\mathbf{d}}, \hat{\mathbf{k}}_i)$ at each point $\mathbf{r}'$ on the surface such that

$$\hat{\mathbf{t}} = \frac{\hat{\mathbf{k}}_i \times \hat{\mathbf{n}}_1}{|\hat{\mathbf{k}}_i \times \hat{\mathbf{n}}_1|} \quad \text{and} \quad \hat{\mathbf{d}} = \hat{\mathbf{k}}_i \times \hat{\mathbf{t}}. \tag{4.13}$$

Because $\hat{\mathbf{t}}$ is normal to the principal plane (the plane containing the incident and surface normal vectors at a point on the surface), it represents the perpendicular (or horizontal) polarization vector. The parallel (or vertical) polarization vector is represented by $\hat{\mathbf{d}}$, which is in the principal plane.

Under the tangent plane approximation, the total field at a point on the rough surface is approximated by the incident field plus the field scattered by an infinite plane tangent to the surface at that point. The surface fields under the tangent plane approximation have been shown to be (Ulaby *et al.*, 1986a)

$$\begin{aligned}\hat{\mathbf{n}}_1 \times \mathbf{E}(\mathbf{r}') \;=\; & E_0 \big\{(1 + r_h)(\hat{\mathbf{a}}\cdot\hat{\mathbf{t}})(\hat{\mathbf{n}}_1 \times \hat{\mathbf{t}}) \\ & - (1 - r_v)(\hat{\mathbf{n}}_1\cdot\hat{\mathbf{n}}_i)(\hat{\mathbf{a}}\cdot\hat{\mathbf{d}})\hat{\mathbf{t}}\big\} e^{ik_1 \hat{\mathbf{k}}_i\cdot\mathbf{r}'} \end{aligned} \tag{4.14}$$

$$\begin{aligned}-\eta_1\hat{\mathbf{n}}_1 \times \mathbf{H}(\mathbf{r}') \;=\; & E_0 \Big\{(1 + r_v)(\hat{\mathbf{a}}\cdot\hat{\mathbf{d}})(\hat{\mathbf{n}}_1 \times \hat{\mathbf{t}}) \\ & + (1 - r_h)(\hat{\mathbf{n}}_1\cdot\hat{\mathbf{n}}_i)(\hat{\mathbf{a}}\cdot\hat{\mathbf{t}})\hat{\mathbf{t}}\big\} e^{ik_1 \hat{\mathbf{k}}_i\cdot\mathbf{r}'}. \end{aligned} \tag{4.15}$$

By using (4.14) and (4.15) in the far-field expression for the scattered field given by (4.12), we obtain

$$\mathbf{E}^s(\mathbf{r}) = ik_1 E_0 \frac{e^{ik_1 r}}{4\pi r}(\hat{\mathbf{v}}_s\hat{\mathbf{v}}_s + \hat{\mathbf{h}}_s\hat{\mathbf{h}}_s)\cdot \int_S \mathbf{U}(h_x, h_y) e^{ik_1(\hat{\mathbf{k}}_i - \hat{\mathbf{k}}_s)\cdot\mathbf{r}'} ds', \quad (4.16)$$

where

$$\begin{aligned}\mathbf{U}(h_x, h_y) = &\\ &\hat{\mathbf{k}}_s \times (\hat{\mathbf{n}}_1 \times \hat{\mathbf{t}})(\hat{\mathbf{a}}\cdot\hat{\mathbf{t}})(1 + r_h) - (\hat{\mathbf{k}}_s \times \hat{\mathbf{t}})(\hat{\mathbf{n}}_1\cdot\hat{\mathbf{k}}_i)(\hat{\mathbf{a}}\cdot\hat{\mathbf{d}})(1 - r_v)\\ &- (\hat{\mathbf{n}}_1 \times \hat{\mathbf{t}})(\hat{\mathbf{a}}\cdot\hat{\mathbf{d}})(1 + r_v) - \hat{\mathbf{t}}(\hat{\mathbf{n}}_1\cdot\hat{\mathbf{k}}_i)(\hat{\mathbf{a}}\cdot\hat{\mathbf{t}})(1 - r_h). \qquad (4.17)\end{aligned}$$

Here, $r_h$ and $r_v$ are the standard Fresnel reflection coefficients for the local incidence and transmission angles, $\theta_{il}$ and $\theta_{tl}$; thus,

$$r_h = \frac{\eta_2 \cos\theta_{il} - \eta_1 \cos\theta_{tl}}{\eta_2 \cos\theta_{il} + \eta_1 \cos\theta_{tl}} \quad (4.18)$$

$$r_v = \frac{\eta_1 \cos\theta_{il} - \eta_2 \cos\theta_{tl}}{\eta_1 \cos\theta_{il} + \eta_2 \cos\theta_{tl}}. \quad (4.19)$$

The validity conditions, based on analytic considerations, for the Kirchhoff (or tangent plane) approximation have been quoted in the literature as (Ulaby *et al.*, 1986a)

$$k_1 l > 6 \quad (4.20)$$

$$R_c > \lambda, \quad (4.21)$$

where $l$ is the surface correlation length, and $R_c$ is the average radius of curvature for the rough surface. Note that $R_c$ depends on both the surface correlation length $l$ and the standard deviation of surface height $s$. The expression relating $R_c$ to $l$ and $s$ is given in terms of the fourth derivative of the surface correlation function $\rho(\xi)$ evaluated at $\xi = 0$ (Ulaby *et al.*, 1986a, pp. 1011–1013):

$$R_c = s\left[\frac{2}{\pi}\left.\frac{\partial^4 \rho(\xi)}{\partial \xi^4}\right|_{\xi=0}\right]^{0.5}. \quad (4.22)$$

Because $\hat{\mathbf{n}}_1$, $\hat{\mathbf{t}}$, and $\hat{\mathbf{d}}$ depend on the position vector $\mathbf{r}'$, the expression for the scattered field under the tangent plane approximation is still difficult to solve. Additional assumptions must be made to obtain an analytic solution for the scattered field. In the next two sections, we will consider two models based on the Kirchhoff approach with additional approximations. Section 4-1.2 discusses the GO model, and Section 4-1.3 discusses the PO model.

### 4-1.2 Geometrical Optics Model

The geometrical optics model for rough surface scattering is obtained by applying the method of stationary phase to evaluate (4.16) in the limit as $k_1 \to \infty$. The resulting scattered field expression contains only contributions from specular points on the surface.

The exponential term in equation (4.16) contains the phase

$$Q = k_1(\hat{\mathbf{k}}_i - \hat{\mathbf{k}}_s)\cdot\mathbf{r}' = q_x x' + q_y y' + q_z h(\mathbf{r}'_\perp), \tag{4.23}$$

where

$$q_x = k_1(\sin\theta_i \cos\phi_i - \sin\theta_s \cos\phi_s) \tag{4.24}$$
$$q_y = k_1(\sin\theta_i \sin\phi_i - \sin\theta_s \sin\phi_s) \tag{4.25}$$
$$q_z = k_1(\cos\theta_i - \cos\theta_s). \tag{4.26}$$

A stationary phase point is defined to be one at which the rate of change of phase is zero. Mathematically, this corresponds to

$$\frac{\partial Q}{\partial x'} = \frac{\partial Q}{\partial y'} = 0. \tag{4.27}$$

By computing the partial derivatives of $Q$ and equating them to zero as required above, at the stationary phase points we find that

$$h_{xs} = \frac{\partial h(\mathbf{r}'_{\perp s})}{\partial x'} = -\frac{q_x}{q_z} \quad \text{and} \quad h_{ys} = \frac{\partial h(\mathbf{r}'_{\perp s})}{\partial y'} = -\frac{q_y}{q_z}. \tag{4.28}$$

By considering only the stationary phase points, the unit vectors $\hat{\mathbf{n}}_1$, $\hat{\mathbf{t}}$, and $\hat{\mathbf{d}}$ are now independent of $\mathbf{r}'$, and are given by

$$\hat{\mathbf{n}}_1 = \frac{|q_z|}{q_z q} k_1(\hat{\mathbf{k}}_i - \hat{\mathbf{k}}_s) \tag{4.29}$$
$$\hat{\mathbf{t}} = \frac{|q_z|}{q_z D}(\hat{\mathbf{k}}_s \times \hat{\mathbf{k}}_i) \tag{4.30}$$
$$\hat{\mathbf{d}} = \frac{|q_z|}{q_z D}\left[\hat{\mathbf{k}}_s - \hat{\mathbf{k}}_i(\hat{\mathbf{k}}_i\cdot\hat{\mathbf{k}}_s)\right], \tag{4.31}$$

where

$$q^2 = q_x^2 + q_y^2 + q_z^2 = 2k_1^2\left[1 - (\hat{\mathbf{k}}_s\cdot\hat{\mathbf{k}}_i)\right] \tag{4.32}$$
$$D = \sqrt{(\hat{\mathbf{k}}_i\cdot\hat{\mathbf{v}}_s)^2 + (\hat{\mathbf{k}}_i\cdot\hat{\mathbf{h}}_s)^2}. \tag{4.33}$$

By evaluating $\mathbf{U}(h_x, h_y)$ at the stationary phase points and moving the result outside the integral, we obtain

$$\mathbf{E}^s(\mathbf{r}) = ik_1 I E_0 \frac{e^{ik_1 r}}{4\pi r}(\hat{\mathbf{v}}_s\hat{\mathbf{v}}_s + \hat{\mathbf{h}}_s\hat{\mathbf{h}}_s)\cdot\mathbf{U}(h_{xs}, h_{ys}), \tag{4.34}$$

where

$$I = \int_S e^{ik_1(\hat{\mathbf{k}}_i - \hat{\mathbf{k}}_s)\cdot\mathbf{r}'} ds'. \tag{4.35}$$

Note that $I$ is independent of the incident and scattered polarizations. The scattering matrix element corresponding to transmission of $\hat{\mathbf{q}}_i$ polarization and reception of $\hat{\mathbf{p}}_s$ polarization is therefore given by

$$S_{pq} = \frac{ik_1}{4\pi} I U_{pq} = \frac{ik_1}{4\pi} I \hat{\mathbf{p}}_s \cdot \mathbf{U}(h_{xs}, h_{ys})\Big|_{\hat{\mathbf{a}}=\hat{\mathbf{q}}_i}. \tag{4.36}$$

By substituting (4.28) through (4.33) into (4.17), we obtain (Ulaby *et al.*, 1986a)

$$U_{vv} = -\frac{q|q_z|}{q_z k_1 D^2}[r_v(\hat{\mathbf{v}}_s\cdot\hat{\mathbf{k}}_i)(\hat{\mathbf{v}}_i\cdot\hat{\mathbf{k}}_s) + r_h(\hat{\mathbf{h}}_s\cdot\hat{\mathbf{k}}_i)(\hat{\mathbf{h}}_i\cdot\hat{\mathbf{k}}_s)] \tag{4.37}$$

$$U_{vh} = -\frac{q|q_z|}{q_z k_1 D^2}[r_v(\hat{\mathbf{v}}_s\cdot\hat{\mathbf{k}}_i)(\hat{\mathbf{h}}_i\cdot\hat{\mathbf{k}}_s) - r_h(\hat{\mathbf{h}}_s\cdot\hat{\mathbf{k}}_i)(\hat{\mathbf{v}}_i\cdot\hat{\mathbf{k}}_s)] \tag{4.38}$$

$$U_{hv} = -\frac{q|q_z|}{q_z k_1 D^2}[r_v(\hat{\mathbf{h}}_s\cdot\hat{\mathbf{k}}_i)(\hat{\mathbf{v}}_i\cdot\hat{\mathbf{k}}_s) - r_h(\hat{\mathbf{v}}_s\cdot\hat{\mathbf{k}}_i)(\hat{\mathbf{h}}_i\cdot\hat{\mathbf{k}}_s)] \tag{4.39}$$

$$U_{hh} = -\frac{q|q_z|}{q_z k_1 D^2}[r_v(\hat{\mathbf{h}}_s\cdot\hat{\mathbf{k}}_i)(\hat{\mathbf{h}}_i\cdot\hat{\mathbf{k}}_s) + r_h(\hat{\mathbf{v}}_s\cdot\hat{\mathbf{k}}_i)(\hat{\mathbf{v}}_i\cdot\hat{\mathbf{k}}_s)]. \tag{4.40}$$

The local angles $\theta_{il}$ and $\theta_{tl}$ are given in terms of the incident and scattered angles by the relations:

$$\cos\theta_{il} = -\frac{q|q_z|}{2k_1 q_z} \tag{4.41}$$

$$\sin\theta_{tl} = \frac{k_1}{k_2}\sin\theta_{il}. \tag{4.42}$$

To compute the Stokes matrix or the Stokes scattering operator for the rough surface, we must find the correlation product $\langle S_{pq} S_{mn}^* \rangle$. Because the only nondeterministic quantity in $S_{pq}$ is $I$, the problem reduces to computing $\langle |I|^2 \rangle$. Tsang has shown (Tsang *et al.*, 1985) that

$$\langle |I|^2 \rangle = \frac{4\pi^2 A q^2}{q_z^4} p(h_{xs}, h_{ys}), \tag{4.43}$$

where $p(h_{xs}, h_{ys})$ is the probability density function for the surface slopes evaluated at $h_x = h_{xs}$ and $h_y = h_{ys}$, and $A$ is the illuminated area in the $x$-$y$ plane. The result in (4.43) is different from Tsang's result by a factor of $q^2/q_z^2$. This difference arises from the fact that Tsang considered $I$ to be the integral over $d\mathbf{r}'_\perp$, where

$$ds' = (\hat{\mathbf{n}}_1 \cdot \hat{\mathbf{z}})^{-1} d\mathbf{r}'_\perp = \frac{q}{|q_z|} d\mathbf{r}'_\perp \quad . \tag{4.44}$$

By using (4.43) with (4.36) results in the correlation product of scattering matrix elements given by

$$\langle S_{pq} S^*_{mn} \rangle = \frac{A|k_1|^2 q^2}{4q_z^4} p(h_{xs}, h_{ys}) U_{pq} U^*_{mn}. \tag{4.45}$$

When the surface height is Gaussian distributed with zero mean, the probability density function is

$$p(h_x, h_y) = \frac{1}{2\pi m^2} \exp\left[-\frac{h_x^2 + h_y^2}{2m^2}\right], \tag{4.46}$$

where $m = s\sqrt{|\rho''(0)|}$ is the *rms surface slope*, $\rho''(0)$ is the second derivative of the surface correlation function evaluated at $\xi = 0$, and $s$ is the standard deviation of the surface height. We have assumed that $\rho$ depends only on the distance from the origin (i.e., the roughness is isotropic). In this case, the correlation product becomes

$$\langle S_{pq} S^*_{mn} \rangle = \frac{A|k_1|^2 q^2}{8\pi q_z^4 m^2} U_{pq} U^*_{mn} \exp\left[-\frac{q_x^2 + q_y^2}{2q_z^2 m^2}\right] \quad . \tag{4.47}$$

Table 4.1 listed the expressions for $m$ which should be used for each of the correlation functions given by (4.1) to (4.3).

The stationary phase approximation used to obtain the GO solution requires an additional restriction to be placed on $s$ (Ulaby *et al.*, 1986a), namely

$$(q_z s)^2 > 10. \tag{4.48}$$

By using the expression for $q_z$ given by (4.26), this condition becomes

$$k_1 s > \frac{\sqrt{10}}{(\cos\theta_i - \cos\theta_s)}. \tag{4.49}$$

For the backscattering case, we should set $\theta_s = \pi - \theta_i$ and $\phi_s = \pi + \phi_i$ in the preceding expressions.

### 4-1.3 Physical Optics Model

For the GO model, we assumed that the variance of the surface height was large compared to the incident wavelength. This restriction is not required with the PO model. Instead, we assume that the rms surface slope and the change in $r_v$ and $r_h$ are both small. If the rms surface slope is small, we can approximate the local surface normal by

$$\hat{\mathbf{n}}_1 \approx -\hat{\mathbf{x}}h_x(\mathbf{r}'_\perp) - \hat{\mathbf{y}}h_y(\mathbf{r}'_\perp) + \hat{\mathbf{z}}, \tag{4.50}$$

except when computing $\hat{\mathbf{t}}$ and $\hat{\mathbf{d}}$, where we assume $\hat{\mathbf{n}}_1 \approx \hat{\mathbf{z}}$. Therefore, the local coordinate unit vectors are $\hat{\mathbf{t}} \approx -\hat{\mathbf{h}}_i$ and $\hat{\mathbf{d}} \approx \hat{\mathbf{v}}_i$. Furthermore, by requiring the change in the Fresnel reflection coefficients to be small, we can approximate them by a series expansion about zero slopes ($h_x = h_y = 0$), and then keep only up to first-order slope terms. The Fresnel reflection coefficients under this approximation are given by (Ulaby *et al.*, 1986a)

$$r_h \approx r_{h0} + r_{h1}(h_x \cos\theta_i + h_y \sin\theta_i) \tag{4.51}$$

$$r_v \approx r_{v0} + r_{v1}(h_x \cos\theta_i + h_y \sin\theta_i) \tag{4.52}$$

$$r_{h0} = \frac{\eta_2\cos\theta_i + \eta_1\cos\theta_t}{\eta_2\cos\theta_i - \eta_1\cos\theta_t} \tag{4.53}$$

$$r_{h1} = r_{h0}\frac{\eta_2\sin\theta_i + \eta_1\sin\theta_t}{\eta_2\cos\theta_i - \eta_1\cos\theta_t} \tag{4.54}$$

$$r_{v0} = \frac{\eta_1\cos\theta_i + \eta_2\cos\theta_t}{\eta_1\cos\theta_i - \eta_2\cos\theta_t} \tag{4.55}$$

$$r_{v1} = -\frac{[\eta_1\sin\theta_i - \eta_2\sin\theta_t - r_{v0}(\eta_1\sin\theta_i + \eta_2\sin\theta_t)]}{\eta_1\cos\theta_i - \eta_2\cos\theta_t}. \tag{4.56}$$

The angles $\theta_i$, $\theta_s$, and $\theta_t$ are the incident, scattered, and transmitted angles with respect to the mean surface. The angles $\theta_i$ and $\theta_s$ are defined by the incident and scattered directions, and $\theta_t$ is given by

$$\sin\theta_t = \frac{k_1}{k_2}\sin\theta_i. \tag{4.57}$$

Recall from (4.16) that the scattering matrix elements are of the form:

$$S_{pq} = \frac{ik_1}{4\pi}\int_S U_{pq} e^{ik_1(\hat{\mathbf{k}}_i - \hat{\mathbf{k}}_s)\cdot\mathbf{r}'} ds', \tag{4.58}$$

where

$$U_{pq} = \hat{\mathbf{p}}_s \cdot \mathbf{U}(h_x, h_y)|_{\hat{\mathbf{a}}=\hat{\mathbf{q}}_i}. \tag{4.59}$$

If $U_{pq}$ is also approximated by a series expansion about zero slopes, by keeping terms up to first order in slope, we obtain (Ulaby *et al.*, 1986a; Ulaby *et al.*, 1988)

$$U_{pq} \approx a_{pq} + b_{pq} h_x + c_{pq} h_y, \tag{4.60}$$

where

$$a_{pq} = \begin{cases} r_{v0}(\cos\theta_i - \cos\theta_s)\cos(\phi_s - \phi_i) & pq = vv \\ r_{h0}(\cos\theta_i \cos\theta_s - 1)\sin(\phi_s - \phi_i) & pq = vh \\ r_{v0}(1 - \cos\theta_i \cos\theta_s)\sin(\phi_s - \phi_i) & pq = hv \\ r_{h0}(\cos\theta_i - \cos\theta_s)\cos(\phi_s - \phi_i) & pq = hh \end{cases} \tag{4.61}$$

$$b_{pq} = Z_{pq}\cos\phi_i \tag{4.62}$$

$$c_{pq} = Z_{pq}\sin\phi_i, \tag{4.63}$$

and

$$Z_{hh} = r_{h0}[\sin\theta_s - \sin\theta_i\cos(\phi_s - \phi_i)] + r_{h1}(\cos\theta_i - \cos\theta_s)\cos(\phi_s - \phi_i) \tag{4.64}$$

$$Z_{vh} = -\sin(\phi_s - \phi_i)[r_{v0}\sin\theta_i\cos\theta_s + r_{h1}(1 - \cos\theta_i\cos\theta_s)] \tag{4.65}$$

$$Z_{hv} = \sin(\phi_s - \phi_i)[r_{v0}\sin\theta_i\cos\theta_s + r_{v1}(1 - \cos\theta_i\cos\theta_s)] \tag{4.66}$$

$$Z_{vv} = r_{v0}[\sin\theta_s - \sin\theta_i\cos(\phi_s - \phi_i)] + r_{v1}(\cos\theta_i - \cos\theta_s)\cos(\phi_s - \phi_i). \tag{4.67}$$

If we let $I_{pq}$ denote the integral factor in (4.58), the correlation of the scattering matrix elements is

$$\langle S_{pq} S^*_{mn}\rangle = \frac{|k_1|^2}{16\pi^2}\langle I_{pq} I^*_{mn}\rangle, \tag{4.68}$$

where

$$\langle I_{pq} I^*_{mn}\rangle = \left\langle \int_{S'}\int_S U_{pq} U^*_{mn} e^{ik_1(\hat{\mathbf{k}}_i - \hat{\mathbf{k}}_s)\cdot(\mathbf{r}-\mathbf{r}')} ds\, ds' \right\rangle. \tag{4.69}$$

By keeping only terms up to first order in surface slope, the product $U_{pq}U^*_{mn}$ becomes

$$U_{pq}U^*_{mn} \approx a_{pq}a^*_{mn} + a_{pq}b^*_{mn}h_{x'} + b_{pq}a^*_{mn}h_x + a_{pq}c^*_{mn}h_{y'} + c_{pq}a^*_{mn}h_y. \tag{4.70}$$

Integration of the first term of (4.70) contributes both *coherent* and *incoherent* scattering terms, which will be denoted as $I_c$ and $I_i$, respectively. Therefore,

$$I_c + I_i = \left\langle \int_{-L}^{L}\int_{-L}^{L} a_{pq}a^*_{mn} e^{ik_1(\hat{\mathbf{k}}_i - \hat{\mathbf{k}}_s)\cdot(\mathbf{r}-\mathbf{r}')} ds\, ds' \right\rangle, \tag{4.71}$$

where $A = (2L)^2$ is the area illuminated by the antenna beam.

Assuming $h(\mathbf{r}_\perp)$ to be a stationary Gaussian random process with zero mean, variance $s^2$, and correlation function $\rho$, we obtain

$$\begin{aligned} I_c + I_i &= a_{pq} a^*_{mn} \int_{-2L}^{2L}\int_{-2L}^{2L} (2L - |u|)(2L - |v|) \\ &\cdot e^{i(q_x u + q_y v)} e^{-q_z^2 s^2 [1-\rho(u,v)]} \, du \, dv. \end{aligned} \tag{4.72}$$

Here, we have used the expression:

$$\left\langle e^{iq_z[h(x,y)-h(x',y')]} \right\rangle = e^{-q_z^2 s^2 [1-\rho(u,v)]} \tag{4.73}$$

for the characteristic function of a Gaussian random process and made the substitutions $u = x - x'$ and $v = y - y'$. Under the assumption of isotropic roughness, $\rho(u,v) = \rho(\xi)$, the series expression:

$$e^{q_z^2 s^2 \rho(\xi)} = \sum_{n=0}^{\infty} \frac{(q_z s)^{2n}}{n!} \rho^n(\xi) \tag{4.74}$$

can be used to simplify the integral in (4.72). Considering only the $n = 0$ term, we obtain for large $L$:

$$I_c = a_{pq} a^*_{mn} (2\pi)^2 A \delta(q_x) \delta(q_y) e^{-q_z^2 s^2}, \tag{4.75}$$

where the product of the delta functions in (4.75) is given by

$$\delta(q_x)\delta(q_y) = \begin{cases} \dfrac{A}{(2\pi)^2} & q_x = q_y = 0 \\ 0 & \text{otherwise} \end{cases}. \tag{4.76}$$

The remaining part of the infinite summation ($n > 0$) contributes to the incoherent intensity. Changing the $u, v$ coordinates into polar coordinates $\xi$, $\alpha$, and evaluating the integral with respect to $\alpha$, gives

$$I_i = 2\pi A a_{pq} a^*_{mn} e^{-q_z^2 s^2} \sum_{n=1}^{\infty} \frac{(q_z s)^{2n}}{n!} \int_0^{\infty} \rho(\xi)^n J_0(q_t \xi) \xi \, d\xi, \tag{4.77}$$

where $q_t^2 = q_x^2 + q_y^2$.

The remaining four terms of (4.70) depend on surface slopes, and also contribute to the incoherent intensity. We shall denote this contribution by $I_s = I_{sx} + I_{sy}$. Using the expressions (Ulaby *et al.*, 1986a):

$$\begin{aligned} \langle h_x e^{iq_z[h(x,y)-h(x',y')]} \rangle &= \langle h_{x'} e^{iq_z[h(x,y)-h(x',y')]} \rangle \\ &= -iq_z s^2 \cos\alpha \frac{\partial \rho(\xi)}{\partial \xi} e^{q_z^2 s^2 [1-\rho(\xi)]} \end{aligned} \tag{4.78}$$

$$\begin{aligned}\langle h_y e^{iq_z[h(x,y)-h(x',y')]}\rangle &= \langle h_{y'} e^{iq_z[h(x,y)-h(x',y')]}\rangle \\ &= -iq_z s^2 \sin\alpha \frac{\partial \rho(\xi)}{\partial \xi} e^{q_z^2 s^2[1-\rho(\xi)]} \quad (4.79)\end{aligned}$$

in (4.69) and (4.70), and performing the $\alpha$ integration, $I_{sx}$ and $I_{sy}$ become

$$I_{sx} = 2\pi A K_x e^{-q_z^2 s^2} \frac{q_x}{q_t} \int_0^\infty \frac{d\rho(\xi)}{d\xi} J_1(q_t\xi) e^{q_z^2 s^2 \rho(\xi)} \xi \, d\xi \quad (4.80)$$

$$I_{sy} = 2\pi A K_y e^{-q_z^2 s^2} \frac{q_y}{q_t} \int_0^\infty \frac{d\rho(\xi)}{d\xi} J_1(q_t\xi) e^{q_z^2 s^2 \rho(\xi)} \xi \, d\xi. \quad (4.81)$$

Again, the roughness has been assumed to be isotropic and Gaussian distributed. The factors $K_x$ and $K_y$ are functions of the polarization-dependent expansion coefficients in (4.61) to (4.63):

$$K_x = q_z s^2 (b_{pq} a^*_{mn} + a_{pq} b^*_{mn}) \quad (4.82)$$

$$K_y = q_z s^2 (c_{pq} a^*_{mn} + a_{pq} c^*_{mn}). \quad (4.83)$$

By combining the three terms $I_c$, $I_i$, and $I_s$, the correlation product of the scattering matrix elements can be written as

$$\langle S_{pq} S^*_{mn}\rangle = \frac{|k_1|^2}{16\pi^2} (I_c + I_i + I_s). \quad (4.84)$$

The first term represents coherent scattering from the surface and is present only in the specular direction with respect to the mean surface. The second term represents incoherent scattering, and the third term represents incoherent scattering due to the surface slopes.

We now consider the three correlation functions given by (4.1) to (4.3). For $\rho_1(\xi)$ and $\rho_3(\xi)$, the integrals in (4.77), (4.80), and (4.81) can be evaluated analytically. With the Gaussian correlation function $\rho_1(\xi)$, we obtain

$$I_i = \pi l^2 A a_{pq} a^*_{mn} e^{-q_z^2 s^2} \sum_{n=1}^{\infty} \frac{(q_z s)^{2n}}{n!n} \exp\left[-\frac{q_t^2 l^2}{4n}\right] \quad (4.85)$$

$$I_{sx} = -\pi A q_x K_x l^2 e^{-q_z^2 s^2} \sum_{n=1}^{\infty} \frac{(q_z s)^{2(n-1)}}{n!n} \exp\left[-\frac{q_t^2 l^2}{4n}\right] \quad (4.86)$$

$$I_{sy} = -\pi A q_y K_y l^2 e^{-q_z^2 s^2} \sum_{n=1}^{\infty} \frac{(q_z s)^{2(n-1)}}{n!n} \exp\left[-\frac{q_t^2 l^2}{4n}\right], \quad (4.87)$$

where we have made use of the definite integral expression:

$$\int_0^\infty \xi^{n+1} e^{-\alpha\xi^2} J_n(\beta\xi) \, d\xi = \frac{\beta^n}{(2\alpha)^{n+1}} \exp\left[-\frac{\beta^2}{4\alpha}\right]. \quad (4.88)$$

A similar evaluation using $\rho_3(\xi)$, leads to expressions identical with those given by (4.85) to (4.87), subsequent to replacing

$$\exp\left[-\frac{q_t^2 l^2}{4n}\right] \text{ with } f(n, lq_t) = \frac{n2^{1-\mu}(lq_t)^\mu}{\Gamma(\mu+1)} K_{-\mu}(lq_t), \tag{4.89}$$

where $\mu = 1.5n - 1$ (Ulaby *et al.*, 1986a). The functions $\Gamma$ and $K_\nu$ are the gamma function and the modified Bessel function (Gradsteyn and Ryzhik, 1980), respectively. For the remaining correlation function, $\rho_2(\xi)$, the integrals in (4.77), (4.80), and (4.81) must be evaluated numerically .

As with the GO model, the PO model is valid only in the range given by (4.20) and (4.21). Another condition limiting the rms slope is usually given in the literature (Ulaby *et al.*, 1986a) as

$$m < 0.25. \tag{4.90}$$

### 4-1.4 Small Perturbation Model

Scattering from a slightly rough surface can be obtained by using the Rayleigh hypothesis to express the transmitted and scattered fields in terms of waves propagating away from the interface. In general, waves traveling in both the upward and downward directions will exist in the region between the minimum and maximum height of the rough surface. However, the Rayleigh hypothesis neglects the scattered waves traveling toward the interface. The transmitted and scattered fields are usually written in terms of a Fourier series expansion (or a Fourier transform), and a power series expansion is obtained in terms of some small quantity such as the surface height variation in wavelengths or the slope of the surface. This technique is known as the small perturbation method, originally used by Rice (Rice, 1951). Additional information on the method is available in the literature (Ishimaru, 1978; Ulaby *et al.*, 1986a).

The small perturbation method can also be formulated by using the *extended boundary condition* (EBC) method (Agarwal, 1977; Nieto-Vesperinas, 1982; Tsang *et al.*, 1985). The two approaches, however, generate the same expansion for the scattered field, so we will follow Tsang *et al.*, (1985) and use the EBC method. First, we must use Huygen's principle and the extinction theorem to formulate diffraction integrals in terms of the surface fields. A series expansion is then used to represent the surface fields, and this expansion is computed up to the *n*th order. We can then use the surface fields to compute the *n*th-order scattered field by using a diffraction integral similar to equation (4.8). In practice, the method becomes difficult above second order, and we will consider here only the first-order solution.

From Tsang *et al.*, (1985), the total field $\mathbf{E}_1(\mathbf{r})$ in region 1 and the transmitted field $\mathbf{E}_2(\mathbf{r})$ in region 2 satisfy

$$\int_{S'} \{i\omega\mu_1 \mathcal{G}_1(\mathbf{r},\mathbf{r}')\cdot[\hat{\mathbf{n}}_1 \times \mathbf{H}_1(\mathbf{r}')] + \nabla \times \mathcal{G}_1(\mathbf{r},\mathbf{r}')\cdot[\hat{\mathbf{n}}_1 \times \mathbf{E}_1(\mathbf{r}')]\}\, ds' = \begin{cases} \mathbf{E}_1(\mathbf{r}) - \mathbf{E}^i(\mathbf{r}) & z > h(\mathbf{r}_\perp) \\ -\mathbf{E}^i(\mathbf{r}) & z < h(\mathbf{r}_\perp) \end{cases} \tag{4.91}$$

$$\int_{S'} \{i\omega\mu_2 \mathcal{G}_2(\mathbf{r},\mathbf{r}')\cdot[\hat{\mathbf{n}}_2 \times \mathbf{H}_2(\mathbf{r}')] + \nabla \times \mathcal{G}_2(\mathbf{r},\mathbf{r}')\cdot[\hat{\mathbf{n}}_2 \times \mathbf{E}_2(\mathbf{r}')]\}\, ds' = \begin{cases} 0 & z > h(\mathbf{r}_\perp) \\ \mathbf{E}_2(\mathbf{r}) & z < h(\mathbf{r}_\perp) \end{cases}, \tag{4.92}$$

where $\hat{\mathbf{n}}_1$ and $\hat{\mathbf{n}}_2$ are the unit vectors normal to the rough surface and pointing into regions 1 and 2, respectively. Also, $\mathcal{G}_1(\mathbf{r},\mathbf{r}')$ and $\mathcal{G}_2(\mathbf{r},\mathbf{r}')$ are the dyadic Green's functions for the two regions. We must now define unit polarization vectors, similar to (4.4) through (4.6), for the incident, scattered, and transmitted fields:

$$\hat{\mathbf{k}}_s^\pm = \hat{\mathbf{x}}\sin\theta_s\cos\phi_s + \hat{\mathbf{y}}\sin\theta_s\sin\phi_s \pm \hat{\mathbf{z}}\cos\theta_s \tag{4.93}$$

$$\hat{\mathbf{h}}_s^\pm = -\hat{\mathbf{x}}\sin\phi_s + \hat{\mathbf{y}}\cos\phi_s \tag{4.94}$$

$$\hat{\mathbf{v}}_s^\pm = \hat{\mathbf{h}}_s^\pm \times \hat{\mathbf{k}}_s^\pm = \pm\hat{\mathbf{x}}\cos\theta_s\cos\phi_s \pm \hat{\mathbf{y}}\cos\theta_s\sin\phi_s - \hat{\mathbf{z}}\sin\theta_s \tag{4.95}$$

$$\hat{\mathbf{k}}_f^\pm = \hat{\mathbf{x}}\sin\theta_f\cos\phi_f + \hat{\mathbf{y}}\sin\theta_f\sin\phi_f \mp \hat{\mathbf{z}}\cos\theta_f \tag{4.96}$$

$$\hat{\mathbf{h}}_f^\pm = -\hat{\mathbf{x}}\sin\phi_f + \hat{\mathbf{y}}\cos\phi_f \tag{4.97}$$

$$\hat{\mathbf{v}}_f^\pm = \hat{\mathbf{h}}_f^\pm \times \hat{\mathbf{k}}_f^\pm = \mp\hat{\mathbf{x}}\cos\theta_f\cos\phi_f \mp \hat{\mathbf{y}}\cos\theta_f\sin\phi_f - \hat{\mathbf{z}}\sin\theta_f, \tag{4.98}$$

where the subscript $f \in \{i,t\}$ denotes either the incident or the transmitted wave. The + and − superscripts denote upward and downward traveling waves, respectively. In addition to these unit vectors, we define the transverse vector wavenumbers $\mathbf{k}_\perp = \hat{\mathbf{x}}k_x + \hat{\mathbf{y}}k_y$ for a general wave and $\mathbf{k}_{\perp i} = \hat{\mathbf{x}}k_{xi} + \hat{\mathbf{y}}k_{yi}$ for the incident wave.

With the above definitions, we can write integral representations for the dyadic Green's functions in regions 1 and 2 (Tsang *et al.*, 1985):

$$\mathcal{G}_1(\mathbf{r},\mathbf{r}') = -\hat{\mathbf{z}}\hat{\mathbf{z}}\frac{\delta(\mathbf{r},\mathbf{r}')}{k_1^2} + \frac{i}{8\pi^2}\int \frac{1}{k_{1z}}\left[\hat{\mathbf{h}}_s^\pm\hat{\mathbf{h}}_s^\pm + \hat{\mathbf{v}}_s^\pm\hat{\mathbf{v}}_s^\pm\right] \cdot e^{i\mathbf{k}_\perp\cdot(\mathbf{r}-\mathbf{r}')} e^{\pm ik_{1z}(z-z')}\, d\mathbf{k}_\perp \tag{4.99}$$

$$\mathcal{G}_2(\mathbf{r},\mathbf{r}') = -\hat{\mathbf{z}}\hat{\mathbf{z}}\frac{\delta(\mathbf{r},\mathbf{r}')}{k_2^2} + \frac{i}{8\pi^2}\int \frac{1}{k_{2z}}\left[\hat{\mathbf{h}}_t^\pm\hat{\mathbf{h}}_t^\pm + \hat{\mathbf{v}}_t^\pm\hat{\mathbf{v}}_t^\pm\right] \cdot e^{i\mathbf{k}_\perp\cdot(\mathbf{r}-\mathbf{r}')} e^{\pm ik_{2z}(z-z')}\, d\mathbf{k}_\perp, \tag{4.100}$$

where the + and − signs refer to the regions $z > z'$ and $z < z'$, respectively. The $z$ components of the vector wavenumbers are given by

$$\begin{aligned} k_{1z} &= k_1 \cos\theta_s &(4.101)\\ k_{1zi} &= k_1 \cos\theta_i &(4.102)\\ k_{2z} &= \sqrt{k_2^2 - k_1^2 \sin^2\theta_s} &(4.103)\\ k_{2zi} &= \sqrt{k_2^2 - k_1^2 \sin^2\theta_i}. &(4.104)\end{aligned}$$

Using (4.99) in (4.91) for $z < h(\mathbf{r}'_\perp)$ and (4.100) in (4.92) for $z > h(\mathbf{r}'_\perp)$, we obtain the extended boundary conditions:

$$\begin{aligned} \mathbf{E}^i(\mathbf{r}) &= \frac{1}{8\pi^2}\int e^{i\mathbf{k}_\perp\cdot\mathbf{r}_\perp} e^{-ik_{1z}z}\frac{k_1}{k_{1z}}\int e^{-i\mathbf{k}_\perp\cdot\mathbf{r}'_\perp} e^{ik_{1z}h(\mathbf{r}'_\perp)}\\ &\cdot\left\{\left[\hat{\mathbf{h}}_s^-\hat{\mathbf{h}}_s^- + \hat{\mathbf{v}}_s^-\hat{\mathbf{v}}_s^-\right]\cdot\mathbf{a}(\mathbf{r}'_\perp) + \left[-\hat{\mathbf{v}}_s^-\hat{\mathbf{h}}_s^- + \hat{\mathbf{h}}_s^-\hat{\mathbf{v}}_s^-\right]\cdot\mathbf{b}(\mathbf{r}'_\perp)\right\}\\ &\cdot d\mathbf{r}'_\perp\, d\mathbf{k}_\perp &(4.105)\\ 0 &= \frac{1}{8\pi^2}\int e^{i\mathbf{k}_\perp\cdot\mathbf{r}_\perp} e^{ik_{2z}z}\frac{k_2}{k_{2z}}\int e^{-i\mathbf{k}_\perp\cdot\mathbf{r}'_\perp} e^{-ik_{2z}h(\mathbf{r}'_\perp)}\\ &\cdot\left\{\frac{k_1}{k_2}\left[\hat{\mathbf{h}}_t^+\hat{\mathbf{h}}_t^+ + \hat{\mathbf{v}}_t^+\hat{\mathbf{v}}_t^+\right]\cdot\mathbf{a}(\mathbf{r}'_\perp) + \left[-\hat{\mathbf{v}}_t^+\hat{\mathbf{h}}_t^+ + \hat{\mathbf{h}}_t^+\hat{\mathbf{v}}_t^+\right]\cdot\mathbf{b}(\mathbf{r}'_\perp)\right\}\\ &\cdot d\mathbf{r}'_\perp\, d\mathbf{k}_\perp, &(4.106)\end{aligned}$$

where the unknown surface field functions, $\mathbf{a}(\mathbf{r}'_\perp)$ and $\mathbf{b}(\mathbf{r}'_\perp)$, are given by the relations:

$$\begin{aligned} \eta_j\hat{\mathbf{n}}_1\times\mathbf{H}_j(\mathbf{r}')\,ds' &= \mathbf{a}(\mathbf{r}'_\perp)\,d\mathbf{r}'_\perp &(4.107)\\ \hat{\mathbf{n}}_1\times\mathbf{E}_j(\mathbf{r}')\,ds' &= \mathbf{b}(\mathbf{r}'_\perp)\,d\mathbf{r}'_\perp. &(4.108)\end{aligned}$$

By noting from (4.107) and (4.108) that $\mathbf{a}(\mathbf{r}'_\perp)$ and $\mathbf{b}(\mathbf{r}'_\perp)$ must be tangential to $\hat{\mathbf{n}}_1$, we obtain two additional vector equations for the surface field unknowns:

$$\hat{\mathbf{z}}\cdot\mathbf{w}(\mathbf{r}'_\perp) = (\hat{\mathbf{x}}\frac{\partial h(\mathbf{r}'_\perp)}{\partial x'} + \hat{\mathbf{y}}\frac{\partial h(\mathbf{r}'_\perp)}{\partial y'})\cdot\mathbf{w}_\perp(\mathbf{r}'_\perp), \quad \mathbf{w}\in\{\mathbf{a},\mathbf{b}\}. \qquad (4.109)$$

Next, we write the surface field unknowns and the surface height function in the form of series expansions in powers of $k_1 s$:

$$\begin{aligned} \mathbf{w}(\mathbf{r}'_\perp) &= \sum_{n=0}^{\infty}\frac{(k_1 s)^{(n)}}{n!}\mathbf{w}^{(n)}(\mathbf{r}'_\perp), \quad \mathbf{w}\in\{\mathbf{a},\mathbf{b}\} &(4.110)\\ h(\mathbf{r}'_\perp) &= \sum_{n=0}^{\infty}\frac{(k_1 s)^{(n)}}{n!}h^{(n)}(\mathbf{r}'_\perp). &(4.111)\end{aligned}$$

We must also expand the exponential terms that depend on the surface height into series expressions;

$$e^{\pm ik_{jz}h(\mathbf{r}'_\perp)} = \sum_{n=0}^{\infty} \frac{1}{n!}\left[\pm ik_{jz}h(\mathbf{r}'_\perp)\right]^n, \quad j \in \{1,2\}. \tag{4.112}$$

By substituting (4.110) and (4.111) into (4.109) and equating the coefficients of equal powers of $k_1 s$, we can obtain recurrence relations for the surface unknown coefficients. To simplify the expressions, we will assume that the surface height function can be approximated by

$$h(\mathbf{r}'_\perp) \approx k_1 s \; h^{(1)}(\mathbf{r}'_\perp). \tag{4.113}$$

The zero-order coefficient of $h(\mathbf{r}'_\perp)$ is equal to zero (i.e. $h^{(0)}(\mathbf{r}'_\perp) = 0$) because we assume that the surface has a mean height of zero. Substituting the series expressions of (4.110), (4.112), and (4.113) into the extended boundary conditions of (4.105) and (4.115), and equating $n$th-order terms, we obtain

$$\hat{\mathbf{z}}\cdot\mathbf{w}^{(0)}(\mathbf{r}'_\perp) = 0 \tag{4.114}$$

$$\hat{\mathbf{z}}\cdot\mathbf{w}^{(n)}(\mathbf{r}'_\perp) = n\left(\hat{\mathbf{x}}\frac{\partial h^{(1)}(\mathbf{r}'_\perp)}{\partial x'} + \hat{\mathbf{y}}\frac{\partial h^{(1)}(\mathbf{r}'_\perp)}{\partial y'}\right)\cdot\mathbf{w}_\perp^{(n-1)}(\mathbf{r}'_\perp). \tag{4.115}$$

Once the surface fields $\mathbf{a}(\mathbf{r}'_\perp)$ and $\mathbf{b}(\mathbf{r}'_\perp)$ are determined up to $n$th order by using (4.105), (4.106), (4.114), and (4.115), we can calculate the scattered field in region 1 by way of

$$\begin{aligned}\mathbf{E}^s(\mathbf{r}) \;=\; & -\frac{1}{8\pi^2}\int e^{i\mathbf{k}_\perp\cdot\mathbf{r}_\perp} e^{ik_{1z}}\frac{k_1}{k_{1z}}\int e^{-i\mathbf{k}_\perp\cdot\mathbf{r}'_\perp} e^{-ik_{1z}h(\mathbf{r}'_\perp)} \\ & \cdot\left\{\left[\hat{\mathbf{h}}_s^+\hat{\mathbf{h}}_s^+ + \hat{\mathbf{v}}_s^+\hat{\mathbf{v}}_s^+\right]\cdot\mathbf{a}(\mathbf{r}'_\perp) + \left[-\hat{\mathbf{v}}_s^+\hat{\mathbf{h}}_s^+ + \hat{\mathbf{h}}_s^+\hat{\mathbf{v}}_s^+\right]\cdot\mathbf{b}(\mathbf{r}'_\perp)\right\} \\ & \cdot d\mathbf{r}'_\perp\, d\mathbf{k}_\perp. \end{aligned} \tag{4.116}$$

The correlation product for the first order scattered field, from Tsang *et al.*, (1985), is given by

$$\langle E_{pq}^{s(1)} E_{mn}^{s(1)*}\rangle = |E_0|^2 \int f_{pq} f_{mn}^* W(|\mathbf{k}_\perp - \mathbf{k}_{\perp i}|)\, d\mathbf{k}_\perp, \tag{4.117}$$

where the polarization dependent factors $f_{pq}$ are

$$f_{vv} = \frac{2k_2^2 k_{1zi}(k_2^2 - k_1^2)}{(k_1^2 k_{2z} + k_2^2 k_{1z})(k_1^2 k_{2zi} - k_2^2 k_{1zi})}\left[k_\rho k_{\rho i} - \frac{k_1^2}{k_2^2} k_{2z} k_{2zi} K_1\right] \tag{4.118}$$

$$f_{vh} = \frac{2k_1 k_{2z} k_{1zi}(k_2^2 - k_1^2)}{(k_1^2 k_{2z} + k_2^2 k_{1z})(k_{2zi} - \quad -k_{1zi})} K_2 \tag{4.119}$$

$$f_{hv} = -\frac{2k_1 k_{2zi} k_{1zi}(k_2^2 - k_1^2)}{(k_{2z} + k_{1z})(k_1^2 k_{2zi} - k_2^2 \quad - k_{1zi})} K_2 \tag{4.120}$$

$$f_{hh} = \frac{2k_{1zi}(k_2^2 - k_1^2)}{(k_{2z} + k_{1z})(k_{2zi} - k_{1zi})} K_1, \tag{4.121}$$

with

$$K_1 = \frac{k_x k_{xi} + k_y k_{yi}}{k_\rho k_{\rho i}} \tag{4.122}$$

$$K_2 = \frac{k_y k_{xi} - k_x k_{yi}}{k_\rho k_{\rho i}}. \tag{4.123}$$

The function $W(|\mathbf{k}_\perp - \mathbf{k}_{\perp i}|)$ is the spectral density of the rough surface, or simply the Fourier transform of the correlation function $\rho(\mathbf{r}_\perp)$, defined by

$$W(\mathbf{k}_\perp) = \frac{s^2}{(2\pi)^2} \int \rho(\mathbf{r}_\perp) e^{i\mathbf{k}_\perp \cdot \mathbf{r}_\perp} \, d\mathbf{r}_\perp. \tag{4.124}$$

From (4.117), we have an expression of the form:

$$\langle E_{pq}^s E_{mn}^{s*} \rangle = |E_0|^2 \int f(k_x, k_y) \, d\mathbf{k}_\perp = |E_0|^2 \int k_1^2 \cos\theta_s f(k_x, k_y) \, d\Omega_s, \tag{4.125}$$

where the superscript denoting the first-order solution has been dropped, and

$$f(k_x, k_y) = f_{pq} f_{mn}^* W(|\mathbf{k}_\perp - \mathbf{k}_{\perp i}|). \tag{4.126}$$

Denoting $\Delta\langle E_{pq}^s E_{mn}^{s*} \rangle$ as the intensity within the narrow spectral bands $\Delta k_x \Delta k_y$ centered at $k_x$ and $k_y$, we obtain (Ulaby *et al.*, 1986a)

$$\Delta\langle E_{pq}^s E_{mn}^{s*} \rangle \approx |E_0|^2 f(k_x, k_y) \Delta k_x \Delta k_y = |E_0|^2 k_1^2 \cos\theta_s f(k_x, k_y) \Delta\Omega_s. \tag{4.127}$$

Thus, the average intensity per unit solid angle is

$$\frac{\Delta\langle E_{pq}^s E_{mn}^{s*} \rangle}{\Delta\Omega_s} = |E_0|^2 k_1^2 \cos\theta_s f(k_x, k_y). \tag{4.128}$$

The average intensity $P$ at a receiver located at a distance $r$ from the illuminated area $A$ is equal to the average intensity per unit solid angle times the solid angle subtended by the receiver, or equivalently

$$P = \frac{\Delta\langle E_{pq}^s E_{mn}^{s*} \rangle}{\Delta\Omega_s} \left( \frac{A\cos\theta_s}{r^2} \right) = \frac{|E_0|^2}{r^2} A k_1^2 \cos^2\theta_s f(k_x, k_y). \tag{4.129}$$

The correlation product of the scattering matrix elements, $\langle S_{pq}S^*_{mn}\rangle$, is therefore given by

$$\langle S_{pq}S^*_{mn}\rangle = \frac{r^2}{|E_0|^2}P = Ak_1^2\cos^2\theta_s f_{pq}f^*_{mn}W(|\mathbf{k}_\perp - \mathbf{k}_{\perp i}|). \qquad (4.130)$$

For a surface with a Gaussian correlation function of the form in (4.1), the spectral density is

$$W(|\mathbf{k}_\perp - \mathbf{k}_{\perp i}|) = \frac{(sl)^2}{4\pi}\exp\left[-\frac{l^2}{4}|\mathbf{k}_\perp - \mathbf{k}_{\perp i}|^2\right], \qquad (4.131)$$

where

$$|\mathbf{k}_\perp - \mathbf{k}_{\perp i}|^2 = k_1^2\left[\sin^2\theta_s + \sin^2\theta_i - 2\sin\theta_s\sin\theta_i\cos(\phi_s - \phi_i)\right]. \qquad (4.132)$$

The validity conditions usually associated with the small perturbation model are (Ulaby *et al.*, 1986a)

$$k_1 s < 0.3 \qquad (4.133)$$

$$m < 0.3 \qquad (4.134)$$

(see Table 4.1 for $m$ in terms of $l$ and $s$). However, a recent paper by Chen and Fung (Chen and Fung, 1988) has shown the need for a small correlation length. On the basis of results given by Chen and Fung, this new condition can be roughly stated as

$$k_1 l < 3.0. \qquad (4.135)$$

However, Chen and Fung used only backscatter data to check validity, so the results may not be accurate for bistatic scattering.

### 4-1.5 Polarization Synthesis for Rough Surface Models

In the previous three subsections (4-1.2 to 4-1.4), first-order scattering models were derived under the geometrical optics, physical optics, and small perturbation approximations. The correlation product of scattering matrix elements, $\langle S_{pq}S^*_{mn}\rangle$, was given for each model by equations (4.45), (4.84), and (4.130), respectively. Because the results were presented in the FSA convention, the average matrices $\mathcal{M}$, $\mathcal{M}_m$, and $\mathcal{C}$ are defined in the FSA convention by

$$\mathcal{M} = \tilde{\mathcal{R}}^{-1}\langle\mathcal{W}\rangle\mathcal{R}^{-1} \tag{4.136}$$

$$\mathcal{M}_m = \tilde{\mathcal{V}}^{-1}\langle\mathcal{W}\rangle\mathcal{V}^{-1} \tag{4.137}$$

$$\mathcal{C} = \langle\mathcal{X}\tilde{\mathcal{X}}^*\rangle, \tag{4.138}$$

where $\mathcal{M}$ is the average Stokes scattering operator, $\mathcal{M}_m$ is the average modified Stokes scattering operator matrix, $\mathcal{C}$ is the average coherency matrix, and

$$\mathcal{W} = \begin{bmatrix} S_{vv}S_{vv}^* & S_{vh}S_{vh}^* & S_{vh}S_{vv}^* & S_{vv}S_{vh}^* \\ S_{hv}S_{hv}^* & S_{hh}S_{hh}^* & S_{hh}S_{hv}^* & S_{hv}S_{hh}^* \\ S_{hv}S_{vv}^* & S_{hh}S_{vh}^* & S_{hh}S_{vv}^* & S_{hv}S_{vh}^* \\ S_{vv}S_{hv}^* & S_{vh}S_{hh}^* & S_{vh}S_{hv}^* & S_{vv}S_{hh}^* \end{bmatrix} \tag{4.139}$$

$$\mathcal{X} = \begin{bmatrix} S_{vv} \\ S_{vh} \\ S_{hv} \\ S_{hh} \end{bmatrix}. \tag{4.140}$$

To compute the scattering coefficient $\sigma_{rt}^0$ as a function of polarization using the techniques of Chapter 2, these average matrices must be converted to the BSA convention. By using (2.113), (2.114), and (2.116) of Chapter 2, the matrices $\overline{\mathcal{M}}$, $\overline{\mathcal{M}}_m$, and $\overline{\mathcal{C}}$ are defined in the BSA convention such that

$$\overline{\mathcal{M}} = \mathcal{T}\mathcal{M} \tag{4.141}$$

$$\overline{\mathcal{M}}_m = \mathcal{T}\mathcal{M}_m \tag{4.142}$$

$$\overline{\mathcal{C}} = \mathcal{T}\mathcal{C}\mathcal{T}, \tag{4.143}$$

where

$$\mathcal{T} = \begin{bmatrix} 1 & 0 & 0 & 0 \\ 0 & 1 & 0 & 0 \\ 0 & 0 & -1 & 0 \\ 0 & 0 & 0 & -1 \end{bmatrix}. \tag{4.144}$$

Corresponding to each of these matrices, the scattering coefficient $\sigma_{rt}^0$ is given by equations (2.75) to (2.77), respectively. For completeness, these expressions are repeated in (4.145) to (4.147) below:

$$\sigma_{rt}^{0}(\psi_r, \chi_r, \psi_t, \chi_t) = \frac{4\pi}{A}\overline{\mathbf{A}}^{r} \cdot \overline{\mathcal{M}}\,\overline{\mathbf{A}}^{t} \tag{4.145}$$

$$\sigma_{rt}^{0}(\psi_r, \chi_r, \psi_t, \chi_t) = \frac{4\pi}{A}\overline{\mathbf{A}}_m^{r} \cdot \overline{\mathcal{M}}_m \overline{\mathbf{A}}_m^{r} \tag{4.146}$$

$$\sigma_{rt}^{0}(\psi_r, \chi_r, \psi_t, \chi_t) = \frac{4\pi}{A}\overline{\mathbf{B}}^{*} \cdot \overline{\mathcal{C}}\,\overline{\mathbf{B}} \tag{4.147}$$

As an example of the polarization responses obtained with the GO, PO, and first order SPM rough surface models, Figures 4.2, 4.3, and 4.4 illustrate typical plots for backscatter from two different surfaces at an incidence angle of $\theta_0 = 20°$ ($\theta_0 = \theta_s = \pi - \theta_i$). For the GO and PO surfaces (Figures 4.2 and 4.3), the standard deviation of surface height and the surface correlation length are $s = 10.0$ cm and $l = 1.0$ m, respectively. At the wavelength of 18.7 cm (L-band) that was used for the plots, these quantities correspond to $k_1 s = 3.36$ and $k_1 l = 33.6$, which are well within the regions of validity stated for both the GO and PO models. Because this surface is outside the region of validity for the SPM model, a different surface is assumed for the example polarization response shown in Figure 4.4 using the SPM model. The surface considered in this case is characterized by $s = 8.0$ mm and $l = 5.0$ cm ($k_1 s = 0.269$ and $k_1 l = 1.68$ for $\lambda = 18.7$ cm).

If we compare the three models considered, the structures of the polarization responses are similar. For the GO model, $\sigma_{rt}^{0}$ is independent of the orientation angle $\psi$, with the maximum co-polarization response occurring at $\chi = 0°$ (linear polarization) and the minimum at $\chi = \pm 45°$ (circular polarization). The cross-polarization response is just the opposite, with the maximum (minimum) being at $\chi = \pm 45°$ ($\chi = 0°$). The response for the PO model, although similar to the response for the GO model, predicts a variation in the co-polarization $\sigma_{rt}^{0}$ with the orientation angle $\psi$. From Figure 4.3, the maximum co-polarization response occurs at vertical ($vv$) polarization. The variation with $\psi$ decreases as the ellipticity angle approaches $\chi = \pm 45°$. The cross-polarization response for the PO model shows no variation with $\psi$, and the maximum and minimum cross-polarization responses occur at the same polarization as that for the GO model, although the absolute levels are slightly different. For the smoother surface, the polarization response with the SPM model (see Figure 4.4) is similar to that of the PO model.

One significant disadvantage of using these models for rough surface scattering is that the cross-polarization scattering matrix elements $S_{vh}$ and $S_{hv}$ are identically zero for backscatter; therefore, the polarization responses do not have the characteristic "pedestal" representing the unpolarized component of the scattered wave. This is evident in Figures 4.2 to 4.4. To overcome this limitation, second order (or higher) rough surface scattering models must be derived. The PO and SPM models can be extended to second order in a manner similar to that presented here for the first-order solutions, in which case the polarization response will show an unpolarized component. Although second-order polarimetric solutions have not been presented using the PO and SPM models, nonpolarimetric results are available in the literature (Tsang *et al.*, 1985; Fung and Pan, 1987; Pan and Fung, 1987). A nonzero cross-polarization response ($\sigma^0_{vh}$ or $\sigma^0_{hv}$) is also present if multiple scattering and shadowing are considered as in (Fung and Eom, 1981).

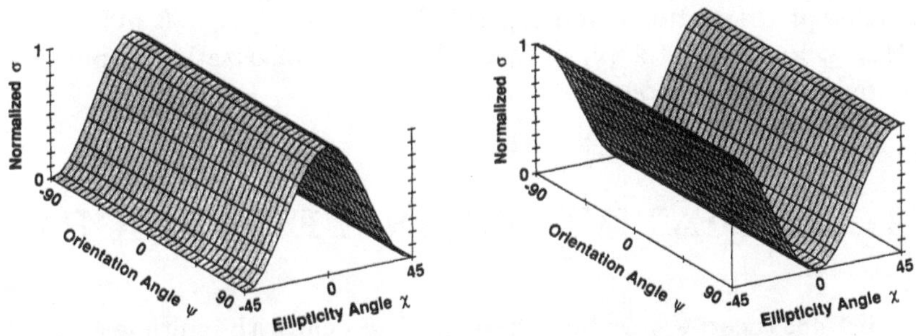


(a) Co-Pol Response (b) Cross-Pol Response

**Figure 4.2** Backscatter polarization response using the GO model ($k_1 s = 3.36, k_1 = 33.6, m = \sqrt{2}(s/l) = 0.1414, \epsilon_{2r} = 5.65 + i1.28, \theta_s = \theta_0 = 20°$) (a) co-polarization response; (b) cross-polarization response.

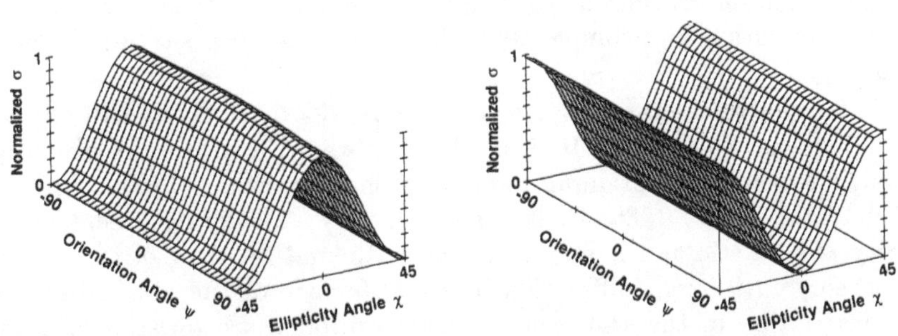


(a) Co-Pol Response (b) Cross-Pol Response

**Figure 4.3** Backscatter polarization response using the PO model with the Gaussian correlation function $\rho_1(\xi)$ ($k_1 s = 3.36, k_1 = 33.6, m = \sqrt{2}(s/l) = 0.1414, \epsilon_{2r} = 5.65 + i1.28, \theta_s = \theta_0 = 20°$) (a) co-polarization response; (b) cross-polarization response.

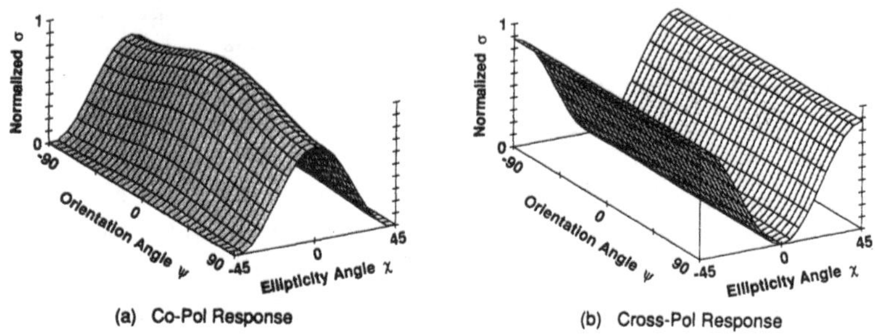


**Figure 4.4** Backscatter polarization response using the SPM model with the Gaussian correlation function $\rho_1(\xi)$ ($k_1 s = 3.36, k_1 = 33.6, m = \sqrt{2}(s/l) = 0.1414, \epsilon_{2r} = 5.65 + i1.28, \theta_s = \theta_0 = 20°$) (a) co-polarization response; (b) cross-polarization response.

# 4-2 RADIATIVE TRANSFER THEORY

Two distinct theories have been developed to deal with multiple scattering problems. One is called *analytical theory* and the other is called *radiative transfer theory* or *transport theory* (Ishimaru, 1978).

The analytical theory starts with the basic differential equations such as Maxwell's equations or the wave equation. In principle, the analytical theory includes all multiple scattering, diffraction, and interference effects. In practice, however, the analytical approach is mathematically complicated and a formulation that completely includes all these effects is impossible to obtain.

The radiative transfer theory, conversely, does not start with the basic differential equations. It deals directly with the transport of energy through a medium containing particles (Chandrasekhar, 1960; Ishimaru, 1978; Tsang *et al.*, 1985; Ulaby *et al.*, 1986b). The development of the theory is heuristic and it may lack the mathematical rigor of the analytical theory. Although diffraction and interference effects are included in the description of the scattering and absorption characteristics of a single particle, radiative transfer theory does not actually include diffraction effects. Radiative transfer theory assumes that there is no correlation between fields, and therefore the addition of power, rather than the addition of fields, holds.

The basic differential equation in radiative transfer theory is the equation of transfer and it is equivalent to Boltzmann's equation used in kinetic theory of gases and in neutron transport theory (Sommerfeld, 1956). If the specific intensity is scalar, as is the case for totally unpolarized light, the radiative transfer equation becomes a single differential equation. For vector electromagnetic fields, we need to use the Stokes parameter for the specific intensities, which leads to four coupled differential equations. To distinguish this from the ***scalar transfer equation***, we call it the ***vector radiative transfer equation*** (Tsang *et al.*, 1985).

The radiative transfer equation is formulated in terms of three constitutive functions: (1) the extinction matrix, which describes the attenuation of the specific intensity due to absorption and scattering; (2) the phase matrix, which characterizes the coupling between the incident and scattered intensities at every point in the medium; and (3) the emission function, which accounts for black body emission by the medium. The first two quantities will be examined and related to the properties of the medium in the sections below. In radar remote sensing problems, the emission contribution is negligibly small and usually ignored.

## 4-2.1 Scalar Radiative Transfer Equation

The radiative transfer equation is formulated in terms of the ***specific intensity*** $I(\mathbf{r}, \hat{\mathbf{s}})$, where $\mathbf{r}$ is a position vector and $\hat{\mathbf{s}}$ denotes the direction of propagation. For totally unpolarized light, $I(\mathbf{r}, \hat{\mathbf{s}})$ is a scalar quantity and is defined in terms of the power $\mathrm{d}P$ (watts) flowing along the $\hat{\mathbf{s}}$ direction within a solid angle $\mathrm{d}\Omega$ through an elementary area $\mathrm{d}A$ as follows (Chandrasekhar, 1960, p. 1; Ulaby *et al.*, 1986, p. 1085):

$$\mathrm{d}P = I(\mathbf{r}, \hat{\mathbf{s}}) \; \cos\theta \; \mathrm{d}A \; \mathrm{d}\Omega \quad , \tag{4.148}$$

where $\theta$ is the angle between the outward normal to $\mathrm{d}A$ and the unit vector $\hat{\mathbf{s}}$. The dimension of $I$ is W-m$^{-2}$sr$^{-1}$.

Let us consider the simple case of a tenuous medium (low density) containing $N$ randomly distributed identical spheres, each characterized by an absorption cross section $\sigma_a$ and a scattering cross section $\sigma_s$. The extinction coefficient due to absorption and scattering by the spheres is

$$\begin{aligned} \kappa_e &= N(\sigma_a + \sigma_s) \\ &= N\sigma_e, \end{aligned} \tag{4.149}$$

where $\sigma_e$ is the extinction cross section of an individual particle. If the background material of the random medium is not air or a pure dielectric, the total extinction coefficient of the medium is

$$\kappa'_e = \kappa_e + \kappa_{ag}, \tag{4.150}$$

where

$$\kappa_{ag} = 2k_0 \mid \text{Im}\sqrt{\epsilon} \mid \tag{4.151}$$

is the power absorption coefficient of the background material, $k_0$ is the wavenumber in free space, and $\epsilon$ is the relative complex dielectric constant of the material.

The equation of transfer is derived by accounting for all the losses and contributions that occur when an incident specific intensity $I(\mathbf{r}, \hat{\mathbf{s}})$ propagates through an elemental cylindrical volume of unit cross section and length d$s$. These include loss terms due to extinction, a source term due to bistatic scattering, and a source term due to emission. The change in $I(\mathbf{r}, \hat{\mathbf{s}})$ over the differential length d$s$ is

$$\frac{\mathrm{d}I(\mathbf{r},\hat{\mathbf{s}})}{\mathrm{d}s} = -\kappa_e I(\mathbf{r},\hat{\mathbf{s}}) - \kappa_{ag} I(\mathbf{r},\hat{\mathbf{s}}) + \int_{4\pi} \mathrm{d}\Omega'\, P(\hat{\mathbf{s}},\hat{\mathbf{s}}')\, I(\mathbf{r},\hat{\mathbf{s}}) + J_e, \tag{4.152}$$

where the first two terms account for extinction by the particles and the background material, the third term accounts for scattering of specific intensity $I(\mathbf{r}, \hat{\mathbf{s}}')$ incident upon the elemental volume from all directions $\hat{\mathbf{s}}'$ into the direction $\hat{\mathbf{s}}$, and $J_e$ is a source function that accounts for self emission by the elemental volume in the direction $\hat{\mathbf{s}}$. The *phase function* $P(\hat{\mathbf{s}}, \hat{\mathbf{s}}')$ specifies the angular distribution of intensity incident from direction $\hat{\mathbf{s}}'$ into all directions including the direction of interest, $\hat{\mathbf{s}}$.

The differential equation given by (4.152) is called the *scalar radiative transfer equation.* At microwave frequencies, the waves propagating through a random medium like vegetation or snow are not totally unpolarized, which requires that we extend the radiative transfer equation to the vector case.

## 4-2.2 Vector Radiative Transfer Equation

The vector analog of (4.152) is given by

$$\frac{\mathrm{d}\mathbf{I}(\mathbf{r},\hat{\mathbf{s}})}{\mathrm{d}s} = -\boldsymbol{\kappa}_e\, \mathbf{I}(\mathbf{r},\hat{\mathbf{s}}) - \kappa_{ag}\, \mathbf{I}(\mathbf{r},\hat{\mathbf{s}}) + \int_{4\pi} \mathrm{d}\Omega' \mathcal{P}(\hat{\mathbf{s}},\hat{\mathbf{s}}')\, \mathbf{I}(\mathbf{r},\hat{\mathbf{s}}'), \tag{4.153}$$

where now the vector specific intensity $\mathbf{I}$ is a $4 \times 1$ column matrix, $\boldsymbol{\kappa}_e$ is a $4 \times 4$ extinction matrix, $\mathcal{P}$ is a $4 \times 4$ phase matrix, and $\kappa_{ag}$ remains unchanged. The emission term has been deleted because in radar remote sensing its contribution is negligibly small in comparison to the other terms in the equation. The definitions of these new matrices follows.

## Vector Specific Intensity

For an elliptically polarized monochromatic plane wave

$$\mathbf{E} = \left(E_v \hat{\mathbf{v}} + E_h \,\hat{\mathbf{h}}\right) e^{i\mathbf{k}\cdot\mathbf{r}} \tag{4.154}$$

propagating in a medium with intrinsic impedance $\eta$, where $\hat{\mathbf{v}}$ and $\hat{\mathbf{h}}$ are the unit vertical and horizontal polarization vectors, the vector specific intensity $\mathbf{I}$ is defined in terms of the modified stokes vector $\mathbf{F}_m$ (see (1.50)) by the relation:

$$\mathbf{I} = \mathbf{F}_m \,/\eta = \begin{bmatrix} I_v \\ I_h \\ U \\ V \end{bmatrix} /\eta = \begin{bmatrix} | E_v |^2 \\ | E_h |^2 \\ 2\,\mathrm{Re}\,(E_v \; E_h^*) \\ 2\,\mathrm{Im}\,(E_v \; E_h^*) \end{bmatrix} \; / \, \eta \; . \tag{4.155}$$

By using the formulation relating the incident field $\mathbf{E}^i$ of a plane wave to the field $\mathbf{E}^s$ of the spherical wave scattered by a particle characterized by a scattering matrix $\mathcal{S}$, as given by (2.15):

$$\begin{bmatrix} E_v^s \\ E_h^s \end{bmatrix} = \frac{e^{ikr}}{r} \begin{bmatrix} S_{vv} & S_{vh} \\ S_{hv} & S_{hh} \end{bmatrix} \begin{bmatrix} E_v^i \\ E_h^i \end{bmatrix} , \tag{4.156}$$

we can easily show that the scattered specific intensity $\mathbf{I}^s$ is related to the incident specific intensity $\mathbf{I}^i$ by

$$\mathbf{I}^s \;(\theta_s,\; \phi_s) \;=\; \frac{1}{r^2} \,\mathcal{L}_m \;(\theta_s,\; \phi_s;\; \theta_i,\; \phi_i;\; \theta_j,\; \phi_j)\; \mathbf{I}^i \;(\theta_i,\; \phi_i), \tag{4.157}$$

where $(\theta_s,\; \phi_s)$ defines the scattered direction, $(\theta_i,\; \phi_i)$ defines the incident direction, $(\theta_j,\; \phi_j)$ defines the particle orientation, and

$$\mathcal{L}_m = \begin{bmatrix} |S_{vv}|^2 & |S_{vh}|^2 & \mathrm{Re}(S_{vh}^* S_{vv}) & -\mathrm{Im}(S_{vh}^* S_{vv}) \\ |S_{hv}|^2 & |S_{hh}|^2 & \mathrm{Re}(S_{hh}^* S_{hv}) & -\mathrm{Im}(S_{hv} S_{hh}^*) \\ 2\mathrm{Re}(S_{vv} S_{hv}^*) & 2\mathrm{Re}(S_{vh} S_{hh}^*) & \mathrm{Re}(S_{vv} S_{hh}^* + S_{vh} S_{hv}^*) & -\mathrm{Im}(S_{vv} S_{hh}^* - S_{vh} S_{hv}^*) \\ 2\mathrm{Im}(S_{vv} S_{hv}^*) & 2\mathrm{Im}(S_{vh} S_{hh}^*) & \mathrm{Im}(S_{vv} S_{hh}^* + S_{vh} S_{hv}^*) & \mathrm{Re}(S_{vv} S_{hh}^* - S_{vh} S_{hv}^*) \end{bmatrix} \tag{4.158}$$

is the modified Mueller matrix, defined according to the forward scattering alignment (FSA) convention.

The relationship given by (4.157) pertains to the situation where the scattering object is regarded as a point target. For a distributed target, we need to define the scattered specific intensity in terms of the net electric field $\mathbf{E}^s$, representing the vector sum of all fields scattered from an illuminated area $A$ subtending a solid angle:

$$d\Omega = \frac{A \cos \theta_s}{r^2} \tag{4.159}$$

at the range $r$ from the surface. The angle $\theta_s$ is the angle between the surface normal and the scattering direction. In this case,

$$\mathbf{I}^s \, d\Omega = \begin{bmatrix} I_v^s \\ I_h^s \\ U^s \\ V^s \end{bmatrix} d\Omega \, / \eta = \begin{bmatrix} | E_v^s |^2 \\ | E_h^s |^2 \\ 2\mathrm{Re}(E_v^s E_h^{s*}) \\ 2\mathrm{Im}(E_v^s E_h^{s*}) \end{bmatrix} / \eta \, . \tag{4.160}$$

In Section 4-3 we will examine techniques for solving the vector radiative transfer equation (4.153) to find the specific intensity $\mathbf{I}^s$ scattered by the random medium under consideration when it is illuminated by an incident plane wave of specific intensity $\mathbf{I}^i$. Once $\mathbf{I}^s$ has been found for the backscattering direction, it can be used to compute the backscattering coefficient $\sigma^0$ of the medium. To show this, let us start with the simple case where both the transmitting and receiving antennas are vertically polarized. The backscattering coefficient is defined by

$$\sigma_{vv}^0 = \frac{4\pi r^2}{A} \frac{| E_v^s |^2}{| E_v^i |^2} \, , \tag{4.161}$$

where $E_v^i$ is the incident field and $r$ is sufficiently large to satisfy the far-field condition. Upon making the following substitutions in (4.161):

$$| E_v^s |^2 = I_v^s d\Omega = I_v^s \frac{A \cos \theta_s}{r^2} \tag{4.162}$$

and

$$| E_v^i |^2 = I_v^i , \tag{4.163}$$

we have

$$\sigma_{pq}^0 = 4\pi \cos \theta_s \, \frac{I_p^s}{I_q^i} \, , \tag{4.164}$$

where $p$ denotes the polarization of the receiving antenna and $q$ denotes the polarization of the transmitting antenna and $p, q = v$ or $h$.

To compute $\sigma^0$ for any arbitrary combination of transmitting and receiving antenna polarizations, a more involved procedure is required, as discussed in Section 4-4.

## Phase Matrix

One of the fundamental assumptions of radiative transfer theory is that in a medium containing a random distribution of particles, the waves scattered from the particles are random in phase, thereby allowing the addition of the waves to be performed incoherently. That is, the Stokes parameters of the mixture are the sum of the respective Stokes parameters of the separate waves (Tsang *et al.*, 1985; p. 127). If the medium contains $N$ identical spherical particles per unit volume, (4.157) ought to be replaced with

$$\mathbf{I}^s\,(\theta_s,\,\phi_s) = \frac{1}{r^2}\,\mathcal{P}\,(\theta_s,\,\phi_s;\,\theta_i,\,\phi_i)\,\mathbf{I}^i\,(\theta_i,\,\phi_i)\,, \tag{4.165}$$

where the phase matrix $\mathcal{P}$ is given by

$$\mathcal{P}\,(\theta_s,\,\phi_s;\,\theta_i,\,\phi_i) = N\,\mathcal{L}_m\,(\theta_s,\,\phi_s;\,\theta_i,\,\phi_i)\ . \tag{4.166}$$

In general, however, the medium may contain particles of different shapes, sizes, and orientations, in which case $\mathcal{P}$ assumes the more general definition:

$$\begin{aligned}\mathcal{P}\,(\theta_s,\,\phi_s;\,\theta_i,\,\phi_i) &= N\,\langle\,\mathcal{L}_m\,(\theta_s,\,\phi_s;\theta_i,\phi_i;\,x_1,\,x_2,\,\ldots,\,x_n)\rangle \\ &= N\int \mathrm{d}x_1\,\mathrm{d}x_2\,\ldots\,\mathrm{d}x_n\,p\,(x_1,\,x_2,\,\ldots,\,x_n) \\ &\quad\cdot\mathcal{L}_m\,(\theta_s,\,\phi_s;\,\theta_i,\,\phi_i;\,x_1,\,x_2,\,\ldots,\,x_n)\end{aligned} \tag{4.167}$$

where $p\,(x_1,\,x_2,\,\ldots,\,x_n)$ is the joint probability density function over the shape, size, and orientation parameters $(x_1,\,x_2,\,\ldots,\,x_n)$ of the particles.

## Extinction Matrix

For a single particle, the optical theorem provides a useful relationship for computing the extinction cross section of the particle in terms of the field scattered in the forward direction, relative to the incident field. If the incident wave is $p$-polarized, where $p = v$ or $h$, the extinction cross section of a particle with orientation direction $(\theta_j, \phi_j)$ is given by

$$\sigma_e^p = \frac{4\pi}{k_0}\,\mathrm{Im}\,[S_{pp}\,(\theta_i,\,\phi_i;\,\theta_i,\,\phi_i;\,\theta_j,\,\phi_j)]\ , \tag{4.168}$$

where $S_{pp}$ is the $pp$ scattering amplitude of the particle, evaluated at $\theta_s = \theta_i$ and $\phi_s = \phi_i$.

The $p$-polarized extinction coefficient of a medium containing a random distribution of particles may be obtained by performing an ensemble average over the particles. Thus,

$$\kappa_e = N \langle \sigma_e^p \rangle . \tag{4.169}$$

For spherical particles, $\kappa_e$ is the same for both polarizations and the extinction matrix is given by the diagonal form:

$$\boldsymbol{\kappa}_e = \begin{bmatrix} \kappa_e & 0 & 0 & 0 \\ 0 & \kappa_e & 0 & 0 \\ 0 & 0 & \kappa_e & 0 \\ 0 & 0 & 0 & \kappa_e \end{bmatrix} . \tag{4.170}$$

Ishimaru and Cheung (1980) have shown that, in the general case, when the particles are nonspherical, the extinction matrix assumes a more complicated structure given by

$$\boldsymbol{\kappa}_e =$$

$$\begin{bmatrix} -2\mathrm{Re}(M_{vv}) & 0 & -\mathrm{Re}(M_{vh}) & -\mathrm{Im}(M_{vh}) \\ 0 & -2\mathrm{Re}(M_{hh}) & -\mathrm{Re}(M_{hv}) & \mathrm{Im}(M_{hv}) \\ -2\mathrm{Re}(M_{hv}) & -2\mathrm{Re}(M_{vh}) & -\mathrm{Re}(M_{vv}+M_{hh}) & \mathrm{Im}(M_{vv}-M_{hh}) \\ 2\mathrm{Im}(M_{hv}) & -2\mathrm{Im}(M_{vh}) & -\mathrm{Im}(M_{vv}-M_{hh}) & -\mathrm{Re}(M_{vv}+M_{hh}) \end{bmatrix} \tag{4.171}$$

where

$$M_{mn} = \frac{i2\pi N}{k_0} \langle S_{mn}(\theta_i, \phi_i, \theta_i, \phi_i, \theta_j, \phi_j \rangle ; \qquad m, n = v, h. \tag{4.172}$$

## Phase Matrix of Simple Objects

With the definitions given in the preceding sections for the phase and extinction matrices, we can, in principle, use a suitable numerical technique to solve the vector radiative transfer equation (4.153) for any specified particle distribution and boundary conditions. In practice, however, to obtain an exact solution for nonspherical particles is difficult beyond first order. If the wavelength is much larger than the particle size, to treat the particles as isotropic point scatterers or as spherical Rayleigh scatterers is reasonable. With these assumptions, we can obtain numerical solutions that take into account multiple scattering effects. In this section, we will present an expression for the phase function of a medium containing isotropic scatterers and expressions for the phase matrix of a medium containing spherical Rayleigh particles. For spheres with radii that are comparable to or larger than the wavelength and for particles of other shapes, such as disks and cylinders, the expressions given in Chapter 3 for the elements of the scattering matrix $\boldsymbol{\mathcal{S}}$ can be used in (4.158) to compute the modified Mueller matrix $\boldsymbol{\mathcal{L}}_m$. Then the elements of $\boldsymbol{\mathcal{L}}_m$ must be averaged in accordance with the specified particle size, shape, or orientation distribution to obtain the phase matrix $\boldsymbol{\mathcal{P}}$, as indicated by (4.167).

## Isotropic Scatterers

Isotropic point scatterers are characterized by a simple scalar phase function, which is independent of both angle and polarization (Chandrasekhar, 1980, p. 6), given by

$$P = \frac{\kappa_s}{4\pi} = \kappa_e \frac{\widetilde{w}}{4\pi} \;, \tag{4.173}$$

where $\widetilde{w} = \kappa_s / \kappa_e$ is the albedo of a single particle. When the expression for $P$ is inserted in the scalar radiative transfer equation given by (4.152), the differential equation can readily be solved numerically, as shown in Chandrasekhar (1960, pp. 15–16).

## Spherical Rayleigh Scatterers

Mie scattering refers to the solution published by Mie in 1908, which provides expressions for the fields scattered by a dielectric sphere of arbitrary size in the form of an infinite series. Expressions for the scattering amplitudes of a spherical particle are given in Section 3-4.2. If the particle radius is small relative to the wavelength such that $ka = 2\pi a \,/\, \lambda << 1$, where $\lambda$ is the wavelength in the dielectric material of the particle, simpler expressions based on the Rayleigh approximation may be used. The phase matrix of a random medium containing spherical Rayleigh scatterers of radius $a$ is given by (Tsang *et al.*, 1985, pp. 155–158):

$$\mathcal{P}(\theta_s,\phi_s;\theta_i,\phi_i) = \begin{bmatrix} P_{11} & P_{12} & P_{13} & 0 \\ P_{21} & P_{22} & P_{23} & 0 \\ P_{31} & P_{32} & P_{33} & 0 \\ 0 & 0 & 0 & P_{44} \end{bmatrix}, \tag{4.174}$$

where

$$\begin{aligned}
P_{11} &= w[\sin^2\theta_s \sin^2\theta_i + \cos^2\theta_s \cos^2\theta_i \cos^2(\phi_s-\phi_i)] \\
&\quad +2\sin\theta_s \sin\theta_i \cos\theta_s \cos\theta_i \cos(\phi_s-\phi_i) \\
P_{12} &= w\cos^2\theta_s \sin^2(\phi_s-\phi_i) \\
P_{13} &= w[\cos\theta_s \sin\theta_s \sin\theta_i \sin(\phi_s-\phi_i) \\
&\quad +\cos^2\theta_s \cos\theta_i \sin(\phi_s-\phi_i)\cos(\phi_s-\phi_i)] \\
P_{21} &= w\cos^2\theta_i \sin^2(\phi_s-\phi_i) \\
P_{22} &= w\cos^2(\phi_s-\phi_i) \\
P_{23} &= -w\cos\theta_s \sin(\phi_s-\phi_i)\cos(\phi_s-\phi_i) \\
P_{31} &= w[-2\sin\theta_s \sin\theta_i \cos\theta_i \sin(\phi_s-\phi_i) \\
&\quad -2\cos\theta_s \cos^2\theta_i \cos(\phi_s-\phi_i)\sin(\phi_s-\phi_i)] \\
P_{32} &= 2w\cos\theta_s \sin(\phi_s-\phi_i)\cos(\phi_s-\phi_i) \\
P_{33} &= w\{\sin\theta_s \sin\theta_i \cos(\phi_s-\phi_i) \\
&\quad +\cos\theta_s \cos\theta_i[\cos^2(\phi_s-\phi_i) - \sin^2(\phi_s-\phi_i)]\} \\
P_{44} &= w[\sin\theta_s \sin\theta_i \cos(\phi_s-\phi_i) + \cos\theta_s \cos\theta_i] \\
w &= \frac{3}{8\pi}\kappa_s
\end{aligned}$$

and $\kappa_s$ is the scattering coefficient given by

$$\begin{aligned}
\kappa_s &= \frac{8\pi}{3} N k^4 a^6 \mid y \mid^2 = 2 f k^4 a^3 \mid y \mid^2 , \\
y &= \frac{\epsilon_s - \epsilon}{\epsilon_s + 2\epsilon} ,
\end{aligned} \tag{4.175}$$

where $f$ is the volume fraction of scatterers, $N$ is the number density, $\epsilon_s$ is the dielectric constant of the sphere material and $\epsilon$ is the dielectric constant of the background material. The absorption coefficient is given by

$$\kappa_a = fk\frac{\epsilon_s''}{\epsilon} \left| \frac{3\epsilon}{\epsilon_s + 2\epsilon} \right|^2 \tag{4.176}$$

and the extinction coefficient $\kappa_e$ is the sum of $\kappa_s$ and $\kappa_a$. The extinction matrix is diagonal with each element equal to $\kappa_e$ (see (4.170)).

## 4-2.3 Boundary Conditions for the Radiative Transfer Equation

When the specific intensity impinges upon the boundary separating two media with different dielectric constants, part of the incident energy will be reflected and the remaining part will be transmitted across the boundary. In this section, we will consider the boundary conditions relating the Stokes parameters of the specific intensities in the two media.

If the boundary is a planar dielectric interface, scattering at the boundary occurs in only the specular direction and does not include a diffuse component. If the boundary is rough, however, the incident wave will be scattered in all directions at the boundary and the incident intensity is coupled into all of the reflected and transmitted directions. In this book, we shall limit our discussion to the smooth-boundary case. For examination of the boundary conditions associated with a rough surface, we refer the interested reader to the article by Fung and Eom (1981), who investigated radar scattering from a layer of Rayleigh particles over a rough surface by using the Kirchhoff approximation to model the scattering by the surface.

For a plane wave incident upon a planar boundary from medium 1 to medium 2, as shown in Fig. 4.5, the Fresnel reflection coefficients, $r_{h12}$ and $r_{v12}$, and the Fresnel transmission coefficients, $t_{h12}$ and $t_{v12}$, are given by

$$r_{v12} = \frac{n_2 \cos\theta_1 - n_1 \cos\theta_2}{n_2 \cos\theta_1 + n_1 \cos\theta_2} \tag{4.177}$$

$$r_{h12} = \frac{n_1 \cos\theta_1 - n_2 \cos\theta_2}{n_1 \cos\theta_1 + n_2 \cos\theta_2} \tag{4.178}$$

$$t_{v12} = \frac{2n_2 \cos\theta_1}{n_2 \cos\theta_1 + n_1 \cos\theta_2} \tag{4.179}$$

$$t_{h12} = \frac{2n_1 \cos\theta_1}{n_1 \cos\theta_1 + n_2 \cos\theta_2}, \tag{4.180}$$

where $n_1$ and $n_2$ are the indices of refraction of media 1 and 2, respectively, $\theta_1$ is the incidence angle, and $\theta_2$ is the refraction angle, which is related to $\theta_1$ by Snell's law:

$$n_2 \sin\theta_2 = n_1 \sin\theta_1 \ . \tag{4.181}$$

We also define the power reflectivity and transmissivity $R_{\beta 12}$ and $T_{\beta 12}$:

$$R_{\beta 12} = |r_{\beta 12}|^2 = 1 - T_{\beta 12} \ , \tag{4.182}$$

$$T_{\beta 12} = \frac{n_2 \cos\theta_2}{n_1 \cos\theta_1} |t_{\beta 12}|^2 \ , \tag{4.183}$$

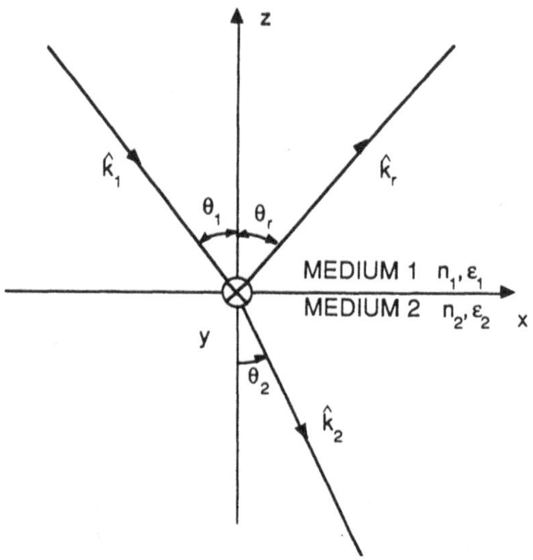


**Figure 4.5** Planar dielectric interface; $\hat{\mathbf{k}}_1$, $\hat{\mathbf{k}}_r$, and $\hat{\mathbf{k}}_2$, are the incident, reflected, and transmitted directions.

with $\beta = v$ or $h$. The reflected specific intensity $I^r_\beta$ in medium 1 is related to the incident intensity $I^i_\beta$ by

$$I^r_\beta = R_{\beta 12} I^i_\beta. \tag{4.184}$$

The transmitted specific intensity is obtained by noting the relationship (Ishimaru, 1978):

$$\frac{I^i_\beta}{\epsilon_1} = \frac{I^r_\beta}{\epsilon_1} + \frac{I^t_\beta}{\epsilon_2} , \tag{4.185}$$

where $\epsilon_1 = n_1^2$ and $\epsilon_2 = n_2^2$. Hence,

$$\begin{aligned} I^t_\beta &= \frac{\epsilon_2}{\epsilon_1}(1 - R_{\beta 12}) I^i_\beta \\ &= \frac{\epsilon_2}{\epsilon_1} T_{\beta 12} I^i_\beta \ . \end{aligned} \tag{4.186}$$

Equations (4.184) and (4.186) relate the first two components of the Stokes vector for the incident, reflected, and transmitted specific intensities at a plane boundary. The relation between the incident and reflected Stokes vector is

$$\mathbf{I}^r = \mathcal{R}_{12}(\theta_1)\mathbf{I}^i \tag{4.187}$$

where $\mathcal{R}_{12}(\theta_1)$ is the reflectivity matrix given by

$$\mathcal{R}_{12}(\theta_1) = \begin{bmatrix} |r_{v12}|^2 & 0 & 0 & 0 \\ 0 & |r_{h12}|^2 & 0 & 0 \\ 0 & 0 & \mathrm{Re}(r_{v12}r^*_{h12}) & -\mathrm{Im}(r_{v12}r^*_{h12}) \\ 0 & 0 & \mathrm{Im}(r_{v12}r^*_{h12}) & \mathrm{Re}(r_{v12}r^*_{h12}) \end{bmatrix}. \tag{4.188}$$

Similarly, the relation between the incident and transmitted Stokes vector is

$$\mathbf{I}^t = \mathcal{T}_{12}(\theta_1)\mathbf{I}^i, \tag{4.189}$$

where $\mathcal{T}_{12}(\theta_1)$ is the transmissivity matrix, given by

$$\mathcal{T}_{12}(\theta_1) = \frac{n_2^3 \cos\theta_2}{n_1^3 \cos\theta_1} \begin{bmatrix} |t_{v12}|^2 & 0 & 0 & 0 \\ 0 & |t_{h12}|^2 & 0 & 0 \\ 0 & 0 & \mathrm{Re}(t_{v12}t^*_{h12}) & -\mathrm{Im}(t_{v12}t^*_{h12}) \\ 0 & 0 & \mathrm{Im}(t_{v12}t^*_{h12}) & \mathrm{Re}(t_{v12}t^*_{h12}) \end{bmatrix}. \tag{4.190}$$

Equation (4.190) applies when $\theta_1$ is less than the critical angle. For $\theta_1$ greater than the critical angle, $\mathcal{T}_{12}(\theta_1) = 0$.

# 4-3 SOLUTIONS FOR THE RADIATIVE TRANSFER EQUATION

The vector radiative transfer equation, in general, does not have analytical solutions, except in some special cases. In this section, we will discuss the approximations and numerical techniques used for solving the radiative transfer equation, namely the iterative solution technique and the discrete ordinate technique (Chandrasekhar, 1960; Ishamaru, 1978; Tsang *et al.*, 1985; Ulaby *et al.*, 1986b). In the iterative approach the radiative transfer equation is cast into an integral form, and then solved iteratively to obtain the zero-, first-, and second-order solutions. In principle, we can obtain an accurate solution by iterating many times. In practice, however, the iteration beyond the second order requires a substantial amount of computer time.

The discrete ordinate eigenanalysis technique provides good numerical solutions for the radiative transfer equation (Ishimaru *et al.*, 1982; Cheung and Ishimaru, 1982) and has been used extensively in the literature. In this technique, the specific intensity and phase matrix are first discretized into a finite number of directions by means of quadrature, and the resulting matrix equation is solved by eigenanalysis.

In addition to these, several other techniques have been applied to solve the radiative transfer equation, including the invariant imbedding method and the finite difference method (Tsang *et al.*, 1985). The major limitation of these numerical techniques is the fact that they are computationally very demanding, particularly if the angular variation of the phase matrix is very complex.

## 4-3.1 Iterative Method

The iterative method is applicable when the random medium is weakly scattering; that is, when the albedo $\tilde{w}$ is small. The procedure starts by computing the zero-order solution, which ignores scattering, except for its contribution to extinction. The zero-order solution then serves as a source function for computing the first-order solution.

Consider the specific intensity incident onto a layer of spherical scatterers, as shown in Fig. 4.6. We assume the background dielectric constant of layer 2 to be the same as that of layer 1. The bottom boundary is flat and the dielectric constant of layer 3 is different from that of layer 2. The top boundary is diffuse, which means that no reflection occurs at the top interface, but the incident intensity will be specularly reflected at the flat interface at $z = -d$. The case for nonspherical scatterer will be considered in Section 4-5.

We separate the specific intensity into upward-going $\mathbf{I}^+(\mu_s, \phi_s, z)$ and downward-going $\mathbf{I}^-(-\mu_s, \phi_s, z)$ components, noting that $\mu_s = \cos\theta_s$ and $\theta_s$ varies between 0 and $\pi/2$.

Inside layer 2, the intensities $\mathbf{I}^+$ and $\mathbf{I}^-$ must satisfy the coupled radiative transfer equations:

$$\frac{\mathrm{d}}{\mathrm{d}z}\mathbf{I}^+(\mu_s, \phi_s, z) = -\frac{\kappa_e}{\mu_s}\mathbf{I}^+(\mu_s, \phi_s, z) + \mathbf{F}^+(\mu_s, \phi_s, z) \tag{4.191}$$

$$-\frac{\mathrm{d}}{\mathrm{d}z}\mathbf{I}^-(-\mu_s, \phi_s, z) = -\frac{\kappa_e}{\mu_s}\mathbf{I}^-(-\mu_s, \phi_s, z) + \mathbf{F}^-(-\mu_s, \phi_s, z), \tag{4.192}$$

where $\kappa_e$ is the extinction coefficient, $\mu_s = \cos\theta_s$, and $-\mu_s = \cos(\pi - \theta_s)$. The extinction coefficient $\kappa_e$ is a scalar quantity because the particles are spheres. The source functions $\mathbf{F}^+(\mu_s, \phi_s, z)$ and $\mathbf{F}^-(-\mu_s, \phi_s, z)$ account for

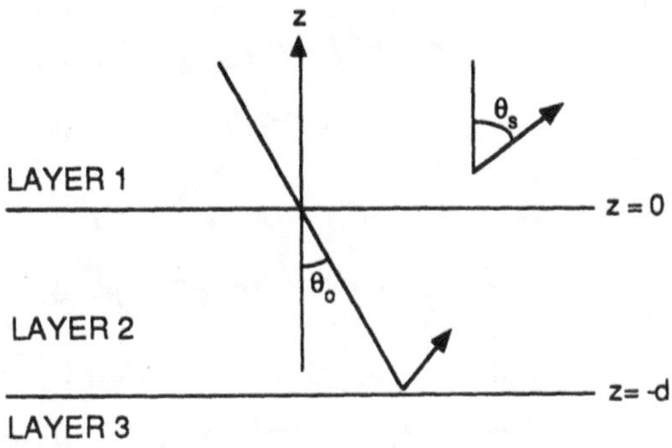


**Figure 4.6** Geometrical configuration of the problem. Background dielectric constant of layer 2 is the same as layer 1.

directing the energy incident upon an elemental volume from all directions into the directions $(\theta_s, \phi_s)$ and $(\pi - \theta_s, \phi_s)$, respectively, and are given by

$$\begin{aligned}\mathbf{F}^+(\mu_s, \phi_s, z) = \frac{1}{\mu_s}\Bigg[&\int_0^{2\pi}\int_0^1 \mathcal{P}(\mu_s, \phi_s; \mu', \phi')\mathbf{I}^+(\mu', \phi', z)\,\mathrm{d}\Omega' \\ &+ \int_0^{2\pi}\int_0^1 \mathcal{P}(\mu_s, \phi_s; -\mu', \phi')\mathbf{I}^-(-\mu', \phi', z)\,\mathrm{d}\Omega'\Bigg]\end{aligned} \tag{4.193}$$

$$\begin{aligned}\mathbf{F}^-(-\mu_s, \phi_s, z) = \frac{1}{\mu_s}\Bigg[&\int_0^{2\pi}\int_0^1 \mathcal{P}(-\mu_s, \phi_s; \mu', \phi')\mathbf{I}^+(\mu', \phi', z)\,\mathrm{d}\Omega' \\ &+ \int_0^{2\pi}\int_0^1 \mathcal{P}(-\mu_s, \phi_s; -\mu', \phi')\mathbf{I}^-(-\mu', \phi', z)\,\mathrm{d}\Omega'\Bigg]\end{aligned} \tag{4.194}$$

where $\mathrm{d}\Omega' = \mathrm{d}\mu'\mathrm{d}\phi' = \sin\theta'\mathrm{d}\theta'\mathrm{d}\phi'$ and $\mathcal{P}(\mu_s, \phi_s; \mu', \phi')$ is the phase matrix of the scatterers.

The boundary conditions are

$$\mathbf{I}^-(-\mu_s, \phi_s, z = 0) = \mathbf{I}_0\delta(\mu_s - \mu_0)\delta(\phi_s - \phi_0) \tag{4.195}$$

$$\mathbf{I}^+(\mu_s, \phi_s, z = -d) = \mathcal{R}_{23}(\mu_s)\mathbf{I}^-(-\mu_s, \phi_s, z = -d), \tag{4.196}$$

where $\mathbf{I}_0$ is the specific intensity incident in medium 1 along the direction $(\theta_0, \phi_0)$ and $\mathcal{R}_{23}(\mu_s)$ is the reflectivity matrix at $z = -d$ at angle $\mu_s$. $\mathcal{R}_{23}$ can be obtained by replacing the subscript 12 with 23 in (4.188). We first cast the differential equations and the boundary conditions into the integral equations:

$$\mathbf{I}^{+}(\mu_s, \phi_s, z) = e^{-\kappa_e(z+d)/\mu_s}\mathbf{I}^{+}(\mu_s, \phi_s, z=-d) + \int_{-d}^{z} e^{-\kappa_e(z-z')/\mu_s}\mathbf{F}^{+}(\mu_s, \phi_s, z')\,dz' \quad (4.197)$$

$$\mathbf{I}^{-}(-\mu_s, \phi_s, z) = e^{\kappa_e z/\mu_s}\mathbf{I}^{-}(-\mu_s, \phi_s, z=0) + \int_{z}^{0} e^{\kappa_e(z-z')/\mu_s}\mathbf{F}^{-}(-\mu_s, \phi_s, z')\,dz'. \quad (4.198)$$

By using the boundary conditions, (4.197) and (4.198) become

$$\mathbf{I}^{+}(\mu_s, \phi_s, z) = e^{-\kappa_e(z+d)/\mu_s}\mathcal{R}_{23}(\mu_0)e^{-\kappa_e d/\mu_0}\mathbf{I}_0\delta(\mu_s-\mu_0)\delta(\phi_s-\phi_0) + e^{-\kappa_e(z+d)/\mu_s}\mathcal{R}_{23}(\mu_s)\int_{-d}^{0} e^{\kappa_e(-d-z')/\mu_s}\mathbf{F}^{-}(-\mu_s, \phi_s, z')\,dz' + \int_{-d}^{z} e^{-\kappa_e(z-z')/\mu_s}\mathbf{F}^{+}(\mu_s, \phi_s, z')\,dz' \quad (4.199)$$

$$\mathbf{I}^{-}(-\mu_s, \phi_s, z) = e^{\kappa_e z/\mu_0}\mathbf{I}_0\delta(\mu_s-\mu_0)\delta(\phi_s-\phi_0) + \int_{z}^{0} e^{\kappa_e(z-z')/\mu_s}\mathbf{F}^{-}(-\mu_s, \phi_s, z')\,dz'. \quad (4.200)$$

In the iterative approach, the solutions are expanded into series of perturbation orders by treating $\tilde{w} = \kappa_s/\kappa_e$ as a small parameter:

$$\mathbf{I}(\mu, \phi, z) = \mathbf{I}^{(0)}(\mu, \phi, z) + \mathbf{I}^{(1)}(\mu, \phi, z) + \mathbf{I}^{(2)}(\mu, \phi, z) + \ldots \quad (4.201)$$

where $-1 < \mu < 1$. Because the scattering is treated as an iterative parameter, the perturbation orders have one-to-one correspondence with the multiple scattering process. The solution $\mathbf{I}^{(0)}$ represents reduced incident intensity, $\mathbf{I}^{(1)}$ represents the single scattering solution, and $\mathbf{I}^{(2)}$ represents the double scattering solution.

### Zero-Order Solution

The zero-order solution is obtained by setting $\mathbf{F}^{+} = 0$ and $\mathbf{F}^{-} = 0$ in (4.199) and (4.200):

$$\mathbf{I}^{(0)-}(-\mu_s, \phi_s, z) = e^{\kappa_e z/\mu_0}\mathbf{I}_0\delta(\mu_s-\mu_0)\delta(\phi_s-\phi_0) \quad (4.202)$$

$$\mathbf{I}^{(0)+}(\mu_s, \phi_s, z) = e^{-\kappa_e(z+d)/\mu_0}\mathcal{R}_{23}(\mu_0)e^{-\kappa_e d/\mu_0}\mathbf{I}_0\delta(\mu_s-\mu_0)\delta(\phi_s-\phi_0) \quad (4.203)$$

The zero-order solution is simply the reduced incident intensity, which attenuates exponentially inside the medium. Scattering is not included, except for its contribution to extinction.

## First-Order Solution

To obtain the first-order solution, we insert the zero-order solutions into (4.193) and (4.194) and compute the zero-order source functions. The first-order solution is then given by

$$\begin{aligned}
\mathbf{I}^{(1)-}(-\mu_s,\phi_s,z) &= e^{\kappa_e z/\mu_0}\, \mathbf{I}_0\, \delta(\mu_s-\mu_0)\delta(\phi_s-\phi_0) \\
&+ \Bigg[\mathcal{P}(-\mu_s,\phi_s;-\mu_0,\phi_0)\left(\frac{e^{\kappa_e z/\mu_s}}{\kappa_2}\right)(1-e^{\kappa_2 z}) \\
&+\mathcal{R}_{23}(\mu_0)\mathcal{P}(-\mu_s,\phi_s;\mu_0,\phi_0)\left(\frac{e^{\kappa_e z/\mu_s-\kappa_e 2d/\mu_0}}{\kappa_1}\right) \\
&\cdot(e^{-\kappa_1 z}-1)\Bigg]\frac{\mathbf{I}_0}{\mu_s}
\end{aligned} \tag{4.204}$$

$$\begin{aligned}
\mathbf{I}^{(1)+}(\mu_s,\phi_s,z) &= e^{-\kappa_e(z+2d)/\mu_0}\mathcal{R}_{23}(\mu_0)\mathbf{I}_0\delta(\mu_s-\mu_0)\delta(\phi_s-\phi_0) \\
&+\mathcal{R}_{23}(\mu_s)e^{-\kappa_e(z+2d)/\mu_s} \\
&\cdot\Bigg[\mathcal{R}_{23}(\mu_0)\mathcal{P}(-\mu_s,\phi_s;\mu_0,\phi_0)\left(\frac{e^{-2\kappa_e d/\mu_0}}{\kappa_1}\right)(e^{\kappa_1 d}-1) \\
&+ \mathcal{P}(-\mu_s,\phi_s;-\mu_0,\phi_0)\left(\frac{1}{\kappa_2}\right)(1-e^{-\kappa_2 d})\Bigg]\left(\frac{\mathbf{I}_0}{\mu_s}\right) \\
&+e^{-\kappa_e z/\mu_s}\Bigg[\mathcal{R}_{23}(\mu_0)\mathcal{P}(\mu_s,\phi_s;\mu_0,\phi_0)\left(\frac{e^{-2\kappa_e d/\mu_0}}{\kappa_2}\right) \\
&\cdot(e^{\kappa_2 d}-e^{-\kappa_2 z}) \\
&+ \mathcal{P}(\mu_s,\phi_s;-\mu_0,\phi_0)\left(\frac{1}{\kappa_1}\right)(e^{\kappa_1 z}-e^{-\kappa_1 d})\Bigg]\left(\frac{\mathbf{I}_0}{\mu_s}\right),
\end{aligned} \tag{4.205}$$

where

$$\kappa_1 \triangleq \kappa_e\,(1/\mu_0+1/\mu_s) \tag{4.206}$$

$$\kappa_2 \triangleq \kappa_e\,(1/\mu_0-1/\mu_s)\ . \tag{4.207}$$

The backscattering coefficient $\sigma^0_{pq}(\theta_0)$ corresponding to a $q$-polarized transmitting antenna and a $p$-polarized receiving antenna, where $q$ and $p = v$ or $h$, is obtained by setting $z = 0$, $\mu_s = \mu_0$ and $\phi_s = \pi + \phi_0$ in (4.205) and then using the relationship:

$$\sigma^0_{pq}(\theta_0) = \frac{4\pi \cos\theta_0 I^s_p(\theta_0, \pi + \phi_0, z = 0)}{I_0}, \tag{4.208}$$

where $I^s_p$ is the first element of $\mathbf{I}^{(1)+}$ for $p = v$ and the second element for $p = h$. For $q = v$, the incident specific intensity is given by $\mathbf{I}_0 = I_0[1\ 0\ 0\ 0]^T$, and for $q = h$ it is given by $\mathbf{I}_0 = I_0[0\ 1\ 0\ 0]^T$, where $T$ denotes the transpose operation.

The first-order solution does not give depolarization in the backscattering direction for spherical scatterers. Therefore,

$$\sigma^0_{vh} = \sigma^0_{hv} = 0 \ . \tag{4.209}$$

### Second-Order Solution

The second-order solution can be obtained by inserting the first-order solution into (4.193) and (4.194), which will give us the first-order source functions $\mathbf{F}^{(1)+}$ and $\mathbf{F}^{(1)-}$. The second-order solution can be expressed in terms of the first-order source functions:

$$\begin{aligned}\mathbf{F}^{(1)+}(\mu_s, \phi_s, z) &= \frac{1}{\mu_s}\left[\int_0^{2\pi}\int_0^1 \mathcal{P}(\mu_s, \phi_s; \mu', \phi')\mathbf{I}^{(1)+}(\mu', \phi', z)\, d\Omega'\right. \\ &\quad \left. + \int_0^{2\pi}\int_0^1 \mathcal{P}(\mu_s, \phi_s; -\mu', \phi')\mathbf{I}^{(1)-}(-\mu', \phi', z)\, d\Omega'\right]\end{aligned} \tag{4.210}$$

$$\begin{aligned}\mathbf{F}^{(1)-}(-\mu_s, \phi_s, z) &= \frac{1}{\mu_s}\left[\int_0^{2\pi}\int_0^1 \mathcal{P}(-\mu_s, \phi'; \mu', \phi')\mathbf{I}^{(1)+}(\mu', \phi', z)\, d\Omega'\right. \\ &\quad \left. + \int_0^{2\pi}\int_0^1 \mathcal{P}(-\mu_s, \phi_s; -\mu', \phi')\mathbf{I}^{(1)-}(-\mu', \phi', z)\, d\Omega'\right]\end{aligned} \tag{4.211}$$

$$\begin{aligned}\mathbf{I}^{(2)-}(-\mu_s, \phi_s, z) &= e^{\kappa_e z/\mu_0}\mathbf{I}_0\delta(\mu_s - \mu_0)\delta(\phi_s - \phi_0) \\ &\quad + \int_z^0 e^{\kappa_e(z-z')/\mu_s}\mathbf{F}^{(1)-}(-\mu_s, \phi_s, z')\, dz'\end{aligned} \tag{4.212}$$

$$\begin{aligned}\mathbf{I}^{(2)+}(\mu_s,\phi_s,z) &= e^{-\kappa_e(z+d)/\mu_0}\mathcal{R}_{23}(\mu_0)e^{-\kappa_e d/\mu_0}\mathbf{I}_0\delta(\mu_s-\mu_0)\delta(\phi_s-\phi_0)\\ &+e^{-\kappa_e(z+d)/\mu_s}\mathcal{R}_{23}(\mu_s)\\ &\cdot\int_{-d}^{0} e^{\kappa_e(-d-z')/\mu_s}\mathbf{F}^{(1)-}(-\mu_s,\phi_s,z')\,dz'\\ &+\int_{-d}^{z} e^{-\kappa_e(z-z')/\mu_s}\mathbf{F}^{(1)+}(\mu_s,\phi_s,z')\,dz'.\end{aligned} \tag{4.213}$$

The second-order solution provides nonzero cross-polarized components in the backscattering direction for spherical scatterers. This indicates that depolarization is a multiple scattering effect. The second-order solution, given by (4.212) and (4.213) was obtained by applying the iterative technique directly to the radiative transfer equation. In an alternative approach used by Kuga *et al.*, (1989), the radiative transfer equation was first expanded into a Fourier series in $\phi$, and then the iterative method was applied separately to each Fourier component.

## 4-3.2 Numerical Technique

In the previous section, we obtained solutions for the radiative transfer equation by using the iterative method. The iterative approach is useful if multiple scattering is small. When multiple scattering is important, we need to solve the radiative transfer equation numerically. The most frequently used approach is the discrete ordinate method.

In this section we will first consider the radiative transfer problem for a plane parallel medium containing spherical scatterers (Ishimaru *et al.*, 1982; Ma *et al.*, 1990; Cheung and Ishimaru, 1982). We assume that the scattering layer has a background dielectric constant different from air and that background absorption is not zero. Instead of separating the specific intensity into upward-going and downward-going specific intensities as we did in the preceding section, we will separate the specific intensity into a reduced incident intensity and a diffuse or incoherent intensity. A new radiative transfer equation is then obtained for the diffuse specific intensity. Next, we expand the specific intensity, the source term, and a modified form of the phase matrix in Fourier series and eliminate the $\phi$ dependence. The resulting equations are then solved by using the discrete ordinate method.

Let us consider a random medium bounded by parallel planes as shown in Fig. 4.7. A vertically or horizontally polarized wave is obliquely incident in medium 1 at incidence angle $\theta_0'$, and for convenience we set $\phi_0 = 0$. The dielectric constants of media 1, 2, and 3 may be different. The scattering layer contains randomly distributed spherical particles with dielectric constant $\epsilon_s$. We first separate the specific intensity into the reduced incident intensity $\mathbf{I}_{ri}$ and the diffuse or incoherent intensity $\mathbf{I}_d$:

$$\mathbf{I} = \mathbf{I}_{ri} + \mathbf{I}_d. \tag{4.214}$$

The reduced incident intensity of the plane wave is given by

$$\mathbf{I}_{ri}(\mathbf{r}, \hat{\mathbf{s}}) = \mathbf{I}_0 e^{-\tau/\mu_0} \delta(\mu - \mu_0)\delta(\phi), \tag{4.215}$$

where $\mu = \cos\theta$, $\mathbf{I}_0$ is the incident specific intensity, and $\tau$ is the optical distance:

$$\begin{aligned} \tau &= (\kappa_e + \kappa_{ag})z \\ &= (\kappa_e + \kappa_{ag})s\mu \\ &= \kappa_e' s\mu. \end{aligned} \tag{4.216}$$

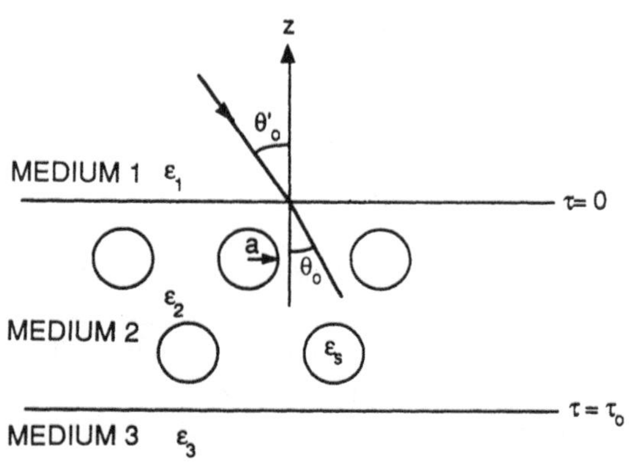


**Figure 4.7** Geometry showing plane parallel media with $\epsilon_1 \neq \epsilon_2 \neq \epsilon_3$.

The quantities $\kappa_e$, $\kappa_{ag}$, and $\kappa_e' = \kappa_e + \kappa_{ag}$ are the extinction coefficient of the scattering particles, the absorption coefficient of the background material, and the total extinction coefficient, respectively.

The diffuse component $\mathbf{I}_d$ is governed by an equation of the form given by (4.153) except for the addition of a new contribution representing conversion of energy from coherent to diffuse as a result of scattering by the particles. Thus,

$$\frac{d\mathbf{I}_d(\mathbf{r}, \hat{\mathbf{s}})}{ds} = -\kappa_e \mathbf{I}_d(\mathbf{r}, \hat{\mathbf{s}}) - \kappa_{ag}\mathbf{I}_d(\mathbf{r}, \hat{\mathbf{s}}) + \int_{4\pi} d\Omega' \mathcal{P}(\hat{\mathbf{s}}, \hat{\mathbf{s}}') \left[\mathbf{I}_d(\mathbf{r}, \hat{\mathbf{s}}') + \mathbf{I}_{ri}(\mathbf{r}, \hat{\mathbf{s}}')\right] \tag{4.217}$$

For convenience, following the procedure used by Ishimaru *et al.* (1982), we define the modified phase matrix:

$$\mathcal{B}(\hat{\mathbf{s}}, \hat{\mathbf{s}}') = \frac{\mathcal{P}(\hat{\mathbf{s}}, \hat{\mathbf{s}}')}{\kappa_e}, \tag{4.218}$$

and assuming that $\mathcal{P}$ and $\kappa_e$ are statistically independent quantities, $\mathcal{B}$ can be rewritten as

$$\mathcal{B} = \frac{1}{\langle \sigma_e \rangle} \begin{bmatrix} \langle |S_{vv}|^2 \rangle & \langle |S_{vh}\rangle|^2 & \mathrm{Re}\langle S_{vv} S_{vh}^* \rangle & -\mathrm{Im}\langle S_{vv} S_{vh}^* \rangle \\ \langle |S_{hv}|^2 \rangle & \langle |S_{hh}|^2 \rangle & \mathrm{Re}\langle S_{hv} S_{hh}^* \rangle & -\mathrm{Im}\langle S_{hv} S_{hh}^* \rangle \\ 2\mathrm{Re}\langle S_{vv} S_{hv}^* \rangle & 2\mathrm{Re}\langle S_{vh} S_{hh}^* \rangle & \mathrm{Re}\langle S_{vv} S_{hh}^* + S_{vh} S_{hv}^* \rangle & -\mathrm{Im}\langle S_{vv} S_{hh}^* - S_{vh} S_{hv}^* \rangle \\ 2\mathrm{Im}\langle S_{vv} S_{hv}^* \rangle & 2\mathrm{Im}\langle S_{vh} S_{hh}^* \rangle & \mathrm{Im}\langle S_{vv} S_{hh}^* + S_{vh} S_{hv}^* \rangle & \mathrm{Re}\langle S_{vv} S_{hh}^* - S_{vh} S_{hv}^* \rangle \end{bmatrix} \tag{4.219}$$

where $\langle \quad \rangle$ denotes the ensemble average over the sphere size distribution, and the elements $S_{ij}$ are the scattering amplitudes of a spherical particle.

Following Sekera (1966), the scattering amplitudes may be expressed as

$$S_{vv} = (l,l)T_v + (r,r)T_h \tag{4.220}$$

$$S_{vh} = -(r,l)T_v + (l,r)T_h \tag{4.221}$$

$$S_{hv} = -(l,r)T_v + (r,l)T_h \tag{4.222}$$

$$S_{hh} = (r,r)T_v + (l,l)T_h \tag{4.223}$$

by using the notation introduced by Chandrasekhar (1960):

$$(l,l) = [(1-\mu^2)(1-\mu'^2)]^{1/2} + \mu\mu' \cos(\phi' - \phi) \tag{4.224}$$

$$(l,r) = -\mu' \sin(\phi' - \phi) \tag{4.225}$$

$$(r,l) = \mu \sin(\phi' - \phi) \tag{4.226}$$

$$(r,r) = \cos(\phi' - \phi) \tag{4.227}$$

$$T_v(\chi) = \frac{A_{rr}(\chi) - \chi A_{ll}(\chi)}{1-\chi^2} \tag{4.228}$$

$$T_h(\chi) = \frac{A_{ll}(\chi) - \chi A_{rr}(\chi)}{1-\chi^2}, \tag{4.229}$$

where

$$\chi = \cos\Theta = [(1-\mu^2)(1-\mu'^2)]^{1/2} \cos(\phi' - \phi) + \mu\mu' \tag{4.230}$$

and where $(\theta', \phi')$ and $(\theta, \phi)$ correspond to the incident and scattered wave directions, respectively, $\mu' = \cos\theta'$, and $\Theta$ is the angle between the incident and scattered wave directions. $A_{ll}$ and $A_{rr}$ are functions of $\Theta$, and they are related to the scattering functions $S_2$ and $S_1$ used by van de Hulst [1957] for the Mie solution:

$$A_{ll} = (i/k)S_2^* \qquad \text{and} \qquad A_{rr} = (i/k)S_1^*. \tag{4.231}$$

In terms of the optical distance $\tau$ and the modified phase matrix $\mathcal{B}$, the equation of transfer becomes

$$\mu \frac{\mathrm{d}\mathbf{I}_d(\tau, \hat{\mathbf{s}})}{\mathrm{d}\tau} = -\mathbf{I}_d(\tau, \hat{\mathbf{s}}) + \frac{\kappa_e}{\kappa_e'} \int_{4\pi} \mathrm{d}\Omega' \mathcal{B}(\hat{\mathbf{s}}, \hat{\mathbf{s}}') \mathbf{I}_d(\tau, \hat{\mathbf{s}}') + \frac{\kappa_e}{\kappa_e'} \mathbf{I}_i(\tau, \hat{\mathbf{s}}) \tag{4.232}$$

where we have introduced the quantity $\mathbf{I}_i(\tau, \hat{\mathbf{s}}')$, which Ishimaru *et al.* (1982) call the incident intensity, to represent the integral:

$$\mathbf{I}_i(\tau, \hat{\mathbf{s}}') = \int \mathrm{d}\Omega' \mathcal{B}(\hat{\mathbf{s}}, \hat{\mathbf{s}}') \mathbf{I}_{ri}(\tau, \hat{\mathbf{s}}). \tag{4.233}$$

The effect of the background absorption is now included in $\kappa_e'$. The term $\kappa_e/\kappa_e'$ is given by

$$\frac{\kappa_e}{\kappa_e'} = \frac{\kappa_e}{\kappa_e + \kappa_{ag}}. \tag{4.234}$$

In a sparse medium, where $\kappa_e = N\sigma_e$, the above equation becomes

$$\frac{\kappa_e}{\kappa_e'} = \frac{\sigma_e}{\sigma_e + \kappa_{ag}/N}. \tag{4.235}$$

This form is similar to the albedo of a single particle, which is defined as

$$w_0 = \frac{\sigma_s}{\sigma_e} = \frac{\sigma_s}{\sigma_s + \sigma_a} \tag{4.236}$$

where $\sigma_s$ and $\sigma_a$ are the scattering and absorption cross sections. If we define the effective albedo of the medium as

$$w_0' = \frac{\sigma_s}{\sigma_e + \kappa_{ag}/N}, \tag{4.237}$$

the effective albedo will decrease as the background absorption increases. The effective albedo is a useful parameter for characterizing scattering in a wet snow medium.

### Incident Specific Intensity for Vertical and Horizontal Polarizations

Because of reflection at the boundaries, we have upward and downward traveling incident waves. Thus,

$$\mathbf{I}_i = \mathbf{I}_i^+ + \mathbf{I}_i^- , \tag{4.238}$$

where $\mathbf{I}_i^+$ is for $\mu > 0$ and $\mathbf{I}_i^-$ for $\mu < 0$. For a *vertically* polarized incident wave, $\mathbf{I}_i^+$ is given by

$$\mathbf{I}_i^+ = \int \mathcal{B}\mathbf{I}_{ri}^+ \, d\Omega' \tag{4.239}$$

$$= C_v \int \mathcal{B} \begin{bmatrix} 1 \\ 0 \\ 0 \\ 0 \end{bmatrix} \delta(\mu' - \mu_0)\delta(\phi')e^{-\tau/\mu_0} \, d\mu' d\phi' \tag{4.240}$$

$$= C_v \begin{bmatrix} B_{11} \\ B_{21} \\ B_{31} \\ B_{41} \end{bmatrix}_{\mu=\mu_0} e^{-\tau/\mu_0} , \tag{4.241}$$

where

$$B_{11} = \frac{\langle |S_{vv}|^2 \rangle}{\langle \sigma_e \rangle},$$

$$B_{21} = \frac{\langle |S_{hv}|^2 \rangle}{\langle \sigma_e \rangle},$$

$$B_{31} = \frac{2\mathrm{Re}\langle S_{vv} S_{hv}^* \rangle}{\langle \sigma_e \rangle},$$

$$B_{41} = \frac{2\mathrm{Im}\langle S_{vv} S_{hv}^* \rangle}{\langle \sigma_e \rangle}. \tag{4.242}$$

In (4.239) $B_{11}$, $B_{21}$, $B_{31}$, and $B_{41}$ are evaluated at $\mu = \mu_0$. The quantity $C_v$ accounts for multiple reflections at the boundaries for a vertically polarized incident wave. For $\beta = v$ or $h$,

$$C_\beta = \frac{T_{\beta 12} I_0}{1 - R_{\beta 23} R_{\beta 21} \exp(-2\tau_0/\mu_0)} \tag{4.243}$$

where $\tau_0 = \kappa_e' d$. The quantities $T_{\beta 12}$, $R_{\beta 21}$, and $R_{\beta 23}$ are the transmissivity and power reflectivity functions defined in (4.182) and (4.183). The reflected incident specific intensity $\mathbf{I}^-$ is

$$\mathbf{I}_i^- = \int \mathcal{B}\mathbf{I}_{ri}^- \, \mathrm{d}\Omega' \tag{4.244}$$

$$= C_v R_{v23} \int \mathcal{B} \begin{bmatrix} 1 \\ 0 \\ 0 \\ 0 \end{bmatrix} \delta(\mu' + \mu_0)\delta(\phi')\mathrm{d}\mu'\mathrm{d}\phi' \exp(-\frac{2\tau_0 - \tau}{\mu_0}) \tag{4.245}$$

$$= C_v R_{v23} \begin{bmatrix} B_{11} \\ B_{21} \\ B_{31} \\ B_{41} \end{bmatrix}_{\mu=-\mu_0} \exp(-\frac{2\tau_0 - \tau}{\mu_0}) \; . \tag{4.246}$$

For a *horizontally* polarized incident wave, $\mathbf{I}_i^+$ is given by

$$\begin{aligned} \mathbf{I}_i^+ &= C_h \begin{bmatrix} B_{12} \\ B_{22} \\ B_{32} \\ B_{42} \end{bmatrix}_{\mu=\mu_0} \exp(-\frac{\tau}{\mu_0}) \\ B_{12} &= \frac{\langle |S_{vh}|^2 \rangle}{\langle \sigma_e \rangle}, \\ B_{22} &= \frac{\langle |S_{hh}|^2 \rangle}{\langle \sigma_e \rangle}, \\ B_{32} &= \frac{2\mathrm{Re}\langle S_{vh} S_{hh}^* \rangle}{\langle \sigma_e \rangle}, \\ B_{42} &= \frac{2\mathrm{Im}\langle S_{vh} S_{hh}^* \rangle}{\langle \sigma_e \rangle} \; . \end{aligned} \tag{4.247}$$

The reflected incident specific intensity for horizontal polarization is

$$\mathbf{I}_i^- = C_h R_{h23} \begin{bmatrix} B_{12} \\ B_{22} \\ B_{32} \\ B_{42} \end{bmatrix}_{\mu=-\mu_0} \exp(-\frac{2\tau_0 - \tau}{\mu_0}) \; . \tag{4.248}$$

Upon substituting $\mathbf{I}_i^+$ and $\mathbf{I}_i^-$ into (4.232), we obtain

$$\begin{aligned} \mu \frac{\mathrm{d}\mathbf{I}_d}{\mathrm{d}\tau} &= -\mathbf{I}_d + \frac{\kappa_e}{\kappa_e'} \int \mathcal{B}\mathbf{I}_d \, \mathrm{d}\Omega' \\ &\quad + \frac{\kappa_e}{\kappa_e'} \left( C_v \mathbf{F}^+ \mathrm{e}^{-\tau/\mu_0} + C_v R_{v23} \mathbf{F}^- \mathrm{e}^{-(2\tau_0 - \tau)/\mu_0} \right) \end{aligned} \tag{4.249}$$

where

$$\mathbf{F}^+ = \begin{bmatrix} B_{11} \\ B_{21} \\ B_{31} \\ B_{41} \end{bmatrix}_{\mu=\mu_0}$$

$$\mathbf{F}^- = \begin{bmatrix} B_{11} \\ B_{21} \\ B_{31} \\ B_{41} \end{bmatrix}_{\mu=-\mu_0} \tag{4.250}$$

for a vertically polarized incident wave, and a similar expression can be written for the case when the incident wave is horizontally polarized.

The boundary condition for the incoherent specific intensity $\mathbf{I}_d$ is that the inward intensities at the boundaries are due to the reflected incoherent specific intensities. Mathematically, we can write

$$\mathbf{I}_d(\tau = 0, -\mu) = \mathcal{R}_{21}(-\mu)\mathbf{I}_d(\tau = 0, \mu) \qquad -1 \leq \mu \leq 0 \tag{4.251}$$

$$\mathbf{I}_d(\tau = \tau_0, -\mu) = \mathcal{R}_{23}(\mu)\mathbf{I}_d(\tau = \tau_0, \mu) \qquad 0 \leq \mu \leq 1 \tag{4.252}$$

where $\mathcal{R}_{21}(\mu)$ and $\mathcal{R}_{23}(\mu)$ are the reflectivity matrices of the upper and lower boundaries (see (4.188)).

We can eliminate the $\phi$ dependence from (4.249) and obtain a new equation in terms of $\tau$ and $\mu$ (Ishimaru *et al.*, 1982; Cheung and Ishimaru, 1982) by expanding $\mathbf{I}_d$, $\mathcal{B}, \mathbf{F}^+$, and $\mathbf{F}^-$ in Fourier series in $\phi$. For the plane wave case, we can show that all Fourier components of the equation of transfer are independent of each other. Also, for the vertically and horizontally polarized incident waves, the first two terms of the Stokes vector are even functions of $\phi$ and the last two terms are odd functions of $\phi$.

In the Fourier expansion technique, we first expand $\mathbf{I}$, $\mathcal{B}, \mathbf{F}^+$, and $\mathbf{F}^-$ in Fourier series. To simplify the notation, we will use $\mathbf{F}$ to represent both $\mathbf{F}^+$ and $\mathbf{F}^-$:

$$\mathbf{I}_d = \sum_{m=0}^{\infty} \mathbf{I}_m^a \cos m\phi + \sum_{m=1}^{\infty} \mathbf{I}_m^b \sin m\phi \tag{4.253}$$

$$\mathcal{B} = \frac{1}{2\pi}\mathcal{B}_0^a + \frac{1}{\pi}\sum_{m=1}^{\infty} \mathcal{B}_m^a \cos m(\phi' - \phi)$$

$$+\frac{1}{\pi}\sum_{m=1}^{\infty} \mathcal{B}_m^b \sin m(\phi' - \phi) \tag{4.254}$$

$$\mathbf{F} = \sum_{m=0}^{\infty} \mathbf{F}_m^a \cos m\phi + \sum_{m=1}^{\infty} \mathbf{F}_m^b \sin m\phi. \tag{4.255}$$

Superscripts $a$ and $b$ denote the even and odd Fourier components. As we see from (4.254), the matrix $\mathcal{B}$ is a function of $(\phi' - \phi)$ only because of the spherical shape of the particles. If the particle is nonspherical, $\mathcal{B}$ becomes a function of $\phi'$ and $\phi$, rather than $(\phi' - \phi)$ (Ishimaru *et al.*, 1982).

The Fourier components, $\mathbf{F}_m^a$ and $\mathbf{F}_m^b$, of the source function $\mathbf{F}$, are given by

$$\mathbf{F}_0^a = \frac{1}{2\pi} \int_0^{2\pi} \mathbf{F}\, d\phi \tag{4.256}$$

$$\mathbf{F}_m^a = \frac{1}{\pi} \int_0^{2\pi} \mathbf{F} \cos m\phi\, d\phi \tag{4.257}$$

$$\mathbf{F}_m^b = \frac{1}{\pi} \int_0^{2\pi} \mathbf{F} \sin m\phi\, d\phi. \tag{4.258}$$

Because $S_{vv}$ and $S_{hh}$ are even function of $(\phi' - \phi)$ and $S_{vh}$ and $S_{hv}$ are odd functions of $(\phi' - \phi)$, $\mathbf{F}_m^a$ has only the first two elements and $\mathbf{F}_m^b$ has only the last two elements:

$$\mathbf{F}_m^a = \begin{bmatrix} F_{m1} \\ F_{m2} \\ 0 \\ 0 \end{bmatrix}, \quad \mathbf{F}_m^b = \begin{bmatrix} 0 \\ 0 \\ F_{m3} \\ F_{m4} \end{bmatrix}. \tag{4.259}$$

Similarly, $\mathcal{B}$ can be separated into even and odd components:

$$\mathcal{B}_m^a = \begin{bmatrix} \mathcal{B}_{1m}^a & 0 \\ 0 & \mathcal{B}_{4m}^a \end{bmatrix} \tag{4.260}$$

$$\mathcal{B}_m^b = \begin{bmatrix} 0 & \mathcal{B}_{2m}^b \\ \mathcal{B}_{3m}^b & 0 \end{bmatrix} \tag{4.261}$$

where $\mathcal{B}_{1m}^a$, $\mathcal{B}_{2m}^b$, $\mathcal{B}_{3m}^a$, and $\mathcal{B}_{4m}^a$ are $2 \times 2$ matrices and are given by

$$\mathcal{B}_{1m}^a = \int_0^{2\pi} \mathcal{B}_1 \cos m(\phi' - \phi)\, d(\phi' - \phi) \tag{4.262}$$

$$\mathcal{B}_{2m}^b = \int_0^{2\pi} \mathcal{B}_2 \sin m(\phi' - \phi)\, d(\phi' - \phi) \tag{4.263}$$

$$\mathcal{B}_{3m}^{b} = \int_0^{2\pi} \mathcal{B}_3 \sin m(\phi' - \phi)\, \mathrm{d}(\phi' - \phi) \tag{4.264}$$

$$\mathcal{B}_{4m}^{a} = \int_0^{2\pi} \mathcal{B}_4 \cos m(\phi' - \phi)\, \mathrm{d}(\phi' - \phi) \tag{4.265}$$

$$\mathcal{B} = \begin{bmatrix} \mathcal{B}_1 & \mathcal{B}_2 \\ \mathcal{B}_3 & \mathcal{B}_4 \end{bmatrix}. \tag{4.266}$$

We know from the iterative solution presented earlier that first-order scattering is proportional to $\mathbf{F}$ and second-order scattering is proportional to $\mathcal{B}\mathbf{F}$. This process yields first- and second-order specific intensities, the first two elements of which are even and the last two elements are odd functions of $(\phi' - \phi)$. If we continued the iterative process, the specific intensity $\mathbf{I}$ should continue to have its first two elements as even functions and its last two elements as odd functions of $(\phi' - \phi)$. Therefore, we write

$$\mathbf{I}_m^a = \begin{bmatrix} I_{mv} \\ I_{mh} \\ 0 \\ 0 \end{bmatrix}, \quad \mathbf{I}_m^b = \begin{bmatrix} 0 \\ 0 \\ U_m \\ V_m \end{bmatrix}. \tag{4.267}$$

By substituting (4.253), (4.254), and (4.256) into the equation of transfer (4.249), integrating with respect to $\phi'$, and making use of orthogonality, we see that the Fourier components for the equation of transfer are all independent of each other. Thus, we can obtain the equation of transfer for each Fourier component in the form given below.

For the Fourier component $m = 0$ and a vertically polarized incident wave, we have

$$\mu \frac{\mathrm{d}\mathbf{I}_{d0}}{\mathrm{d}\tau} = -\mathbf{I}_{d0} + \frac{\kappa_e}{\kappa_e'} \int_{-1}^{1} \mathcal{L}_0 \mathbf{I}_{d0}\, \mathrm{d}\mu' + \frac{\kappa_e}{\kappa_e'} C_v \mathbf{F}_0^+ \mathrm{e}^{-\tau/\mu_0}$$
$$+ \frac{\kappa_e}{\kappa_e'} C_v R_{v23} \mathbf{F}_0^- \mathrm{e}^{-(2\tau_0 - \tau)/\mu_0} \tag{4.268}$$

where

$$\mathbf{I}_{d0} = \begin{bmatrix} I_{01} \\ I_{02} \end{bmatrix}, \qquad \mathcal{L}_0 = \mathcal{B}_{10}^{a},$$

$$\mathbf{F}_0^+ = \begin{bmatrix} B_{11} \\ B_{21} \end{bmatrix}_{\mu=\mu_0}, \qquad \mathbf{F}_0^- = \begin{bmatrix} B_{11} \\ B_{21} \end{bmatrix}_{\mu=-\mu_0}. \tag{4.269}$$

For $m > 0$ ,

$$\begin{aligned}\mu\frac{\mathrm{d}\mathbf{I}_{dm}}{\mathrm{d}\tau} &= -\mathbf{I}_{dm} + \frac{\kappa_e}{\kappa_e'}\int_{-1}^{1}\mathcal{L}_m\mathbf{I}_{dm}\,\mathrm{d}\mu' \\ &\quad + \frac{\kappa_e}{\kappa_e'}C_v\mathbf{F}_m^{+}\mathrm{e}^{-\tau/\mu_0} \\ &\quad + \frac{\kappa_e}{\kappa_e'}C_v R_{v23}\mathbf{F}_m^{-}\mathrm{e}^{-(2\tau_0-\tau)/\mu_0} \qquad (4.270)\end{aligned}$$

where

$$\mathbf{I}_{dm} = \begin{bmatrix} I_{mv} \\ I_{mh} \\ U_m \\ V_m \end{bmatrix}, \qquad \mathcal{L}_m = \begin{bmatrix} \mathcal{B}_{1m}^{a} & \mathcal{B}_{2m}^{b} \\ -\mathcal{B}_{3m}^{b} & \mathcal{B}_{4m}^{a} \end{bmatrix},$$

$$\mathbf{F}_m^{+} = \begin{bmatrix} B_{11} \\ B_{21} \\ B_{31} \\ B_{41} \end{bmatrix}_{\mu=\mu_0}, \qquad \mathbf{F}_m^{-} = \begin{bmatrix} B_{11} \\ B_{21} \\ B_{31} \\ B_{41} \end{bmatrix}_{\mu=-\mu_0}. \qquad (4.271)$$

The equation of transfer given in (4.268) and (4.270) together with the boundary conditions given in (4.251) and (4.252) constitute the complete mathematical formulation of the problem. No analytical solution of the integral-differential equation is available, but we can solve the equation of transfer numerically by several techniques. Once $\mathbf{I}_{d0}$ and $\mathbf{I}_{dm}$ are found from (4.268) and (4.270), the total scattered incoherent specific intensities at locations above the upper boundary ($\tau = 0$) and below the lower boundary ($\tau = \tau_0$) are given by

$$\mathbf{I}^s = \mathcal{T}_{21}\left\{\sum_{m=o}^{\infty}\begin{bmatrix} I_{mv} \\ I_{vh} \\ 0 \\ 0 \end{bmatrix}\cos m\phi + \sum_{m=1}^{\infty}\begin{bmatrix} 0 \\ 0 \\ U_m \\ V_m \end{bmatrix}\sin m\phi\right\} \quad \text{at} \quad \tau = 0, \qquad (4.272)$$

$$\mathbf{I}^s = \mathcal{T}_{23}\left\{\sum_{m=o}^{\infty}\begin{bmatrix} I_{mv} \\ I_{mh} \\ 0 \\ 0 \end{bmatrix}\cos m\phi + \sum_{m=1}^{\infty}\begin{bmatrix} 0 \\ 0 \\ U_m \\ V_m \end{bmatrix}\sin m\phi\right\} \quad \text{at} \quad \tau = \tau_0, \qquad (4.273)$$

where $\mathcal{T}_{21}$ and $\mathcal{T}_{23}$ are the transmissivity matrices given by (4.190). If the incidence angle is larger than the critical angle, $\mathcal{T}_{ij} = 0$ and there will be no transmitted intensity. In the special case of normal incidence, all Fourier components of $\mathbf{F}$ and $\mathbf{I}_d$ are zero except $m = 0$ and $m = 2$ (Cheung and Ishimaru, 1982).

The above formulation is for spherical Mie particles, but the same equations can be used for spherical Rayleigh particles. For the Rayleigh particles, we need to replace $\mathcal{B}$ as given in (4.219) by $(1/\kappa_e)\mathcal{P}$, where $\mathcal{P}$ is the Rayleigh phase function given in (4.174). Fourier series expansion of the Rayleigh phase function in $(\phi' - \phi)$ will give nonzero terms for $m = 0$, 1, and 2. In the radiative transfer equation, therefore, we only need to compute the Fourier components up to $m = 2$ for the Rayleigh phase function at any incidence angle.

Among the numerical techniques available for solving equations of the form given by (4.268) and (4.270) is the discrete ordinate method, which has been used extensively in the literature (Ishimaru, 1978; Tsang *et al.*, 1985). In this method, the specific intensity and phase function are discretized into a finite number of directions and the radiative transfer equation is converted into a system of ordinary differential equations with constant coefficients. The solution can be obtained by the eigenanalysis method.

# 4-4 POLARIZATION SYNTHESIS

In this section, we shall examine the relationship between the specific intensity $\mathbf{I}^s$ backscattered from a random medium and the backscattering coefficient of the medium for any specified transmitting and receiving antenna polarization configuration. For the principal linear polarization cases ($vv, vh, hv$, and $hh$), the procedure is straightforward. To compute $\sigma^0_{rt}$ corresponding to an $r$-polarized receiving antenna and a $t$-polarized transmitting antenna, we must first compute the scattered intensity:

$$\mathbf{I}^s = \begin{bmatrix} I_v^s \\ I_h^s \\ U^s \\ V^s \end{bmatrix} \tag{4.274}$$

for an incident plane wave with specific intensity:

$$\mathbf{I}^i = I_0 \begin{bmatrix} 1 \\ 0 \\ 0 \\ 0 \end{bmatrix} \quad \text{if } t = v, \tag{4.275}$$

or

$$\mathbf{I}^i = I_0 \begin{bmatrix} 0 \\ 1 \\ 0 \\ 0 \end{bmatrix} \qquad \text{if } t = h \ . \tag{4.276}$$

Equation (4.164) can then be used to compute the backscattering coefficient $\sigma^0_{rt}$. For example, if (4.274) represents the scattered specific intensity for an incident $v$-polarized wave, we have

$$\sigma^0_{vv} = 4\pi \cos\theta_s \frac{I^s_v}{I_0} \tag{4.277}$$

$$\sigma^0_{hv} = 4\pi \cos\theta_s \frac{I^s_h}{I_0} \ . \tag{4.278}$$

Similarly, if (4.274) represents the scattered specific intensity for an incident $h$- polarized wave, we have

$$\sigma^0_{vh} = 4\pi \cos\theta_s \frac{I^s_v}{I_0} \tag{4.279}$$

$$\sigma^0_{hh} = 4\pi \cos\theta_s \frac{I^s_h}{I_0} \ . \tag{4.280}$$

To compute $\sigma^0_{rt}$ for $rt \notin \{vv, vh, hv, hh\}$, a more general procedure must be used. In describing this procedure, we will consider two slightly different cases. The first case is when the incident and scattered specific intensities are related by some known transformation matrix $\mathcal{H}$, such that

$$\mathbf{I}^s = \mathcal{H}\, \mathbf{I}^i \ . \tag{4.281}$$

This form is present, for example, in the first-order radiative transfer solution given by (4.204) and (4.205). The second case is when the scattered specific intensity can be computed for a given incident specific intensity, but an explicit expression for the transformation matrix $\mathcal{H}$ is unavailable. This situation arises when a numerical solution to the radiative transfer equation is obtained, or the scattered specific intensity is directly measured for a given set of polarization configurations.

First, consider the case in which $\mathcal{H}$ is known. The incident and scattered specific intensities $\mathbf{I}^i$ and $\mathbf{I}^s$ are related to the modified Stokes vectors $\mathbf{F}^t_m$ and $\mathbf{F}^r_m$ by the expressions:

$$\mathbf{I}^i = \frac{1}{\eta}\, \mathbf{F}^t_m \tag{4.282}$$

$$\mathbf{I}^s = \frac{1}{\eta\, d\Omega}\, \mathbf{F}^r_m \ , \tag{4.283}$$

where $d\Omega$ is the solid angle subtended by the illuminated area $A$. By using $d\Omega$ as defined in (4.159) and substituting (4.282) and (4.283) into (4.281) we obtain the expression:

$$\mathbf{F}_m^r = \frac{A \cos \theta_s}{r^2} \mathcal{H} \, \mathbf{F}_m^t \ . \tag{4.284}$$

After converting (4.284) to the BSA convention and comparing the result with (2.45), we find that the modified Mueller matrix $\overline{\mathcal{L}}_m$ and the modified Stokes scattering operator $\overline{\mathcal{M}}_m$ are given by

$$\overline{\mathcal{L}}_m = A \cos \theta_s \, \mathcal{T} \, \mathcal{H} \tag{4.285}$$

$$\overline{\mathcal{M}}_m = A \cos \theta_s (\mathcal{V} \, \tilde{\mathcal{V}})^{-1} \, \overline{\mathcal{H}} \, \overline{A}_m^t \ , \tag{4.286}$$

and, by using the polarization synthesis techniques of Chapter 2, $\sigma_{rt}^0$ can be written in terms of $\overline{\mathcal{M}}_m$ as

$$\sigma_{rt}^0 = 4\pi \ \theta_s \overline{A}_m^r \cdot \left(\mathcal{V}\tilde{\mathcal{V}}\right)^{-1} \overline{\mathcal{H}} \, \overline{A}_m^t \ , \tag{4.287}$$

where

$$\overline{\mathcal{H}} = \mathcal{T} \, \mathcal{H} \tag{4.288}$$

$$(\mathcal{V} \, \tilde{\mathcal{V}})^{-1} = \begin{bmatrix} 1 & 0 & 0 & 0 \\ 0 & 1 & 0 & 0 \\ 0 & 0 & 1/2 & 0 \\ 0 & 0 & 0 & -1/2 \end{bmatrix} \ . \tag{4.289}$$

We now consider the second case in which only the scattered intensity can be computed (or directly measured) for a given incident specific intensity. Although $\mathcal{H}$ is not explicitly known, it can be determined by computing (or measuring) the scattered specific intensity for a linearly independent set of incident polarizations. A convenient set of polarizations contains vertical, horizontal, 45° linear, and left-hand circular. The specific intensities for these polarizations, respectively, are given by

$$\mathbf{I}_v^i = \begin{bmatrix} 1 \\ 0 \\ 0 \\ 0 \end{bmatrix}, \mathbf{I}_h^i = \begin{bmatrix} 0 \\ 1 \\ 0 \\ 0 \end{bmatrix}, \mathbf{I}_{45}^i = \begin{bmatrix} 1/2 \\ 1/2 \\ 1 \\ 0 \end{bmatrix}, \text{and } \mathbf{I}_L^i = \begin{bmatrix} 1/2 \\ 1/2 \\ 0 \\ 1 \end{bmatrix} . \tag{4.290}$$

The corresponding scattered specific intensities are denoted by $\mathbf{I}_v^s$, $\mathbf{I}_h^s$, $\mathbf{I}_{45}^s$, and $\mathbf{I}_L^s$.

The first and second columns of $\mathcal{H}$ are obtained directly by substituting $\mathbf{I}_v^i$ and $\mathbf{I}_h^i$ from (4.290) into (4.281):

$$\mathbf{I}_v^s = \mathcal{H}\,\mathbf{I}_r^v = \begin{bmatrix} h_{11} \\ h_{21} \\ h_{31} \\ h_{41} \end{bmatrix}, \quad \mathbf{I}_h^s = \mathcal{H}\,\mathbf{I}_h^i = \begin{bmatrix} h_{12} \\ h_{22} \\ h_{32} \\ h_{42} \end{bmatrix}. \tag{4.291}$$

However, to obtain the third and fourth columns of $\mathcal{H}$, a convenient approach is to define two additional intensities composed of linear combinations of $\mathbf{I}_v^i, \mathbf{I}_h^i, \mathbf{I}_{45}^i$, and $\mathbf{I}_L^i$. These intensities are given by

$$\mathbf{I}_1^i = \mathbf{I}_{45}^i - \frac{1}{2}\left(\mathbf{I}_v^i + \mathbf{I}_h^i\right) = \begin{bmatrix} 0 \\ 0 \\ 1 \\ 0 \end{bmatrix} \tag{4.292}$$

$$\mathbf{I}_2^i = \mathbf{I}_L^i - \frac{1}{2}\left(\mathbf{I}_v^i + \mathbf{I}_h^i\right) = \begin{bmatrix} 0 \\ 0 \\ 0 \\ 1 \end{bmatrix}. \tag{4.293}$$

By substituting these intensities into (4.281), we obtain

$$\mathbf{I}_1^s = \mathcal{H}\,\mathbf{I}_1^i = \begin{bmatrix} h_{31} \\ h_{32} \\ h_{33} \\ h_{34} \end{bmatrix}, \quad \mathbf{I}_2^s = \mathcal{H}\,\mathbf{I}_2^i = \begin{bmatrix} h_{41} \\ h_{42} \\ h_{43} \\ h_{44} \end{bmatrix}. \tag{4.294}$$

Therefore, the matrix $\mathcal{H}$ can be written as

$$\mathcal{H} = \left[\mathbf{I}_v^s,\ \mathbf{I}_h^s,\ \mathbf{I}_1^s,\ \mathbf{I}_2^s\right], \tag{4.295}$$

where

$$\begin{aligned}
\mathbf{I}_1^s &= \mathcal{H}\,\mathbf{I}_1^i \\
&= \mathcal{H}\left[\mathbf{I}_{45}^i - \frac{1}{2}\left(\mathbf{I}_v^i + \mathbf{I}_h^i\right)\right] \\
&= \mathbf{I}_{45}^s - \frac{1}{2}\left(\mathbf{I}_v^s + \mathbf{I}_h^s\right) \\
\mathbf{I}_2^s &= \mathcal{H}\,\mathbf{I}_2^i \\
&= \mathcal{H}\left[\mathbf{I}_L^i - \frac{1}{2}\left(\mathbf{I}_v^i + \mathbf{I}_h^i\right)\right] \\
&= \mathbf{I}_L^s - \frac{1}{2}\left(\mathbf{I}_v^s + \mathbf{I}_h^s\right).
\end{aligned}$$

Because $\mathcal{H}$ is now known, the backscattering coefficient for the arbitrary polarization configuration $rt$ can be determined by using (4.287) to (4.289).

# 4-5 VEGETATION CANOPY MODEL

Two models will be considered as illustrations of radiative transfer theory. Section 4-5.1 presents a model for microwave backscatter from a single vegetation layer of height $d$ over a ground surface. Section 4-5.2 discusses a backscatter model for a two-layer forest canopy. This model consists of a two layers of vegetation over a ground surface. The top layer contains an appropriate combination of branches, leaves, and needles, and the bottom layer contains the trunks. Section 4-5.3 presents some numerical simulations in which these models are applied to model measured backscatter data from various canopies.

## 4-5.1 Radiative Transfer Model for a Single Vegetation Layer Over a Soil Surface

A first-order polarimetric solution to the problem of scattering by a uniform layer of vegetation may be formulated by using the iterative radiative transfer method discussed in Section 4-3.1. As shown in Fig. 4.8, the canopy is considered to consist of a single horizontal vegetation layer of height $d$ over a dielectric ground surface. The layer is assumed to have a diffuse upper boundary and to be statistically homogeneous and continuous in the horizontal direction. The layer consists of $K$ constituent classes such as branches, leaves, and needles, each with a corresponding number density of $N_k$ particles per unit volume. The joint probability density function $p_k(s_1, s_2; \theta_j, \phi_j)$ describes the distribution in size and orientation of constituent class $k$. Here, $(s_1, s_2)$ and $(\theta_j, \phi_j)$ are the size and orientation parameters of class $k$.

The total backscattered intensity $\mathbf{I}_t^s(\mu_0, \phi_0)$ may be related to the intensity $\mathbf{I}_0$ incident upon the canopy through the transformation matrix $\boldsymbol{\mathcal{T}}_t(\mu_0, \phi_0)$ by the equation:

$$\mathbf{I}_t^s(\mu_0, \phi_0) = \boldsymbol{\mathcal{T}}_t(\mu_0, \phi_0)\mathbf{I}_0(-\mu_0, \phi_0). \tag{4.296}$$

$\boldsymbol{\mathcal{T}}_t(\mu_0, \phi_0)$ will be called the $4 \times 4$ total canopy backscattering transformation matrix.

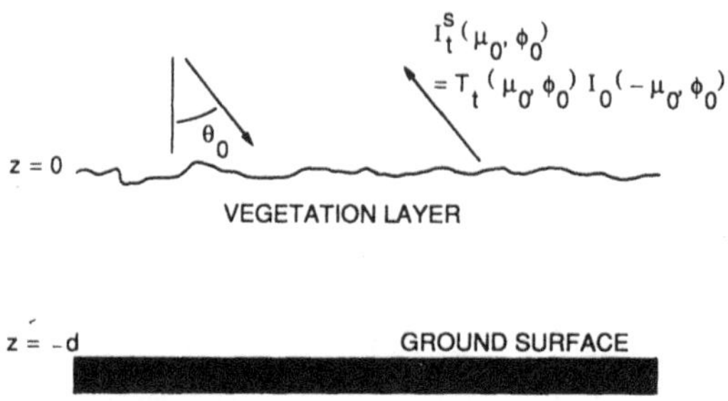


**Figure 4.8** Model for a single layer of vegetation over a ground surface.

To solve for the total canopy backscattering transformation matrix, the problem is divided into two parts. In the first part, a solution is found with the ground surface treated as a specular dielectric interface. This assumption is reasonable as long as the intensity scattered by the surface is dominated by its coherent component in the specular direction. If the surface is very rough or its mean slope is not zero relative to the vertical direction, this assumption will not hold. The scattered intensity for a vegetation layer over a specular surface is

$$\mathbf{I}^s(\mu_0, \phi_0) = \mathcal{T}_c(\mu_0, \phi_0)\mathbf{I}_0(-\mu_0, \phi_0), \tag{4.297}$$

where $\mathcal{T}_c(\mu_0, \phi_0)$ is the transformation matrix relating the incident and scattered intensities in the absence of direct backscatter from the ground surface.

In the second part of the problem, direct backscatter from the ground surface is accounted for by expressing the backscattered intensity as

$$\mathbf{I}^s_g(\mu_0, \phi_0) = \mathcal{T}_g(\mu_0, \phi_0)\mathbf{I}_0(-\mu_0, \phi_0), \tag{4.298}$$

where $\mathcal{T}_g(\mu_0, \phi_0)$ is the transformation matrix that accounts for propagation through the canopy down to the ground surface, backscatter by the ground surface, and propagation again through the canopy back to the radar. The expression for $\mathcal{T}_g$ is given by

$$\mathcal{T}_g(\mu_0, \phi_0) = e^{-\boldsymbol{\kappa}_e^+ d/\mu_0}\mathcal{G}(\mu_0)e^{-\boldsymbol{\kappa}_e^- d/\mu_0}, \tag{4.299}$$

where $\boldsymbol{\kappa}_e^+$ and $\boldsymbol{\kappa}_e^-$ are the $4 \times 4$ extinction matrices of the vegetation layer for upward propagation in the direction $(\mu_0, \phi_0)$ and downward propagation in the direction $(-\mu_0, \phi_0)$, respectively. $\mathcal{G}(\mu_0)$ is the ground backscattering matrix given by

$$\mathcal{G}(\mu_0) = \frac{1}{\cos\,\theta_0}\,\overline{\mathcal{M}}_m, \tag{4.300}$$

where $\overline{\mathcal{M}}_m$ is the modified Stokes scattering operator, which describes ground backscatter and is given in (4.142).

Combining (4.297) and (4.298) yields the total backscattered intensity in terms of the total canopy backscattering transformation matrix $\mathcal{T}_t$:

$$\begin{aligned}\mathbf{I}_t^s(\mu_0,\phi_0) &= [\mathcal{T}_c(\mu_0,\phi_0)+\mathcal{T}_g(\mu_0,\phi_0)]\,\mathbf{I}_0(-\mu_0,\phi_0) &(4.301)\\ &= \mathcal{T}_t(\mu_0,\phi_0)\mathbf{I}_0(-\mu_0,\phi_0). &(4.302)\end{aligned}$$

To determine the extinction matrices, we first have to compute the $2\times 2$ matrix $\mathcal{M}^e$, the elements of which are defined as

$$\mathcal{M}^e_{mn} = \sum_{k=1}^{K} \frac{i2\pi N_k}{k_0}\langle S_{mnk}(\theta_i,\phi_i;\theta_i,\phi_i;\theta_j,\phi_j)\rangle_k; \qquad m,n=v,h, \tag{4.303}$$

where the summation over $k$ represents an addition over the constituent classes within the canopy, $N_k$ is the number density per unit volume of each constituent, $\langle\ \rangle_k$ represents statistical averages over the size and orientation distributions of class $k$, and $S_{mnk}$ is the component of the scattering matrix corresponding to polarization $mn$ for class $k$. The extinction matrix is then given by

$$\boldsymbol{\kappa}_e =$$

$$\begin{bmatrix} -2\mathrm{Re}M^e_{vv} & 0 & -\mathrm{Re}(M^e_{vh}) & -\mathrm{Im}(M^e_{vh}) \\ 0 & -2\mathrm{Re}M^e_{hh} & -\mathrm{Re}(M^e_{hv}) & \mathrm{Im}(M^e_{hv}) \\ -2\mathrm{Re}M^e_{hv} & -2\mathrm{Re}M^e_{vh} & -\mathrm{Re}(M^e_{vv}+M^e_{hh}) & -\mathrm{Im}(M^e_{vv}-M^e_{hh}) \\ 2\mathrm{Im}M^e_{hv} & -2\mathrm{Im}M^e_{vh} & -\mathrm{Im}(M^e_{vv}-M^e_{hh}) & -\mathrm{Re}(M^e_{vv}+M^e_{hh}) \end{bmatrix} \tag{4.304}$$

The quantity $e^{\boldsymbol{\kappa}_z/\mu}$ is defined by Tsang *et al.* (1985, p. 243) as

$$e^{\boldsymbol{\kappa}_z/\mu} = \mathcal{E}(\mu,\phi)\mathcal{D}(\mu,\phi;-z/\mu)\mathcal{E}^{-1}(\mu,\phi). \tag{4.305}$$

The $4\times 4$ eigenmatrix $\mathcal{E}(\mu,\phi)$ is given by

$$\mathcal{E}(\theta,\phi) = \begin{bmatrix} 1 & b_2^* & b_2 & |b_2|^2 \\ |b_1|^2 & b_1 & b_1^* & 1 \\ 2\mathrm{Re}(b_1) & 1+b_1b_2^* & 1+b_1^*b_2 & 2\mathrm{Re}(b_2) \\ -2\mathrm{Im}(b_1) & -i(1-b_1b_2^*) & i(1-b_1^*b_2) & 2\mathrm{Im}(b_2) \end{bmatrix}, \tag{4.306}$$

where

$$b_1 = \frac{2M^e_{hv}}{M^e_{vv}-M^e_{hh}+r} \tag{4.307}$$

$$b_2 = \frac{2M^e_{vh}}{-M^e_{vv} + M^e_{hh} - r} \tag{4.308}$$

$$r = \{(M^e_{vv} - M^e_{hh})^2 + 4M^e_{hv}M^e_{vh}\}^{1/2}. \tag{4.309}$$

$\mathcal{D}(\mu, \phi; -z/\mu)$ is a diagonal matrix with elements:

$$[\mathcal{D}(\mu, \phi; -z/\mu)]_{ii} = \mathrm{e}^{-\beta_i(\mu,\phi)z/\mu}, \tag{4.310}$$

where $\beta_i(\mu, \phi)$ represents the $i$th eigenvalue of $\boldsymbol{\kappa}_e(\mu, \phi)$. These eigenvalues are given by

$$\beta(\hat{\mathbf{s}}) = \begin{bmatrix} \beta_1(\theta,\phi) \\ \beta_2(\theta,\phi) \\ \beta_3(\theta,\phi) \\ \beta_4(\theta,\phi) \end{bmatrix} = -\mathrm{Re}(M^e_{vv} + M^e_{hh}) + \frac{1}{2}\begin{bmatrix} -r - r^* \\ -r + r^* \\ r - r^* \\ r + r^* \end{bmatrix} \tag{4.311}$$

$$= \begin{bmatrix} 2\mathrm{Im}K_1 \\ iK_2^* - iK_1 \\ iK_1^* - iK_2 \\ 2\mathrm{Im}K_2 \end{bmatrix}, \tag{4.312}$$

where

$$K_1 = k_0 - \frac{i}{2}[M^e_{vv} + M^e_{hh} + r] \tag{4.313}$$

$$K_2 = k_0 - \frac{i}{2}[M^e_{vv} + M^e_{hh} - r]. \tag{4.314}$$

For particles that do not depolarize, $M^e_{hv} = M^e_{vh} = 0$ and $\mathcal{E}$ becomes

$$\mathcal{E} = \begin{bmatrix} 1 & 0 & 0 & 0 \\ 0 & 0 & 0 & 1 \\ 0 & 1 & 1 & 0 \\ 0 & -i & i & 0 \end{bmatrix}. \tag{4.315}$$

By using the first-order iterative solution for the radiative transfer equations (4.199) to (4.200) and specializing these to the backscatter case, where $\mu = \mu_0$ and $\phi = \phi_0 + \pi$, the matrix $\mathcal{T}_c(\mu_0, \phi_0)$ takes the form:

$$\begin{aligned} \mathcal{T}_c(\mu_0, \phi_0) = & \frac{1}{\mu_0}\mathrm{e}^{-\boldsymbol{\kappa}_e^+ d/\mu_0}\mathcal{R}(\mu_0)\mathcal{E}(-\mu_0, \phi_0+\pi)\mathcal{A}_1\mathcal{E}^{-1}(\mu_0, \phi_0)\mathcal{R}(\mu_0)\mathrm{e}^{-\boldsymbol{\kappa}_e^- d/\mu_0} \\ & + \frac{1}{\mu_0}\mathcal{E}(\mu_0, \phi_0 + \pi)\mathcal{A}_2\mathcal{E}^{-1}(\mu_0, \phi_0)\mathcal{R}(\mu_0)\mathrm{e}^{-\boldsymbol{\kappa}_e^- d/\mu_0} \\ & + \frac{1}{\mu_0}\mathrm{e}^{-\boldsymbol{\kappa}_e^+ d/\mu_0}\mathcal{R}(\mu_0)\mathcal{E}(-\mu_0, \phi_0 + \pi)\mathcal{A}_3\mathcal{E}^{-1}(-\mu_0, \phi_0) \\ & + \frac{1}{\mu_0}\mathcal{E}(\mu_0, \phi_0 + \pi)\mathcal{A}_4\mathcal{E}^{-1}(-\mu_0, \phi_0), \end{aligned} \tag{4.316}$$

where $\mathcal{R}(\mu_0)$ is the reflectivity matrix of the specular surface given by (4.188).

The four terms in (4.316) represents scattering mechanisms (1), (2a), (2b), and (4) respectively illustrated in Fig. 4.9. For a canopy symmetric in $\phi$, $k_e^+ = k_e^-$ and the $4 \times 4$ $\mathcal{A}_n$ matrices are given by

$$[\mathcal{A}_1(\mu_0, \phi_0 + \pi; \mu_0, \phi_0)]_{ij} = \frac{1 - \exp\{-[\beta_i(-\mu_0, \phi_0 + \pi)/\mu_0 + \beta_j(\mu_0, \phi_0)/\mu_0]d\}}{\beta_i(-\mu_0, \phi_0 + \pi)/\mu_0 + \beta_j(\mu_0, \phi_0)/\mu_0} \cdot [\mathcal{E}^{-1}(-\mu_0, \phi_0 + \pi)\mathcal{P}(-\mu_0, \phi_0 + \pi; \mu_0, \phi_0)\mathcal{E}(\mu_0, \phi_0)]_{ij} \tag{4.317}$$

$$[\mathcal{A}_2(\mu_0, \phi_0 + \pi; \mu_0, \phi_0)]_{ij} = \begin{cases} \left. \begin{array}{l} d e^{-\beta_i(\mu_0)d/\mu_0} \cdot [\mathcal{E}^{-1}(\mu_0, \phi_0 + \pi) \\ \cdot \mathcal{P}(\mu_0, \phi_0 + \pi; \mu_0, \phi_0)\mathcal{E}(\mu_0, \phi_0)]_{ii} \end{array} \right\}, & i = j \\ \left. \begin{array}{l} \dfrac{\exp[-\beta_j(\mu_0, \phi_0)d/\mu_0] - \exp[-\beta_i(\mu_0, \phi_0 + \pi)d/\mu_0]}{\beta_i(\mu_0, \phi_0 + \pi)/\mu_0 - \beta_j(\mu_0, \phi_0)/\mu_0} \\ \cdot [\mathcal{E}^{-1}(\mu_0, \phi_0 + \pi)\mathcal{P}(\mu_0, \phi_0 + \pi; \mu_0, \phi_0)\mathcal{E}(\mu_0, \phi_0)]_{ij} \end{array} \right\}, & i \neq j \end{cases} \tag{4.318}$$

$$[\mathcal{A}_3(\mu_0, \phi_0 + \pi; \mu_0, \phi_0)]_{ij} = \begin{cases} \left. \begin{array}{l} d e^{-\beta_i(\mu_0)d/\mu_0} \cdot [\mathcal{E}^{-1}(-\mu_0, \phi_0 + \pi) \\ \cdot \mathcal{P}(-\mu_0, \phi_0 + \pi; -\mu_0, \phi_0)\mathcal{E}(-\mu_0, \phi_0)]_{ii} \end{array} \right\}, & i = j \\ \left. \begin{array}{l} \dfrac{\exp[-\beta_i(-\mu_0, \phi_0 + \pi)d/\mu_0] - \exp[-\beta_j(-\mu_0, \phi_0)d/\mu_0]}{-\beta_i(-\mu_0, \phi_0 + \pi)/\mu_0 + \beta_j(-\mu_0, \phi_0)/\mu_0} \\ \cdot [\mathcal{E}^{-1}(-\mu_0, \phi_0 + \pi)\mathcal{P}(-\mu_0, \phi_0 + \pi; -\mu_0, \phi_0)\mathcal{E}(-\mu_0, \phi_0)]_{ij} \end{array} \right\}, & i \neq j \end{cases} \tag{4.319}$$

$$[\mathcal{A}_4(\mu_0, \phi_0 + \pi; \mu_0, \phi_0)]_{ij} = \frac{1 - \exp[-(\beta_i(\mu_0, \phi_0 + \pi)/\mu_0 + \beta_j(-\mu_0, \phi_0)/\mu_0)d]}{\beta_i(\mu_0, \phi_0 + \pi)/\mu_0 + \beta_j(-\mu_0, \phi_0)/\mu_0} \cdot [\mathcal{E}^{-1}(\mu_0, \phi_0 + \pi)\mathcal{P}(\mu_0, \phi_0 + \pi; -\mu_0, \phi_0)\mathcal{E}(-\mu_0, \phi_0)]_{ij} \; . \tag{4.320}$$

The phase matrices $\mathcal{P}$ are given by

$$\mathcal{P}(\mu_s, \phi_s; \mu_i, \phi_i) = \sum_{k=1}^{K} N_k \iiiint p_k(s_1, s_2; \theta_p, \phi_p) \cdot \mathcal{L}_k(\mu_s, \phi_s; \mu_i, \phi_i; \theta_p, \phi_p)\, ds_1\, ds_2\, d\theta_p\, d\phi_p, \tag{4.321}$$

where the summation over $k$ represents an addition over the canopy constituent classes, $N_k$ is the number density per unit volume of class $k$, $p_k(s_1, s_2; \theta_p, \phi_p)$ is the joint probability density function describing the distribution in size and orientation of class $k$ and $(\mu_s, \phi_s)$ and $(\mu_i, \phi_i)$ represent the propagation directions of the incident and scattered intensities. The Mueller matrix $\mathcal{L}_k(\mu_s, \phi_s; \mu_i, \phi_i; \theta_p, \phi_p)$ relates the incident and scattered intensities for a single particle with an orientation defined by $(\theta_p, \phi_p)$.

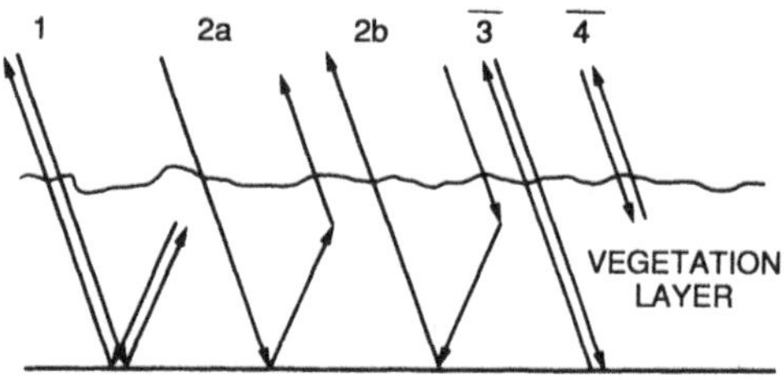


**Figure 4.9** First order backscattering terms for a vegetation layer over a specular surface.

Equation (4.316) may be combined with (4.299) to account for term 3 of Fig. 4.9, thereby yielding the complete expression for $\mathcal{T}_t(\mu_0, \phi_0)$. The matrix $\mathcal{T}_t(\mu_0, \phi_0)$ may be used to compute the linearly polarized total canopy backscattering coefficients:

$$\sigma^0_{vv} = 4\pi \cos\theta_0 \left[\mathcal{T}_t(\mu_0, \phi_0)\right]_{11} \tag{4.322}$$

$$\sigma^0_{hh} = 4\pi \cos\theta_0 \left[\mathcal{T}_t(\mu_0, \phi_0)\right]_{22} \tag{4.323}$$

$$\sigma^0_{hv} = 4\pi \cos\theta_0 \left[\mathcal{T}_t(\mu_0, \phi_0)\right]_{21} \tag{4.324}$$

$$\sigma^0_{vh} = 4\pi \cos\theta_0 \left[\mathcal{T}_t(\mu_0, \phi_0)\right]_{12} \tag{4.325}$$

or, by applying wave synthesis techniques, the matrix may be used to compute the backscattering coefficient for any transmitting and receiving polarization combination.

## 4-5.2 Radiative Transfer Model for a Forest Canopy

A first-order radiative transfer model for simulating microwave backscatter from tree canopies has been proposed by Ulaby *et al.*, (1988, 1990). This model, known as the *Michigan microwave canopy scattering* model

(MIMICS), models a forest canopy as shown in Fig. 4.10. Here, the canopy consists of two distinct horizontal vegetation layers over a dielectric ground surface. The top (crown) layer consists of the tree crowns and may be comprised of an appropriate combination of leaves, needles, and branches. The top has height $d$, is statistically homogeneous, and is continuous in the horizontal direction. The bottom (trunk) layer has height $H_t$ and consists of the tree trunks.

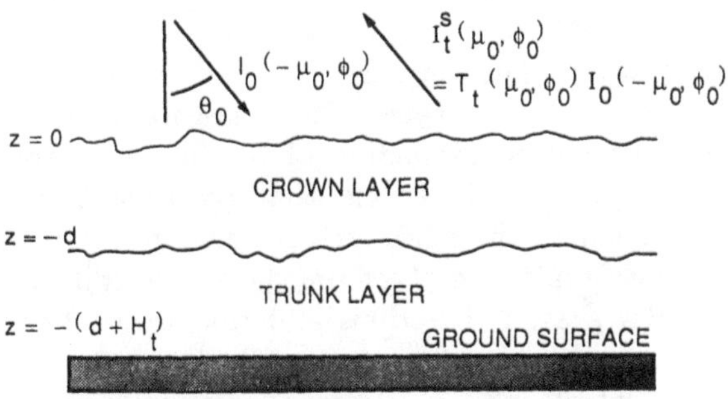


**Figure 4.10** Forest canopy model.

Here, the problem is again solved in two parts. First, the problem is solved with the ground surface treated as a specular interface. This yields an expression for the canopy transformation matrix $\mathcal{T}_c(\mu_0, \phi_0)$. Then, direct backscatter from the ground surface is taken into account as in (4.298). Here, however, $\mathcal{T}_g(\mu_0, \phi_0)$ is given by

$$\mathcal{T}_g(\mu_0, \phi_0) = e^{-\boldsymbol{\kappa}_c^+ d/\mu_0} e^{-\boldsymbol{\kappa}_t^+ H_t/\mu_0} \mathcal{G}(\mu_0) e^{-\boldsymbol{\kappa}_t^- H_t/\mu_0} e^{-\boldsymbol{\kappa}_c^- d/\mu_0}, \quad (4.326)$$

where $\boldsymbol{\kappa}_c^+$ and $\boldsymbol{\kappa}_c^-$ are the extinction matrices of the crown layer for upward and downward propagation, respectively, $\boldsymbol{\kappa}_t^+$ and $\boldsymbol{\kappa}_t^-$ are the extinction matrices of the trunk layer, and $\mathcal{G}(\mu_0)$ is the ground backscattering matrix.

To solve for the canopy transformation matrix, the radiative transfer equations are set up separately in each layer and boundary conditions are applied at $z = 0$, $z = -d$, and $z = -(d + H_t)$. In the crown layer,

$$\mathbf{I}_c^+(\mu, \phi, z) = e^{-\boldsymbol{\kappa}_c^+(z+d)/\mu} \mathbf{I}_c^+(\mu, \phi, -d) + \int_{-d}^{z} e^{-\boldsymbol{\kappa}_c^+(z-z')/\mu} \mathcal{F}_c^+(\mu, \phi, z')\, dz' \quad (4.327)$$

$$\mathbf{I}_c^-(-\mu, \phi, z) = e^{\boldsymbol{\kappa}_c^- z/\mu} \mathbf{I}_c^-(-\mu, \phi, 0) + \int_{z}^{0} e^{\boldsymbol{\kappa}_c^-(z-z')/\mu} \mathcal{F}_c^-(-\mu, \phi, z')\, dz'. \quad (4.328)$$

Similarly, the equations:

$$\begin{aligned} \mathbf{I}_t^+(\mu,\phi,z) &= \mathrm{e}^{-\boldsymbol{\kappa}_t^+[z+(d+H_t)]/\mu}\mathbf{I}_t^+[\mu,\phi,-(d+H_t)] \\ &+ \int_{-(d+H_t)}^{z} \mathrm{e}^{-\boldsymbol{\kappa}_t^+(z-z')/\mu}\boldsymbol{\mathcal{F}}_t^+(\mu,\phi,z')\,\mathrm{d}z' \quad (4.329) \\ \mathbf{I}_t^-(-\mu,\phi,z) &= \mathrm{e}^{\boldsymbol{\kappa}_t^-(z+d)/\mu}\mathbf{I}_t^-(-\mu,\phi,-d) \\ &+ \int_{z}^{-d} \mathrm{e}^{\boldsymbol{\kappa}_t^-(z-z')/\mu}\boldsymbol{\mathcal{F}}_t^-(-\mu,\phi,z')\,\mathrm{d}z' \quad (4.330) \end{aligned}$$

apply in the trunk layer. Furthermore, we shall consider only two types of scattering within the trunk layer. These are forward scattering from the trunks that gives rise to extinction of the forward propagating field and specular scattering from the trunks that gives rise to a ground-trunk interaction contribution to the field backscattered from the canopy. This means that the phase matrix in the trunk layer is significant only in the forward and specular scatter directions and that direct backscatter form the trunk layer may be neglected.

If the air-crown interface ($z = 0$) and the crown-trunk interface ($z = -d$) are considered as diffuse boundaries then the intensities across these interfaces are continuous. This assumption yields the boundary conditions:

$$\begin{aligned} \mathbf{I}_c^-(-\mu,\phi,0) &= \mathbf{I}_0\delta(\mu-\mu_0)\,\delta(\phi-\phi_0) \quad (4.331) \\ \mathbf{I}_c^+(\mu,\phi,-d) &= \mathbf{I}_t^+(\mu,\phi,-d) \quad (4.332) \\ \mathbf{I}_t^-(-\mu,\phi,-d) &= \mathbf{I}_c^-(-\mu,\phi,-d)\,. \quad (4.333) \end{aligned}$$

Furthermore, at the trunk-ground interface where $z = -(d+H_t)$, we have

$$\mathbf{I}_t^+[\mu,\phi,-(d+H_t)] = \boldsymbol{\mathcal{R}}(\mu)\,\mathbf{I}_t^-[-\mu,\phi,-(d+H_t)]\,, \quad (4.334)$$

where $\boldsymbol{\mathcal{R}}(\mu)$ is the reflectivity matrix of the specular surface.

Making use of these boundary conditions and setting $\boldsymbol{\mathcal{F}}_t^\pm = \boldsymbol{\mathcal{F}}_c^\pm = 0$ in (4.327) through (4.330) gives us the zero-order solution for the specific intensities in the crown and trunk layer:

$$\begin{aligned} \mathbf{I}_{c0}^-(-\mu,\phi,z) &= \mathrm{e}^{\boldsymbol{\kappa}_c^- z/\mu}\mathbf{I}_0\delta(\mu-\mu_0)\delta(\phi-\phi_0) \quad (4.335) \\ \mathbf{I}_{t0}^-(-\mu,\phi,z) &= \mathrm{e}^{\boldsymbol{\kappa}_t^-(z+d)/\mu}\mathrm{e}^{-\boldsymbol{\kappa}_c^- d/\mu}\mathbf{I}_0\delta(\mu-\mu_0)\delta(\phi-\phi_0) \quad (4.336) \\ \mathbf{I}_{t0}^+(\mu,\phi,z) &= \mathrm{e}^{-\boldsymbol{\kappa}_t^+(z+(d+H_t))/\mu}\boldsymbol{\mathcal{R}}(\mu)\,\mathrm{e}^{-\boldsymbol{\kappa}_t^- H_t/\mu}\mathrm{e}^{-\boldsymbol{\kappa}_c^- d/\mu} \\ &\cdot\mathbf{I}_0\delta(\mu-\mu_0)\delta(\phi-\phi_0) \quad (4.337) \\ \mathbf{I}_{c0}^+(\mu,\phi,z) &= \mathrm{e}^{-\boldsymbol{\kappa}_c^+(z+d)/\mu}\mathrm{e}^{-\boldsymbol{\kappa}_t^+ H_t/\mu}\boldsymbol{\mathcal{R}}(\mu)\,\mathrm{e}^{-\boldsymbol{\kappa}_t^- H_t/\mu}\mathrm{e}^{-\boldsymbol{\kappa}_c^- d/\mu} \\ &\cdot\mathbf{I}_0\delta(\mu-\mu_0)\delta(\phi-\phi_0). \quad (4.338) \end{aligned}$$

To find the first-order solution, we need to solve for $\mathbf{I}_c^+(\mu,\phi,z)$ and evaluate it at $z=0$. By using (4.327) through (4.330) and applying boundary conditions, we find

$$\begin{aligned}
\mathbf{I}_c^+(\mu,\phi,z) \;=\;& e^{-\boldsymbol{\kappa}_c^+(z+d)/\mu}\mathcal{R}'(\mu,\phi)\cdot\Big[e^{-\boldsymbol{\kappa}_c^- d/\mu}\mathbf{I}_c^-(-\mu,\phi,0) \\
&+\int_{-d}^{0} e^{-\boldsymbol{\kappa}_c^-(d+z')/\mu}\mathcal{F}_c^-(-\mu,\phi,z')\,dz'\Big] \\
&+\int_{-d}^{z} e^{-\boldsymbol{\kappa}_c^+(z-z')/\mu}\mathcal{F}_c^+(\mu,\phi,z')\,dz' \\
&+e^{-\boldsymbol{\kappa}_c^+(z+d)/\mu}\Big[e^{-\boldsymbol{\kappa}_t^+ H_t/\mu}\mathcal{R}(\mu) \\
&\cdot\int_{-(d+H_t)}^{-d} e^{-\boldsymbol{\kappa}_t^-[z'+(d+H_t)]/\mu}\mathcal{F}_t^-(-\mu,\phi,z')\,dz' \\
&+\int_{-(d+H_t)}^{-d} e^{\boldsymbol{\kappa}_t^+(z'+d)/\mu}\mathcal{F}_t^+(\mu,\phi,z')\,dz'\Big]
\end{aligned} \tag{4.339}$$

where

$$\mathcal{R}'(\mu,\phi) = e^{-\boldsymbol{\kappa}_t^+ H_t/\mu}\cdot\mathcal{R}(\mu)\cdot e^{-\boldsymbol{\kappa}_t^- H_t/\mu}. \tag{4.340}$$

The matrix $\mathcal{R}'(\mu,\phi)$ accounts for extinction in the trunk layer and reflection at the specular surface. We now use the zero-order solution in (4.193) and (4.194) to compute the source functions $\mathcal{F}_{t0}^+$, $\mathcal{F}_{t0}^-$, $\mathcal{F}_{c0}^+$, and $\mathcal{F}_{c0}^-$. These source functions are substituted into (4.339) and the resulting expression is evaluated at $z=0$. To consider backscatter, we take $\mu=\mu_0$ and $\phi=\phi_0+\pi$. This gives the first-order solution:

$$\mathbf{I}^s = \mathcal{T}_c(\mu_0,\phi_0)\mathbf{I}_0, \tag{4.341}$$

where

$$\begin{aligned}
\mathcal{T}_c(\mu_0,\phi_0) =\;& \frac{1}{\mu_0}e^{-\boldsymbol{\kappa}_c^+ d/\mu_0}\mathcal{R}'(\mu_0,\phi_0+\pi)\mathcal{E}_c(-\mu_0,\phi_0+\pi) \\
&\cdot\mathcal{A}_1\mathcal{E}_c^{-1}(\mu_0,\phi_0)\mathcal{R}'(\mu_0,\phi_0)e^{-\boldsymbol{\kappa}_c^- d/\mu_0} \\
&+\frac{1}{\mu_0}\mathcal{E}_c(\mu_0,\phi_0+\pi)\mathcal{A}_2\mathcal{E}_c^{-1}(\mu_0,\phi_0)\mathcal{R}'(\mu_0,\phi_0)e^{-\boldsymbol{\kappa}_c^- d/\mu_0} \\
&+\frac{1}{\mu_0}e^{-\boldsymbol{\kappa}_c^+ d/\mu_0}\mathcal{R}'(\mu_0,\phi_0+\pi)\mathcal{E}_c(-\mu_0,\phi_0+\pi)\mathcal{A}_3\mathcal{E}_c^{-1}(-\mu_0,\phi_0) \\
&+\frac{1}{\mu_0}\mathcal{E}_c(\mu_0,\phi_0+\pi)\mathcal{A}_4\mathcal{E}_c^{-1}(-\mu_0,\phi_0)
\end{aligned}$$

(continued next pg.)

$$
\begin{aligned}
&+\frac{1}{\mu_0} e^{-\boldsymbol{\kappa}_c^+ d/\mu_0} e^{-\boldsymbol{\kappa}_t^+ H_t/\mu_0} \mathcal{R}(\mu_0) \mathcal{E}_t(-\mu_0, \phi_0 + \pi) \\
&\cdot \mathcal{E}_t^{-1}(-\mu_0, \phi_0) e^{-\boldsymbol{\kappa}_c^- d/\mu_0} \\
&+\frac{1}{\mu_0} e^{-\boldsymbol{\kappa}_c^+ d/\mu_0} \mathcal{E}_t(\mu_0, \phi_0 + \pi) \\
&\cdot \mathcal{A}_6 \mathcal{E}_t^{-1}(\mu_0, \phi_0) \mathcal{R}(\mu_0) e^{-\boldsymbol{\kappa}_t^- H_t/\mu_0} e^{-\boldsymbol{\kappa}_c^- d/\mu_0}
\end{aligned}
\tag{4.342}
$$

where $\epsilon_t$ and $\epsilon_c$ are the $4 \times 4$ eigenmatrices of the trunk and crown layers, respectively.

The six terms in (4.342) correspond to the first six scattering mechanisms illustrated in Fig. 4.11. For a canopy symmetric in $\phi$, $k_c^+ = k_c^-$, $k_t^+ = k_t^-$, and the $4 \times 4$ $\mathcal{A}_n$ matrices are given by

$$
\begin{aligned}
&[\mathcal{A}_1(\mu_0, \phi_0 + \pi; \mu_0, \phi_0)]_{ij} = \\
&\qquad \frac{1 - \exp[-(\beta_i^c(-\mu_0, \phi_0 + \pi)/\mu_0 + \beta_j^c(\mu_0, \phi_0)/\mu_0) d]}{\beta_i^c(-\mu_0, \phi_0 + \pi)/\mu_0 + \beta_j^c(\mu_0, \phi_0)/\mu_0} \\
&\qquad \cdot [\mathcal{E}_c^{-1}(-\mu_0, \phi_0 + \pi) \mathcal{P}_c(-\mu_0, \phi_0 + \pi; \mu_0, \phi_0) \mathcal{E}_c(\mu_0, \phi_0)]_{ij}
\end{aligned}
\tag{4.343}
$$

$$
[\mathcal{A}_2(\mu_0, \phi_0 + \pi; \mu_0, \phi_0)]_{ij} =
\begin{cases}
\left.\begin{array}{l} d e^{-\beta_i^c(\mu_0) d/\mu_0} \cdot [\mathcal{E}_c^{-1}(-\mu_0, \phi_0 + \pi) \\ \cdot \mathcal{P}_c(-\mu_0, \phi_0 + \pi; -\mu_0, \phi_0) \mathcal{E}_c(-\mu_0, \phi_0)]_{ii} \end{array}\right\} & i = j \\[2ex]
\left.\begin{array}{l} \dfrac{\exp[-\beta_j^c(\mu_0, \phi_0) d/\mu_0] - \exp[-\beta_i^c(\mu_0, \phi_0 + \pi) d/\mu]}{\beta_i^c(\mu_0, \phi_0 + \pi)/\mu - \beta_j^c(\mu_0, \phi_0)/\mu_0} \\ \cdot [\mathcal{E}_c^{-1}(\mu_0, \phi_0 + \pi) \mathcal{P}_c(\mu_0, \phi_0 + \pi; \mu_0, \phi_0) \mathcal{E}_c(\mu_0, \phi_0)]_{ij} \end{array}\right\} & i \neq j
\end{cases}
\tag{4.344}
$$

$$
[\mathcal{A}_3(\mu_0, \phi_0 + \pi; \mu_0, \phi_0)]_{ij} =
\begin{cases}
\left.\begin{array}{l} d e^{-\beta_i^c(\mu_0) d/\mu_0} \cdot [\mathcal{E}_c^{-1}(-\mu_0, \phi_0 + \pi) \\ \cdot \mathcal{P}_c(-\mu_0, \phi_0 + \pi; -\mu_0, \phi_0) \mathcal{E}_c(-\mu_0, \phi_0)]_{ii} \end{array}\right\} & i = j \\[2ex]
\left.\begin{array}{l} \dfrac{\exp[-\beta_i^c(-\mu_0, \phi_0 + \pi) d/\mu_0] - \exp[-\beta_j^c(-\mu_0, \phi_0) d/\mu_0]}{-\beta_i^c(-\mu_0, \phi_0 + \pi)/\mu_0 + \beta_j^c(-\mu_0, \phi_0)/\mu_0} \\ \cdot [\mathcal{E}_c^{-1}(-\mu_0, \phi_0 + \pi) \mathcal{P}_c(-\mu_0, \phi_0 + \pi; -\mu_0, \phi_0) \mathcal{E}_c(-\mu_0, \phi_0)]_{ij} \end{array}\right\} & i \neq j
\end{cases}
\tag{4.345}
$$

$$[\mathcal{A}_4(\mu_0, \phi_0 + \pi; \mu_0, \phi_0)]_{ij} = \frac{1 - \exp[-(\beta_i^c(\mu_0, \phi_0 + \pi)/\mu + \beta_j^c(-\mu_0, \phi_0)/\mu_0)d]}{\beta_i^c(\mu_0, \phi_0 + \pi)/\mu_0 + \beta_j^c(-\mu_0, \phi_0)/\mu_0} \cdot [\mathcal{E}_c^{-1}(\mu_0, \phi_0 + \pi)\mathcal{P}_c(\mu_0, \phi_0 + \pi; -\mu_0, \phi_0)\mathcal{E}_c(-\mu_0, \phi_0)]_{ij} \quad (4.346)$$

$$[\mathcal{A}_5(\mu_0, \phi_0 + \pi; \mu_0, \phi_0)]_{ij} = \begin{cases} \left.\begin{array}{l} H_t e^{-\beta_i^t(\mu_0)H_z/\mu_0} \cdot [\mathcal{E}_t^{-1}(-\mu_0, \phi_0 + \pi) \\ \cdot \mathcal{P}_t(-\mu_0, \phi_0 + \pi; -\mu_0, \phi_0)\mathcal{E}_t(-\mu_0, \phi_0)]_{ii} \end{array}\right\} & i = j \\ \left.\begin{array}{l} \dfrac{\exp[-\beta_i^t(-\mu_0, \phi_0 + \pi)H_t/\mu] - \exp[-\beta_j^t(-\mu_0, \phi_0)H_t/\mu_0]}{-\beta_i^t(-\mu_0, \phi_0 + \pi)/\mu_0 + \beta_j^t(-\mu_0, \phi_0)/\mu_0} \\ \cdot [\mathcal{E}_t^{-1}(-\mu_0, \phi_0 + \pi)\mathcal{P}_t(-\mu_0, \phi_0 + \pi; -\mu_0, \phi_0)\mathcal{E}_t(-\mu_0, \phi_0)]_{ij} \end{array}\right\} & i \neq j \end{cases} \quad (4.347)$$

$$[\mathcal{A}_6(\mu_0, \phi_0 + \pi; \mu_0, \phi_0)]_{ij} = \begin{cases} \left.\begin{array}{l} H_t e^{-\beta_i^t(\mu_0)H_z/\mu_0} \cdot [\mathcal{E}_t^{-1}(\mu_0, \phi_0 + \pi) \\ \cdot \mathcal{P}_t(\mu_0, \phi_0 + \pi; \mu_0, \phi_0)\mathcal{E}_t(\mu_0, \phi_0)]_{ii} \end{array}\right\} & i = j \\ \left.\begin{array}{l} \dfrac{\exp[-\beta_j^t(\mu_0, \phi_0)H_t/\mu_0] - \exp[-\beta_i^t(\mu_0, \phi_0 + \pi)H_t/\mu_0]}{\beta_i^t(\mu_0, \phi_0 + \pi)/\mu_0 - \beta_j^t(\mu_0, \phi_0)/\mu_0} \\ \cdot [\mathcal{E}_t^{-1}(\mu_0, \phi_0 + \pi)\mathcal{P}_t(\mu_0, \phi_0 + \pi; \mu_0, \phi_0)\mathcal{E}_t(\mu_0, \phi_0)]_{ij} \end{array}\right\} & i \neq j \end{cases} \quad (4.348)$$

where $\beta_m^t$ and $\beta_m^c$ are the $n$th eigenvalues of $k_t^t$ and $k_c^t$, respectively.

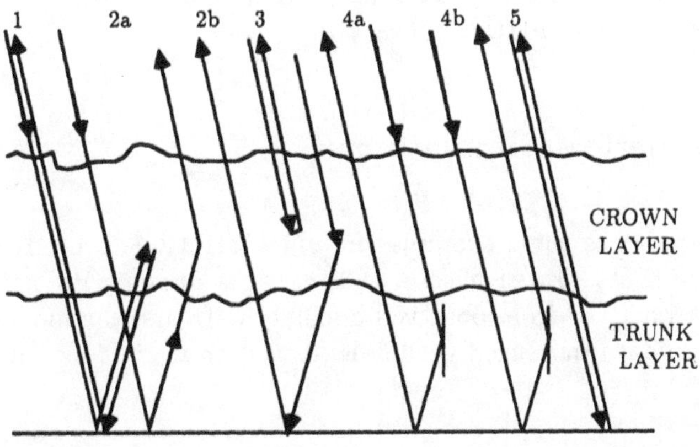


**Figure 4.11** First-order scattering terms for the backscattering case.

In the crown layer, the extinction and phase matices may be computed by applying (4.303) and (4.321). In the trunk layer, one of two cases may apply. If the canopy has an area density of $N_t$ trees per square meter and the trunks have an average height equal to the height of the trunk layer $H_t$, the elements of the extinction matrix are given by

$$M^t_{mn} = \frac{i2\pi\left(\frac{N_t}{H_t}\right)}{k_0}\langle S_{mn}\rangle \tag{4.349}$$

and the phase matrix is

$$\begin{aligned}\mathcal{P}_t(\theta_s,\phi_s;\theta_i,\phi_i) &= \frac{N_t}{H_t}\iiiint p_t(H,D;\theta_t,\phi_t)\\ &\quad\cdot\mathcal{L}(\theta_s,\phi_s;\theta_i,\phi_i;\theta_t,\phi_t)\,\mathrm{d}H\,\mathrm{d}D\,\mathrm{d}\theta_t\,\mathrm{d}\phi_t\end{aligned} \tag{4.350}$$

where $H$ and $D$ are the trunk height and diameter parameters, $\theta_t$ and $\phi_t$ are the trunk orientation parameters, and $(\theta_i,\phi_i)$ and $(\theta_s,\phi_s)$ are the angles of incidence and scattering, respectively.

Alternatively, the trunk layer may consist of some larger branches as well as the tree trunks. In this case,

$$M^t_{mn} = \frac{i2\pi\,(N_{tv})}{k_0}\langle S_{mn}\rangle \tag{4.351}$$

and

$$\begin{aligned}\mathcal{P}_t(\theta_s,\phi_s;\theta_i,\phi_i) = N_{tv}\iiiint p_t(H,D;\theta_t,\phi_t)\\ \cdot\mathcal{L}(\theta_s,\phi_s;\theta_i,\phi_i;\theta_t,\phi_t)\,\mathrm{d}H\,\mathrm{d}D\,\mathrm{d}\theta_t\,\mathrm{d}\phi_t\end{aligned} \tag{4.352}$$

where $N_{tv}$ is the combined number of trunks and branches per unit volume.

Equation (4.326) may be combined with (4.342) to obtain the total canopy transformation matrix $\mathcal{T}_t(\mu_0,\phi_0)$.

### 4-5.3 Numerical Simulations

This section presents some examples in which MIMICS is used to model radar measurements. In each case, the model is driven with parameters that have been derived from observed ancillary data and the model output is compared to data measured with truck-based or aircraft-mounted radar systems.

Figures 4.12 through 4.17 illustrate a modeling analysis of a walnut orchard. The scatterometer data presented here were recorded in Fresno county, California during the summer of 1987 with a truck-mounted scatterometer system as part of the Eos synergism study (Dobson *et al.*, 1988, McDonald *et al.*, 1989, 1990). As part of this study, an extensive set of ancillary data was collected. Data describing individual canopy constituents as well as gross canopy architecture were recorded and adapted for input to MIMICS.

Table 4.2 summarizes some of the gross canopy characteristics. The trunk layer consists of all tree trunks along with some of the larger branches. The crown layer consists of three branch size classes along with the leaves. The branch and ground dielectric constants were determined by averaging several values measured with dielectric probes during the course of the experiment.

**Table 4.2** Gross Canopy Characteristics

| |
|---|
| Trunk layer thickness = 1.7 m |
| Crown layer thickness = 2.5 m |
| Ground dielectric constant = 25 $+i$ 2.5 |
| Trunk dielectric constant = 45 $+i$ 11.2 |
| Primary branch dielectric constant = 34 $+i$ 8.5 |
| Secondary branch dielectric constant = 30 $+i$ 7.5 |

Table 4.3 summarizes the size and orientation characteristics of the branch classes. The class that constitutes the trunk layer includes the trunks and all branches with diameters greater than 4.0 cm. The remaining branches were divided into three size classes distributed throughout the crown layer. The primary branch class consists of all branches with diameters less than 4.0 cm and greater than 0.9 cm. The secondary branch class includes those branches whose diameter is less than 0.9 cm and greater than 0.4 cm. The high order stems constitute the smallest size class and have diameters less than 0.4 cm.

In general, the larger branches tend to have mostly vertical orientations whereas the smaller branches exhibit no preferred orientation. Consequently, the forms of the orientation distributions listed in Table 4-5.3 were inferred. Figure 4.12 shows plots of the orientation PDFs. The branches in the trunk layer were assigned a $\cos^6\theta$ distribution such that the mean value of $\theta$ is 0. The primary branches were assigned a distribution of $\sin^4 2\theta$ such that $\theta$ has a mean value of 45°. The secondary branches and higher order stems were assigned spherical distribution functions. This implies

**Table 4.3** Canopy Branch Classes

| Constituent Class | Trunk Branches | Crown Branches Primary | Crown Branches Secondary | Crown Branches Stems |
|---|---|---|---|---|
| Max. Diam. (cm) | — | 4.0 | 0.9 | 0.4 |
| Min. Diam. (cm) | 4.0 | 0.9 | 0.4 | — |
| Ave. Diam. (cm) | 7.3 | 1.9 | 0.6 | 0.1 |
| Ave. Length (cm) | 92.8 | 35.8 | 10.9 | 5.0 |
| Density (#/m$^3$) | 0.13 | 1.55 | 1.41 | 900. |
| Orientation PDF | $\cos^6\theta$ | $\sin^4\theta$ | spherical | spherical |

that these branches are oriented such that the branch axes are uniformly distributed on a spherical surface.

Table 4.4 summarizes the characteristics of the leaves in this canopy. Here, a spherical orientation function was used such that the normal to the leaf surface is distributed uniformly over a spherical surface. The leaf density together with the leaf diameter and crown height specify a canopy leaf area index of 3.2.

**Table 4.4** Leaf Characteristics

| |
|---|
| Leaf density = 652 leaves/m$^3$ |
| Leaf dielectric constant = $36.5 + i\,11.3$ |
| Average leaf diameter = 5.0 cm |
| Average leaf thickness = 0.02 cm |
| Leaf area index = 3.2 |
| Leaf orientation PDF = spherical |

These data were used to run MIMICS as a function of radar look angle for a frequency of 1.62 GHz. Dielectric cylinders were used to model the trunks and branches, and disks were used to model the leaves. Figure 4.13 shows a comparison of modeled and measured scatterometer data for three polarization configurations over the range $40° \leq \theta \leq 55°$. This figure illustrates very good agreement between MIMICS and the scatterometer data. The two like-polarized backscatter configurations exhibit similar amplitudes with $hh$ being slightly higher than $vv$ in both the measured and the MIMICS-generated data. The cross-polarized backscatter is about 5 dB lower than the like-polarized response.

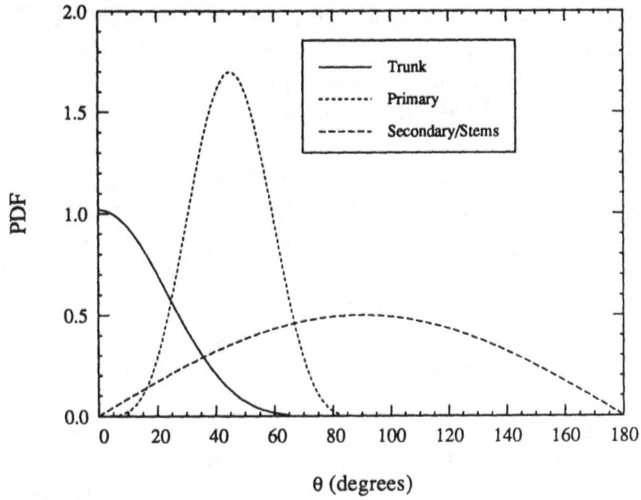


**Figure 4.12** Branch orientation probability distribution functions (PDFs).

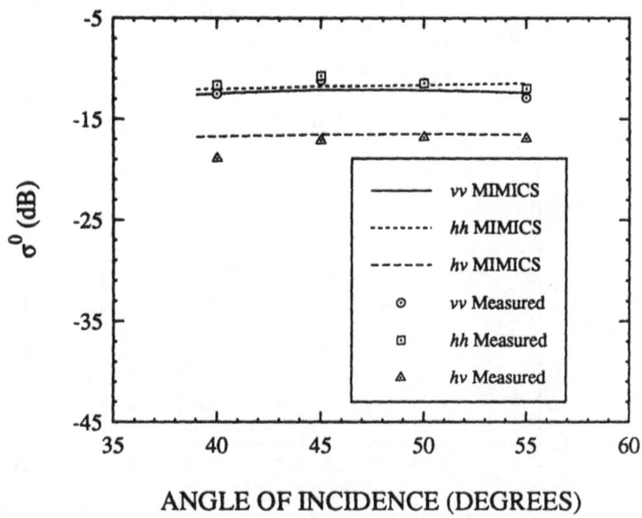


**Figure 4.13** Comparison of MIMICS with measured data for a walnut canopy.

Figures 4.14 and 4.15 show the relative contributions of the three most dominant contributions to the total like-polarized backscatter. The crown-ground mechanism corresponds to the sum of terms (2a) and (2b) in Fig. 4.11. Similarly, the trunk-ground mechanism corresponds to the sum of terms (4a) and (4b), and the direct crown mechanism corresponds to term (3). Whereas the same three mechanisms dominate the total $hh$ and $vv$ polarized backscatter, MIMICS predicts that the crown-ground and trunk-ground terms are the more significant terms for the $hh$ backscatter and the direct crown backscatter is the most significant term for $vv$ backscatter.

Figure 4.16 shows the major contributors to the cross-polarized backscatter. Here, those mechanisms generated by the crown layer are most important. This is expected because the constituents of the crown layer are oriented such that they depolarize more than the constituents in the trunk layer.

Observations of the diurnal behavior of relative dielectric constant of soil and vegetation over a three-day period allow canopy backscatter to be modeled as a function of time (McDonald *et al.*, 1990). Figure 4.17 shows the measured and modeled backscatter for $vv$ polarization at 1.62 GHz. The dips in canopy backscatter observed in the early afternoon for each of the study days may be attributed to changes in the dielectric constant of the crown layer branches.

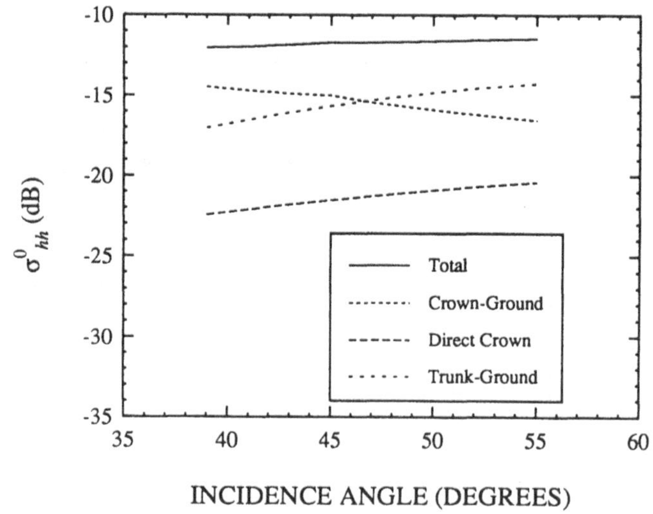


**Figure 4.14** Components of $hh$ polarization backscatter.

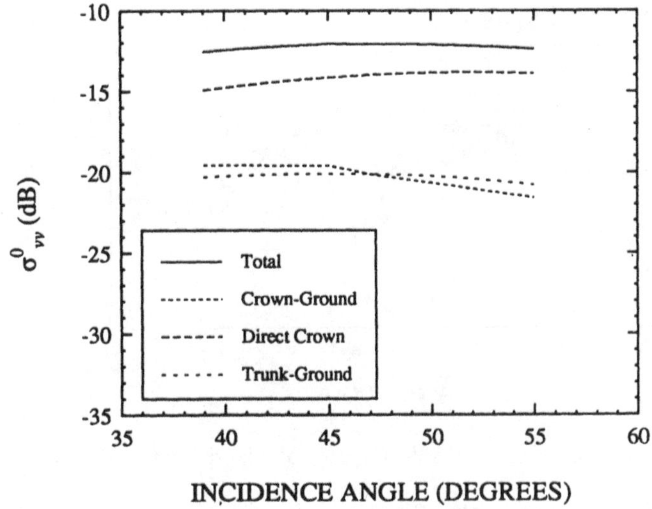


**Figure 4.15** Components of $vv$ polarization backscatter.

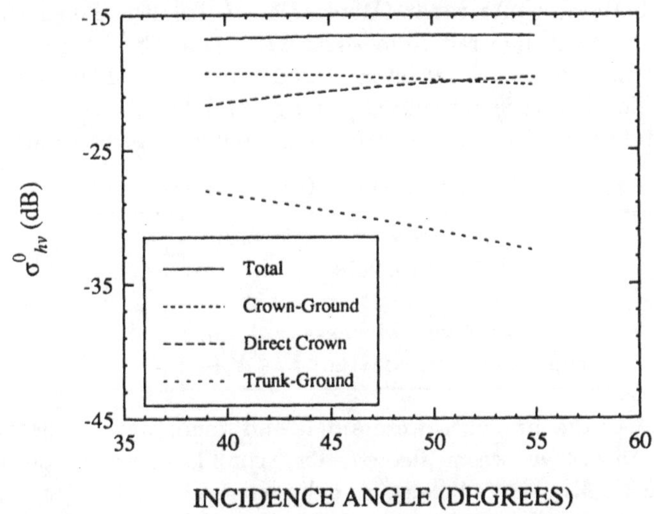


**Figure 4.16** Components of $hv$ polarization backscatter.

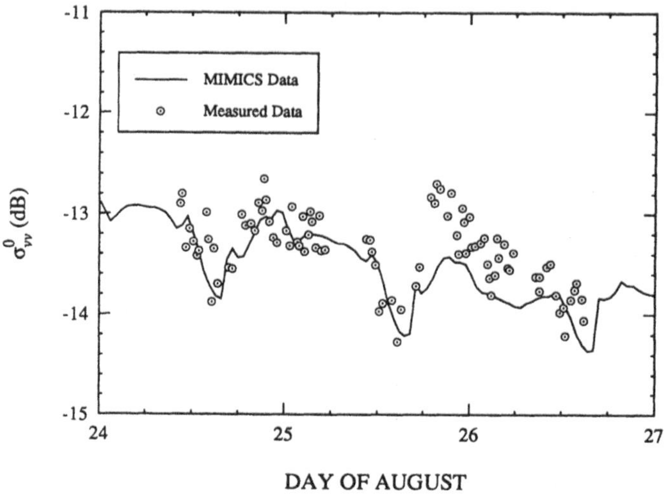


**Figure 4.17** Measured and predicted *vv* polarized backscatter.

Figure 4.18 shows a modeling analysis of extinction through a stand of thin leaf alders together with some measured extinction data. These data were recorded near Fairbanks, Alaska, in March of 1988 with an aircraft-mounted synthetic aperture radar (Dobson *et al.*, 1989). The measured extinction values were inferred from the SAR imagry by comparing the backscatter from a trihedral reflector placed under the alder canopy to that from an identical reflector placed in an open field. Table 4.5 lists the groundtruth data recorded for this stand that were used to drive MIMICS.

**Table 4.5** Alder Stand Characteristics

| |
|---|
| Canopy density = 0.90 trees/m$^2$ |
| Trunk layer thickness = 6.84 m |
| Trunk diameter = 7.0 cm |
| Crown layer thickness = 0.0 m |
| Trunk dielectric constant = 4.4 $+i$ 1.2 |

Here, because the branches were small and there were no leaves on the trees, the crown layer was neglected. The trunk layer was modeled as a collection of vertically oriented dielectric cylinders 6.84 m tall. MIMICS was run as a function of radar incidence angle for C- and X-band frequencies of 5.30 and 9.60 GHz, respectively. Figure 4.18 shows that MIMICS predicts the senstivity of extinction to incidence angle reasonably well.

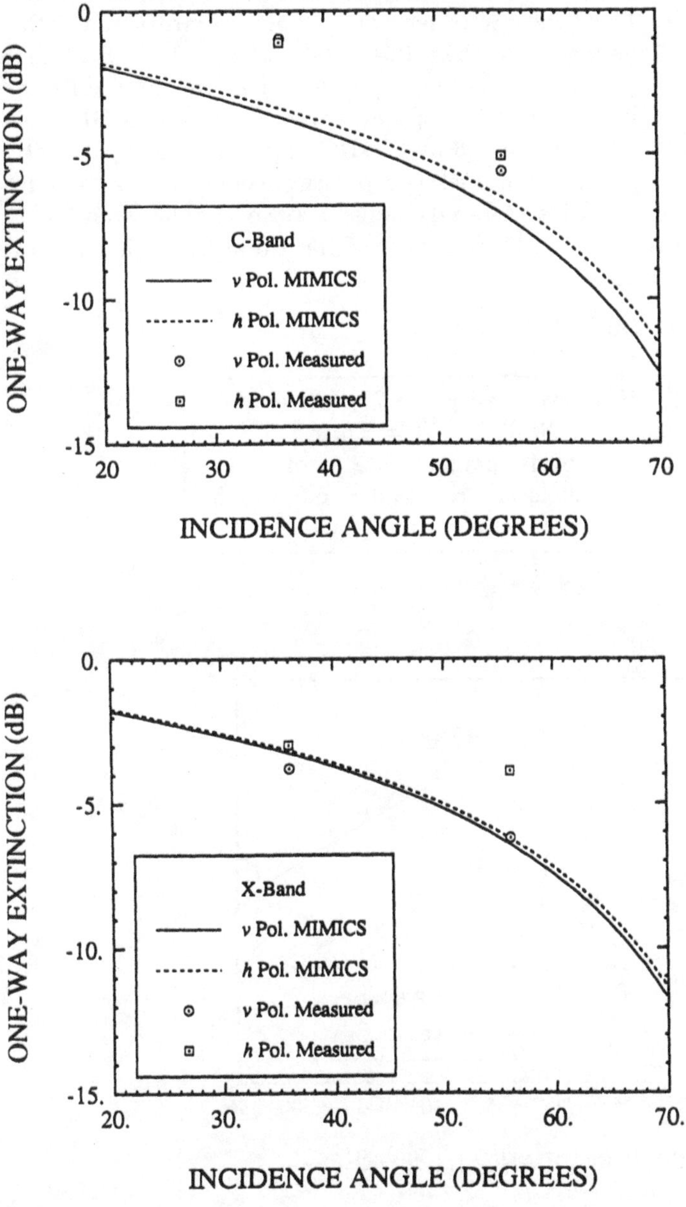


**Figure 4.18** Extinction loss of an Alder canopy.

Figure 4.19 illustrates a phase difference modeling analysis for a number of corn fields. These data were recorded by an aircraft-mounted SAR during the SIR-B mission in October 1984 (Ulaby *et al.*, 1987). Each point on the scatter plot represents the measured polarization phase difference $(\phi_{HH} - \phi_{VV})$ for a different corn canopy at a frequency of 1.28 GHz. The solid line shows the value predicted by MIMICS for the canopy parameters listed in Table 4-5.3. These parameters represent average values computed over the observed corn fields. A rapid change in $\Delta\phi$ occurs between $\theta = 24°$ and $\theta = 30°$. This is caused by the effect of the Brewster angle on the reflection from the corn stalks.

**Table 4.6** Corn Canopy Parameters

| Corn Canopy Parameters |
|---|
| Canopy density = 8.77 stalks/m$^2$ |
| Stalk height = 2.5 m |
| Stalk diameter = 2.5 cm |
| Stalk dielectric constant = $6.5 + i\ 0.5$ |
| Soil dielectric constant = $15 + i\ 2$ |

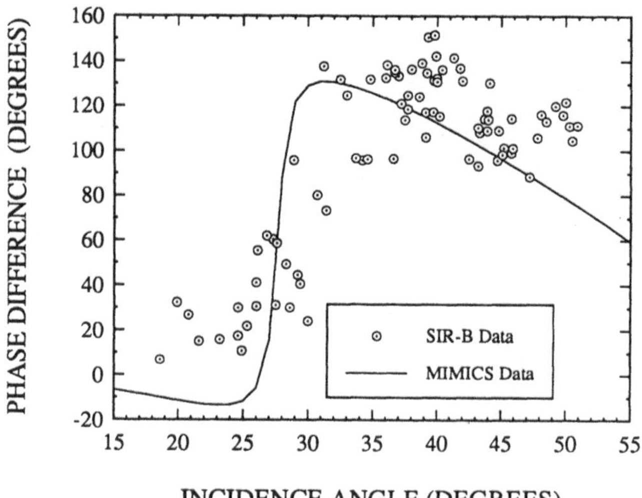


**Figure 4.19** Corn canopy polarization phase difference.

# 4-6 DEFICIENCIES OF THE RADIATIVE TRANSFER THEORY

The radiative transfer equation is based on the assumption that there is no correlation between the fields scattered by the different particles in the random medium, and the addition of power rather than the addition of fields holds. In the formulation, we also assumed that the scatterers are randomly and sparsely distributed and there is no correlation between them. These assumptions are valid in many practical cases including that of a rain cloud. However, the traditional radiative transfer formulation may not be valid for densely distributed media, and fails to account for the backscattering enhancement effect, which may be significant under certain circumstances (Kuga and Ishimaru, 1984; Ishimaru and Kuga, 1982; van Albada and Lagendijk, 1985; Wolf and Maret, 1985).

Backscattering enhancement is a phenomena in which a strong increase in intensity can be observed in the backward direction when the random medium is illuminated by a plane wave, caused by the constructive interference of two waves propagating in the opposite directions in the random medium. The intensity enhancement may be twice as large as the unenhanced intensity, and the angular width is much smaller than $1^\circ$. The effect is most observable if the size parameter $ka$ is between 1 and 10, and the medium volume density is more than 1%. Similar enhancement has also been reported for very large $ka$ and sparsely distributed particles (Kuga and Ishimaru, 1989). Because the radiative transfer theory does not include phase correlation between waves, the backscattering enhancement effect cannot be explained by the radiative transfer theory. The importance of backscattering enhancement has been recognized in optics and solid-state physics, (van Albada and Lagendijk, 1985; Wolf and Maret, 1985), but its significance in microwave and millimeter-wave remote sensing is not well understood. In an effort to correct for this deficiency, Tsang and Ishimaru (1985, 1987) recently introduced a modified form of the radiative transfer theory, which they call the *radiative wave theory*.

If the particles are sparsely distributed, the extinction coefficient is linearly proportional to the number density. This is a valid assumption if the volume density does not exceed about 1%. Beyond 1%, however, the extinction coefficient is no longer linearly proportional to the number density. The *normalized* extinction coefficient, defined as the actual extinction coefficient divided by the average total extinction per unit volume (or $\langle N\sigma_e \rangle$), decreases sharply when the volume fraction of particles becomes more than 1%, and the amount of decrease is related to the size parameter, dielectric constant and volume fraction of particles. Extensive experimental and theoretical studies on the extinction coefficient in a dense medium have been conducted in recent years (Kuga and Ishimaru, 1982; Tsang *et al.*, 1985). The available theoretical models include Twersky's model the perturbation solution with hold-correction, the quasicrystalline approximation (QCA) with Percus-Yevick (PY) pair correlation function, and the quasicrystalline approximation with coherent potential (QCA-CP), (Tsang *et al.*, 1985). Twersky's model is simple, but it is applicable only for small particles. The formula based on the hold-correction is valid if the volume franction is much less than 10%. For higher concentrations, QCA and QCA-CP with the PY pair correlation function have been shown to be effective (Tsang *et al.*, 1985).

One important example of a dense medium is ground snow. Snow has a volume density of 10 to 40% and the dielectric constant of ice is much larger than that of the background medium. When the radiative transfer equation is used at microwave and millimeter wave frequencies, modifying the radiative transfer equation to include the pair-correlation effect is necessary (Wen *et al.*, 1989).

# References

Broschat, S.L. "Numerical studies of the Phase Perturbation Technique for Random Rough Surface Scattering," Ph.D. dissertation, University of Washington, Seattle, January 1989.

Chandrasekhar, S., *Radiative Transfer,* Clarendon, Oxford, 1950 and Dover, New York, 1960.

Chen, M.F., and A.K. Fung, "A Numerical Study of the Regions of Validity of the Kirchhoff and Small-Perturbation Rough Surface Scattering Models," *Radio Science*, Vol. 23, No. 2, pp. 163-170, March-April 1988.

Cheung, R.L.-T., and A. Ishimaru, "Transmission, Backscattering, and depolarization of Waves in Randomly Distributed Spherical Particles," *Applied Optics,* Vol. 21, No. 20, pp. 3792–3798, 1982.

Dobson, M.C., E.S. Kasischke, and K.C. McDonald, "Microwave Extinction by Boreal Forest Canopies in Winter," *Proc. IGARSS '89 Symposium*, Vancouver, British Columbia, July 1989.

Dobson, M.C., K. McDonald, F.T. Ulaby, and J.F. Paris, "Diurnal Patterns in Multifrequency, Multipolarization Backscattering by a Walnut Orchard," *Proc. IGARSS '88 Symposium*, Edinburgh, Scotland, Vol. III, p. 1755, 1988.

Fung, A.K., and H.J. Eom, "A Theory of Wave Scattering from an Inhomogeneous Layer with an Irregular Interface," *IEEE Trans. Antennas and Propagation,* Vol. AP-29, pp. 899–910, 1981.

Fung, A.K., and H.J. Eom, "Multiple Scattering and Depolarization by a Randomly Rough Kirchhoff Surface," *IEEE Trans. Antennas and Propagation,* Vol. AP-29, No. 3, pp. 463–471, May 1981.

Fung, A.K., and G.W. Pan, "A Scattering Model for Perfectly Conducting Random Surfaces: I. Model Development," *Int. J. Remote Sensing*, Vol. 8, No. 11, pp. 1579–1593, 1987.

Gradsteyn, I.S., and I.M. Ryzhik, *Tables of Integrals, Series, and Products*, San Diego, CA: Academic Press, 1980.

Ishimaru, A., *Wave Propagation and Scattering in Random Media,* Vols. I and II, Academic Press, New York, 1978.

Ishimaru , A., and R.L.-T. Cheung, "Multiple Scattering Effects on Wave Propagation due to Rain," *Ann. Telecommunication,* Vol. 35, pp. 373–379, 1980.

Ishimaru, A., and Y. Kuga, "Attenuation Constant of Coherent Field in a Dense Distribution of Particles," *J. Opt. Soc. Am.,* Vol. 72, pp. 1317–1320, 1982.

Ishimaru, A., R. Woo, J.W. Armstrong, and D.C. Blackman, "Multiple Scattering Calculations of Rain Effects," *Radio Science,* Vol. 17, No. 6, pp. 1425–1433, 1982.

Kuga, Y., and A. Ishimaru, "Retroreflectance from a dense distribution of spherical particles," *Journal of the Optical Society of America,* Vol. A-1, pp. 831–835, 1984.

Kuga, Y., and A. Ishimaru, "Backscattering Enhancement by Randomly Distributed Very Large Particles," *Applied Optics,* Vol. 28, pp. 2165–2169, 1989.

Ma, Q., A. Ishimaru, P. Phu, and Y. Kuga, "Transmission, Bidirectional Reflectance and Depolarization of an Optical Wave for a Single Leaf," *Radio Science,* 1990.

McDonald, K.C., M.C. Dobson, and F.T. Ulaby, "Using MIMICS to Model Microwave Backscatter from Tree Canopies," *Proc. IGARSS '89 Symposium,* Vancouver, British Columbia, July 1989.

McDonald, K.C., M.C. Dobson, and F.T. Ulaby, "Using MIMICS to Model L-band Multi-Angle and Multi-Temporal Backscatter from a Walnut Orchard," *IEEE Trans. on Geoscience and Remote Sensing,* Vol. GE-28, 1990.

Pan, G.W., and A.K. Fung, "A Scattering Model for Perfectly Conducting Random Surfaces: II. Range of Validity," *Int. J. Remote Sensing*, Vol. 8, No. 11, pp. 1595–1605, 1987.

Rice, S.O., "Reflection of Electromagnetic Waves from Slightly Rough Surfaces," *Communications in Pure and Applied Mathematics*, Vol. 4, pp. 361–378, 1951.

Sekera, Z., "Scattering Matrices and Reciprocity Relationships for Various Representations of the State of Polarization," *J. Opt. Soc. Am.,* Vol. 56, pp. 1732–1740, 1966.

Sommerfeld, A., *Thermodynamics of Statistical Mechanics,* Academic Press, New York, 1956.

Thorsos, E.I., and D.R. Jackson, "The Validity of the Perturbation Approximation for Rough Surface Scattering using a Gaussian Roughness Spectrum," *J. Acoust. Soc. Am.*, pp. 261–277, July 1989.

Tsang, L., and A. Ishimaru, "Backscattering Enhancement of Random Discrete Scatterers," *J. Opt. Soc. Am.,* Vol. A-1, pp. 836–839, 1984.

Tsang, L., and A. Ishimaru, "Radiative Wave and Cyclical Transfer Equations for Dense Nontenuous Media," *J. Opt. Soc. Am.,* Vol. A-2, pp. 2187–2193, 1985.

Tsang, L., and A. Ishimaru, "Radiative Wave Equations for Vector Electromagnetic Propagation in Dense Nontenuous Media," *J. of Electromagnetic Waves and Applications,* Vol. 1, No. 1, pp. 52–72, 1987.

Tsang, L., J.A. Kong, and R.T. Shin,*Theory of Microwave Remote Sensing,* Wiley-Interscience, New York, 1985.

Ulaby, F.T., D. Held, M.C. Dobson, K.C. McDonald, and T.B.A. Senior, "Relating Polarization Phase Difference of SAR Signals to Scene Properties", *IEEE Trans. Geoscience and Remote Sensing*, Vol. GE-25, pp. 83–92, 1987.

Ulaby, F.T., R.K. Moore, and A.K. Fung, *Microwave Remote Sensing: Active and Passive,* Vol. II,Artech House, Norwood, MA, 1986a, Chapter 12.

Ulaby, F.T., R.K. Moore, and A.K. Fung, *Microwave Remote Sensing,* Vol. III, Artech House, Norwood, MA, 1986b, Chapter 13.

Ulaby, F.T., K. Sarabandi, K.C. McDonald, M.W. Whitt and M.C. Dobson, *Michigan Microwave Canopy Scattering Model (MIMICS),* The University of Michigan Radiation Laboratory, Report No. 022486-T-1, July 1988.

Ulaby, F.T., K. Sarabandi, K.C. McDonald, M.W. Whitt and M.C. Dobson, "Michigan Microwave Canopy Scattering Model (MIMICS)," *Int. J. Remote Sensing,* 1990.

van Albada, M., and A. Lagendijk, "Observation of Weak Localization of Light in a Random Medium," *Phys. Rev. Lett.,* Vol. 55, pp. 2692–2695, 1985.

van de Hulst, *Light Scattering by Small Particles,* Wiley-Interscience, New York, 1957.

Watson, J.G., and J.B. Keller, "Rough Surface Scattering via the Smoothing Method," *J. Acoust. Soc. Am.*, Vol. 75, No. 6, pp. 1705–1708, June 1984.

Wen, B., L. Tsang, D.P. Winebrenner, and A. Ishimaru, "Dense Medium Radiative Transfer Theory: Comparison with Experiment and Application to Microwave Remote Sensing and Polarimetry," *IEEE Trans. Geoscience and Remote Sensing,* Vol. GE-28, 1990.

Watson, J.G., and J.B. Keller, "Rough Surface Scattering via the Smoothing Method," *J. Acoust. Soc. Am.*, Vol. 75, No. 6, pp. 1705–1708, June 1984.

Winebrenner, D.P. "A Surface Field Phase-Perturbation Technique for Scattering from Rough Surfaces," Ph.D. dissertation, University of Washington, Seattle, 1985.

Wolf, P.E., and G. Maret, "Weak Localization and Coherent Backscattering of Photons in Disordered Media," *Phys. Rev. Lett.,* Vol. 55, pp. 2696–2699, 1985.

# Chapter 5

# Polarimetric Scatterometer Systems and Measurements

Michael W. Whitt, Fawwaz T. Ulaby, and Kamal Sarabandi

Until recently, the primary function of scatterometer systems has been to make absolute measurements of the backscattering coefficient $\sigma^0$ of distributed targets, usually as a function of the incidence angle $\theta$. Often, the measurements are made for various configurations of receiving and transmitting antenna polarizations — such as $hh$[1], $hv, vv, RR, RL$ and $LL$ — but, in all cases, the quantity measured by the scatterometer is the magnitude of the received signal and no phase information is recorded. In contrast with such a magnitude-only scatterometer, a polarimetric scatterometer is configured to measure the complete scattering matrix $\overline{\mathcal{S}}$ of a target by measuring the magnitude and phase of the scattered electric field for all linear polarization combinations.[2] The attractive feature of the polarimetric approach lies in the fact that given a matrix $\overline{\mathcal{S}}$, we can use the polarization

[1] The first letter signifies the polarization of the receiving antenna and the second letter signifies the polarization of the transmitting antenna, with antenna polarization always defined for both transmitting and receiving antennas as if the antenna were in a transmitting mode; the letters $h, v, R$, and $L$ denote horizontal, vertical, right circular, and left circular, respectively.

[2] The backscatter alignment (BSA) convention will be used throughout this chapter. The overbar denotes that the quantity is defined in the BSA convention.

synthesis technique (Section 2-3) to compute the scattering coefficient of the target for any desired receiving-transmitting polarization configuration.

The principles of operation of a magnitude-only scatterometer and its attendant design and calibration problems have been the subject of many articles and book chapters (for example, Ulaby *et al.*,1986b, Chapters 10 and 14). Hence, we shall give only a brief review of these topics in this chapter, primarily to provide the background necessary for discussion of those aspects of a scatterometer design and operation that pertain to its use in making polarimetric measurements of the backscattering cross section of point and distributed targets.

# 5-1 SCATTEROMETER SYSTEM CONFIGURATION

The generic block diagram shown in Fig. 5.1 illustrates the basic components of any radar system used in a monostatic mode, wherein the transmitting and receiving antennas are collocated or the same antenna is used for both transmission and reception. The system consists of a transmitter, a receiver, a signal processor, and either one or two antennas. If a single antenna is used for both transmission and reception, a *transmit-receive switch* (TR switch) is used to switch the antennas alternately between transmitter and receiver. With low-power radars, a microwave circulator may be used instead to provide the necessary isolation between the transmitter and receiver sections.

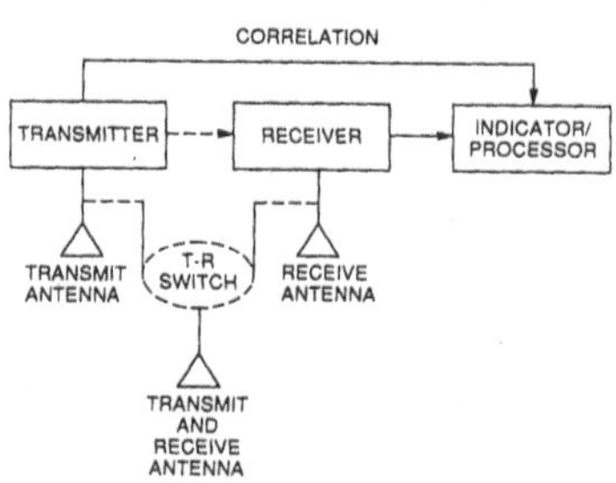


**Figure 5.1** Basic elements of a radar system.

The signal produced by the transmitter is coupled into space through the transmitting antenna. The signal then propagates out to the target, part of the signal reflects or scatters back towards the radar, and part of the scattered wave is coupled by the receiving antenna into the receiver circuit. The receiver amplifies and processes the scattered signal and then passes it on to an indicator of some kind, which may also process the signal in various ways to extract different characteristics of interest.

Radar systems are used for a variety of civilian and military applications. Our concern in this chapter, however, is with the *scatterometer*, which is a specific type of radar used to measure the scattering properties of distributed targets, such as terrain surfaces, although it can be used to measure the *radar cross section* (RCS) of point targets as well. Until recently, the scatterometer configurations most commonly used were operated in the CW-doppler, the FM-CW, or the pulse radar mode, each of which will be described briefly below. These systems were used primarily to measure only the magnitude of the scattered signal. With the introduction of the Hewlett Packard family of Automatic Vector Network Analyzers (HP 8510, HP 8753, and HP 8720) in recent years (as well as similar systems by other manufacturers), making accurate measurements of both the magnitude and phase of the scattered signal is now possible by using the signal processing and error correction capabilities of the network analyzer. Because this new approach is rapidly becoming the standard technique for RCS measurements, we shall devote Section 5-2 to a discussion of its principal features.

### 5-1.1 CW Radar

The continuous-wave (CW) radar is the simplest type of system to build, but it is seldom used because it has some serious limitations. The transmitter consists of a radio frequency (RF) oscillator that produces a continuous, single-frequency signal, and the receiver is essentially a detector with a dc output voltage that is proportional to the power of the received RF signal (Fig. 5.2). For a point target at a range $r$ from the radar antennas, the power received at the receive-antenna terminals is given by the radar equation:

$$P_r = \frac{P_t \, G_t \, G_r \, \lambda^2}{(4\pi)^3 \, r^4} \, \sigma \,, \tag{5.1}$$

where $G_t$ and $G_r$ are the gains of the transmitting and receiving antennas in the direction of the target, $\lambda$ is the wavelength, and $\sigma$ is the RCS of the target. In general, $\sigma$ depends on the target size, shape, and material composition as well as the illumination direction and the polarization

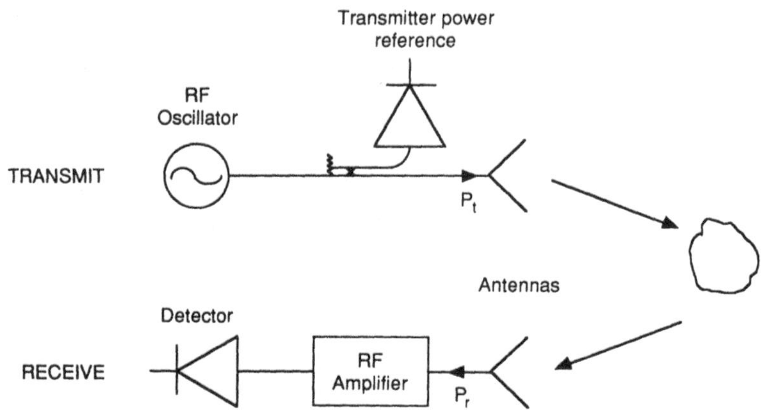


**Figure 5.2** Block diagram of a basic CW radar system.

configurations of the transmitting and receiving antennas.

Because the transmitted signal is unmodulated, a CW radar cannot discriminate between signals received from targets at different ranges. Consequently, the radar cannot measure the range between it and the observed target, nor can it isolate the signal reflected from the target from any stray signals, including that which is coupled directly from the transmitting antenna to the receiving antenna. Moreover, because the output is in the form of a dc voltage, the receiver will exhibit poor noise performance; the noise spectrum generated by the receiver components generally varies as $1/f$.

To overcome the limitations of the CW radar, we can use pulse modulation or phase modulation to control the timing of the transmitted signal. Frequency modulation, which is a specific form of phase modulation, has been particularly useful in this regard. Many other variations on this theme are also used, including the chirp radar mode, which incorporates the features of both pulse and frequency modulations by transmitting short pulses with a carrier frequency that is linearly swept across a certain bandwidth over the duration of the pulse (Farnett *et al.*, 1970).

Returning to the single-frequency CW radar, its signal-to-noise performance can be improved significantly if we precede the detector with a high-gain, low-noise intermediate frequency (IF) amplifier. This is accomplished by using an auxiliary oscillator as shown in Fig. 5.3 to up-convert the frequency of a sample of the transmitted signal from $f_1$ to $(f_1 + f_2)$, and then mixing the output signal with the received signal at frequency $f_1$ to produce an IF signal at $f_2$. Typically, $f_1$ is in the GHz range and $f_2$ is in the kHz or MHz range.

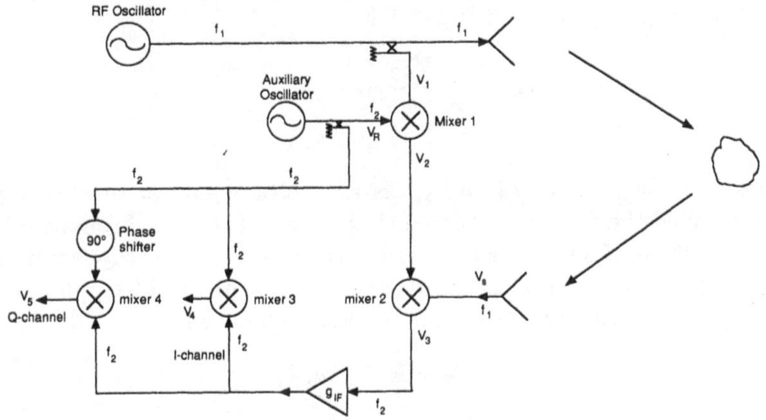


**Figure 5.3** Block diagram of a radar system that uses in-phase and quadrature-phase channels to measure the phase of the received signal $V_s$.

The voltage of the signal sampled through the directional coupler is

$$V_1 = V_{10} \cos \omega_1 t, \tag{5.2}$$

where $\omega_1 = 2\pi f_1$ is the angular frequency of the signal generated by the RF oscillator. Upon *beating* $V_1$ in mixer 1 with $V_R$, the signal from the auxiliary oscillator, and then rejecting the difference frequency $(f_1 - f_2)$, we have

$$\begin{aligned} V_2 &= \frac{V_{R0} V_{10}}{2} \cos[(\omega_1 + \omega_2)t] \\ &= V_{20} \cos[(\omega_1 + \omega_2)t], \end{aligned} \tag{5.3}$$

where $V_{R0}$ is the amplitude of the signal generated by the auxiliary reference oscillator.

The received signal is denoted as

$$V_S = V_{S0} \cos(\omega_1 t + \theta_S), \tag{5.4}$$

where $V_{S0}$ is its magnitude and $\theta_S$ is defined relative to the phase of $V_2$, which is taken here as the reference. The difference frequency output of mixer 2 is given by

$$V_3 = \frac{V_{20} V_{S0}}{2} \cos(\omega_2 t - \theta_S), \tag{5.5}$$

and, after amplification by the IF amplifier and mixing with $V_R$ in mixer 3, the dc output of mixer 3 is given by

$$\begin{aligned} V_4 &= \frac{V_{R0}}{4} g_{\mathrm{IF}} V_{20} V_{S0} \cos\theta_S \\ &= K V_{S0} \cos\theta_S , \end{aligned} \tag{5.6}$$

where $K = (V_{R0}^2 g_{\mathrm{IF}} V_{10})/8$ and $g_{\mathrm{IF}}$ is the voltage gain of the IF amplifier. This is called the ***in-phase channel*** (I-channel) of the synchronous detector.

The output of mixer 4 is the ***quadrature-phase channel*** (Q-channel). The input signals of mixer 4 are similar to those of mixer 3 except that $V_R$ is shifted by 90° prior to the mixing process. This leads to

$$V_5 = K V_{S0} \sin\theta_S . \tag{5.7}$$

The constant $K$ can be determined through calibration by observing a target with known scattering properties, and the voltages $V_4$ and $V_5$ can be combined to determine both $V_{S0}$ and $\theta_S$. Thus,

$$V_{S0} = \frac{1}{K}\sqrt{V_4^2 + V_5^2} \tag{5.8}$$

and

$$\theta_S = \tan^{-1} \frac{V_5}{V_4} . \tag{5.9}$$

Hence, when used in conjunction with a synchronous detector, the frequency up-conversion–down-conversion technique not only results in improved signal-to-noise performance, but also provides the means to measure both the magnitude and the phase of the received signal. Nonetheless, such a single-frequency CW radar is still incapable of measuring the target's range because the radar has no resolving power in the range direction. This limitation can be overcome, however, by using a stepped-frequency technique, which involves measuring the radar return as a function of frequency over a bandwidth $B$.

The measurements can be made at discrete frequencies by stepping the transmitter frequency in fixed increments over the total bandwidth $B$. The Fourier transform of the frequency-domain response is a time-domain response with a range resolution:

$$\Delta\, r_{\min} = \frac{c}{2B} , \tag{5.10}$$

where $c$ is the velocity of light. This result is discussed in Section 5-2 in connection with the use of the automatic vector network analyzer to measure the scattering amplitude of radar targets.

## 5-1.2 CW-Doppler Radar

The CW-doppler radar uses its motion relative to the target in order to produce a doppler shift in the return signal. The system is usually mounted on a moveable platform to provide the motion. A simplified block diagram of a typical system is shown in Fig. 5.4. A single-frequency oscillator transmits a signal, which, upon return, is doppler shifted in frequency by an amount proportional to the velocity of the radar relative to the target. This return signal passes through a circulator into the input port of a single-ended mixer. Transmitter leakage through the circulator is mixed with the return signal, producing a frequency equal to the doppler shift at the output of the mixer. This signal is usually in the kHz range, where the $1/f$ noise is low. After amplification, the signal-to-noise ratio (SNR) is significantly better than that for a comparable CW system. Although the CW-doppler system provides a convenient method of improving the SNR, range information is still unavailable from the radar.

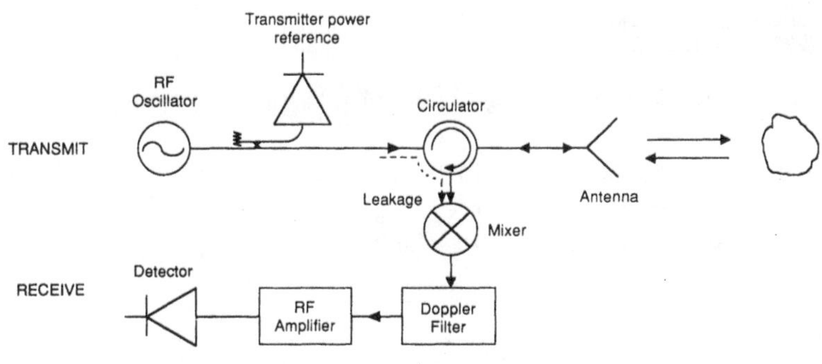


**Figure 5.4** Block diagram of a CW-doppler radar system.

## 5-1.3 FM-CW Radar

The FM-CW radar is a frequency modulated, continuous-wave system with inherent ranging capability (Skolnik, 1980, pp. 81–92; Saunders, 1970, Ulaby *et al.*, 1986a, pp. 512–517). A simplified block diagram illustrating the major components of a typical FM-CW radar is shown in Fig. 5.5. A sweep generator is used to frequency-modulate the transmitted signal, usually in a saw-toothed or triangle form. The backscattered signal from the target, at the RF $f_r(t)$, is then received and fed into a mixer where it is mixed with a sample of the current transmitted signal at the frequency $f_{\mathrm{LO}}(t)$, the local oscillator (LO) frequency. Signals at both the sum and difference frequencies of the received and transmitted signals are generated at the mixer output. The sum frequency is rejected by a bandpass filter, and the difference frequency signal is amplified and then detected by an IF processor (a spectrum analyzer or narrowband detector) or a true rms voltmeter.

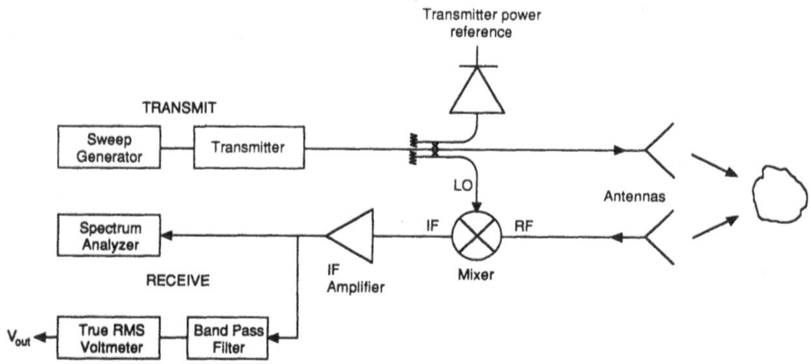


**Figure 5.5** Block diagram of a typical FM-CW radar system.

An illustration of how the system operates for a point target return at a range $r$ is shown in Fig. 5.6. Except during the small segment of time that the modulating waveform changes slope, the IF is given by

$$f_{\mathrm{IF}} = \frac{4r}{c} f_m B , \tag{5.11}$$

where

$$\begin{aligned} f_{\mathrm{IF}} &= \text{intermediate frequency at mixer output (Hz)}, \\ r &= \text{range to point target (m)}, \\ c &= \text{speed of light (m/s)}, \\ B &= \text{modulation bandwidth (Hz)}, \\ f_m &= \text{modulation rate (Hz)}. \end{aligned}$$

Usually, in order to simplify the receiver, the IF bandpass filter is fixed at a particular center frequency. To measure the backscattered power at a range $r$, the modulation rate $f_m$ is adjusted to place the mixer output frequency within the passband of the filter for the particular range of interest. Thus, different ranges can be selected by adjusting the modulation rate and calculating $r$ from the above equation.

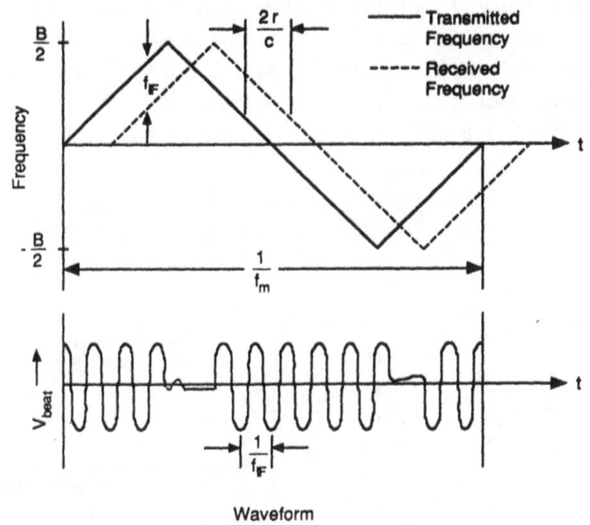


**Figure 5.6** The upper part shows the RF of the transmitted wave and of the wave received from a point target as a function of time, and the lower part shows the voltage of the difference frequency at the mixer output, also as a function time.

The resolution of an FM-CW radar is determined by the antenna beamwidth in the azimuth direction and by the IF filter bandwidth in the range direction, provided that the bandwidth is sufficiently large. Because the range is linearly related to the IF,

$$\Delta r = r \frac{\Delta f_{\mathrm{IF}}}{f_{\mathrm{IF}}} , \tag{5.12}$$

where $\Delta f_{\mathrm{IF}}$ is the IF filter bandwidth. However, the range resolution cannot be made arbitrarily small by decreasing $\Delta f_{\mathrm{IF}}$. The minimum resolution possible is determined by the modulation bandwidth:

$$\Delta r_{\mathrm{min}} = \frac{c}{2B} . \tag{5.13}$$

By making $\Delta r$ larger than $\Delta r_{\mathrm{min}}$, more independent samples can be obtained to reduce signal fading (Ulaby and Dobson, 1989, pp. 69–75).

If the radar is directed at a point target at a range $r$ and the IF filter totally encompasses the return from the target (Fig. 5.7), the output voltage measured by the true rms voltmeter will be given by

$$V_{\mathrm{out}} = g\sqrt{P_r} , \tag{5.14}$$

where $g$ is a voltage gain constant that incorporates the gains and losses of all components between the receiving antenna terminals and the rms voltmeter, and $P_r$ is the received power given by (5.1). The IF filter serves to isolate the desired return (i.e., the target return) from spurious returns, including those caused by internal reflections within the radar system.

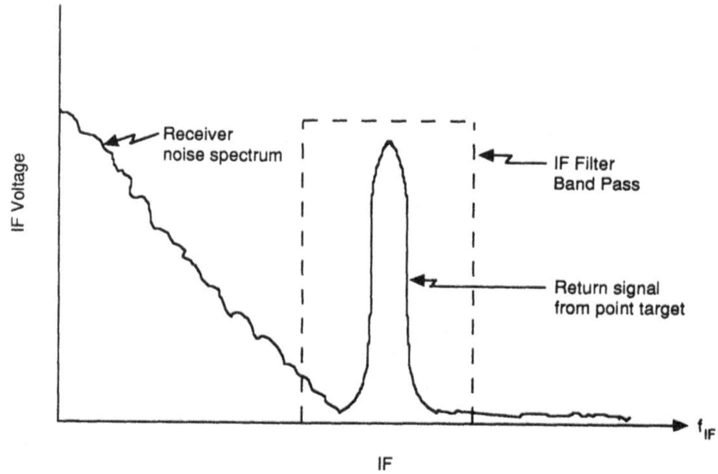


**Figure 5.7** Sketch of the IF voltage on a spectrum analyzer.

## 5-1.4 Pulse Radar

Another common radar system with ranging capability is the pulse radar. A block diagram of this type of system is shown in Fig. 5.8. The timing of the system is controlled by a synchronizer that sends trigger pulses to the modulator and the display. The modulator generates high-power pulses of a particular shape and duration, which are used to modulate an RF oscillator in the pulsed transmitter. The transmitted signal is directed to the antenna and, upon return from the target, the signal is mixed down to a suitable IF and amplified before detection. The detected signal is then displayed.

Range information for a pulsed radar is determined by measuring the time delay between the beginning of the transmitted pulse and the beginning of the received pulse. For a point target, this time delay will be

$$T_d = \frac{2r}{c}, \tag{5.15}$$

where $r$ is the range, and $c$ is the speed of light. As in the FM-CW radar, azimuth resolution is determined by the antenna beamwidth. The range resolution depends on the integration time $\tau$ of the receiver, given by

$$\Delta r = \frac{c\tau}{2}. \tag{5.16}$$

The minimum range resolution available is determined by the pulse length $\tau_p$ of the transmitted waveform, given by

$$\Delta r_{\min} = \frac{c\tau_p}{2}. \tag{5.17}$$

The radar equation given by (5.1) applies to the pulse radar case, but now $P_t$ represents the peak power of the transmitted pulse, and $P_r$ is equal to zero except over the duration of the received pulse. The radar equation may be expressed in terms of the average transmitted power $\overline{P}_t$ by noting that

$$\overline{P}_t = P_t \tau_p f_p, \tag{5.18}$$

where $f_p$ is the *pulse repetition frequency* (PRF). To match the radar receiver to the received pulse, the receiver bandwidth $B$ is usually selected such that

$$B \approx \frac{1}{\tau_p}. \tag{5.19}$$

Insertion of (5.19) into (5.17) leads to (5.13), which demonstrates that an FM-CW radar with a modulation bandwidth $B$ has the same range resolution capability as a pulse radar configuration with matched receiver bandwidth $B$.

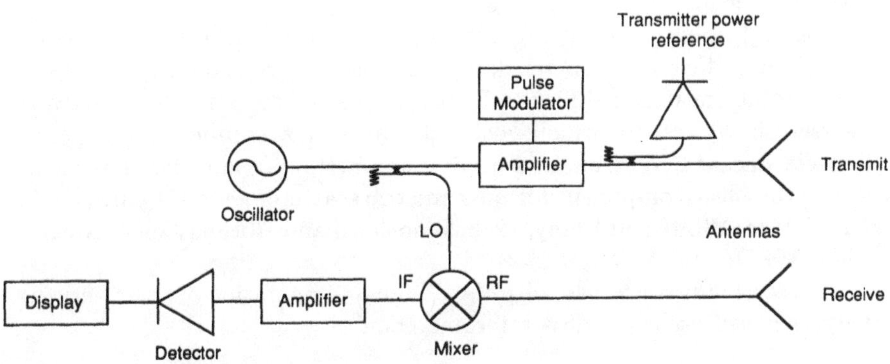


**Figure 5.8** Block diagram of basic coherent pulse radar.

# 5-2 NETWORK ANALYZER BASED POLARIMETRIC SCATTEROMETERS

In general, polarimetric scatterometers must have measurement capabilities that are not usually required with other radars. To measure the scattering matrix, they must be able to transmit and receive two orthogonal polarizations (usually vertical, $v$, and horizontal, $h$) and to measure both the magnitude and phase of the received signals. Because they are used specifically to measure the scattering properties of distributed targets, some bandwidth is also desirable to reduce fading by adding independent frequency samples and to provide ranging capability. With range information available, the radar returns from different locations within a particular random medium (for example, a vegetation canopy) can be separated. This simplifies the correlation of measurements with actual scattering mechanisms occurring within the medium.

Another very important difference between polarimetric scatterometers (as well as traditional scatterometers) and other radars is the range at which targets are to be measured. Although scatterometers may be operated from aircraft or satellite platforms, those to be considered in this chapter will be short-range scatterometers that typically measure targets only up to ranges of about 20 m. This corresponds to a two-way time delay of roughly 133 ns. Because the target is at such a close range, leakage signals that are multiply reflected due to mismatch within the system can appear at the target range as noise with a level well above that of thermal noise. This problem, which is specific to short-range radars, must be considered in any scatterometer design.

Although these additional requirements and problems can be addressed by modifying the scatterometer designs discussed in Section 5-1, the design of such modifications is difficult. In recent years, however, vector network analyzers have become widely available through a number of companies (Hewlett Packard, Wiltron, and other manufacturers), and they have been used as the basic component for polarimetric scatterometer systems (Ulaby *et al.*, 1988a; Whitt and Ulaby, 1988; Blanchard and Rochier, 1987; Riegger *et al.*, 1987).

This section is concerned with the design and operation of basic network analyzer-based polarimetric scatterometers.

## 5-2.1 Network Analyzer Principles of Operation

In general, network analyzers measure the magnitude and phase characteristics of linear networks relative to some standard or reference. This is accomplished by making both transmission and reflection measurements to obtain specific characteristics of the network (see Fig. 5.9). For transmission measurements, the signal transmitted through the test network is compared to the incident signal which serves as the reference. For reflection measurements, the signal reflected from the input port is compared with the incident signal. The output port, in this case, is usually terminated with a matched load to remove reflections that travel back through the network and contribute to the reflected signal.

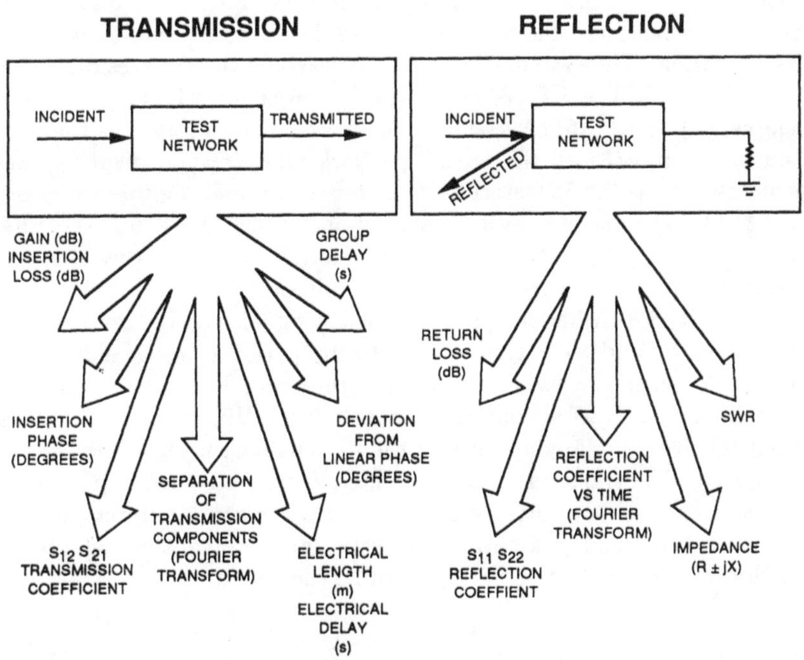


**Figure 5.9** Transmission and reflection measurements (from *HP 8510 Network Analyzer Operating and Programming Manual*, Hewlett-Packard Company, Santa Rosa, CA, 1984).

A basic vector network analyzer system consists of (1) an RF source, (2) an RF to IF converter, (3) an IF signal detector and analog-to-digital

(A/D) converter, and (4) a digital microprocessor and display. In the context of Fig. 5.9, the RF source is used to supply the incident signal to the test network. The transmitted or reflected RF signal and a sample of the incident signal are then mixed down to the IF range, maintaining the magnitude and phase relationships between the two signals. The IF signals are then detected, converted to digital form, and processed to obtain information about the test network. If a frequency-swept incident signal is used, the result is a measurement of the frequency response of the network, and Fourier transform techniques can be used to obtain the time-domain response.

As an example, Fig. 5.10 illustrates the basic components of an HP 8510 vector network analyzer system used for making transmission-reflection measurements. In general, the principles are the same for other vector network analyzer systems (HP 8753, HP 8720, Wiltron 360, and similar systems by other manufacturers). The Hewlett Packard equipment will be stressed throughout this chapter because the authors are more familiar with its operation. The system of Fig. 5.10 consists of three pieces of equipment: (1) an HP 8340A or HP 8341A synthesized sweeper, which serves as the RF source; (2) an HP 8511A frequency converter, which serves as the RF to IF converter; and (3) an HP 8510A network analyzer and display-processor, which serves as the IF detector, A/D converter, and digital processor. The HP 8511A frequency converter is only one of several IF test sets that can be configured to operate as a scatterometer. Because it represents the most general form, however, we will use it to illustrate the basic principles.

The HP 8511A has four ports to which signals can be applied. Of these ports, the $a_1$ and $a_2$ ports are used to input the reference signal, and the $b_1$ and $b_2$ ports are used to input the measurement signals. In the basic transmission-reflection configuration (see Fig. 5.10), a sample of the signal from the RF source is applied to the $a_1$ reference port, while the $a_2$ port is unused. The transmitted and reflected signals are then applied to the $b_1$ and $b_2$ measurement ports, respectively. The network analyzer can now give the magnitude and phase of the reflected and transmitted signals relative to that of the incident signal by forming the ratios:

$$S_{11} = \frac{b_1}{a_1} \quad \text{and} \quad S_{21} = \frac{b_2}{a_1}, \tag{5.20}$$

respectively. Other characteristics of the network, as shown in Fig. 5.9, can also be determined.

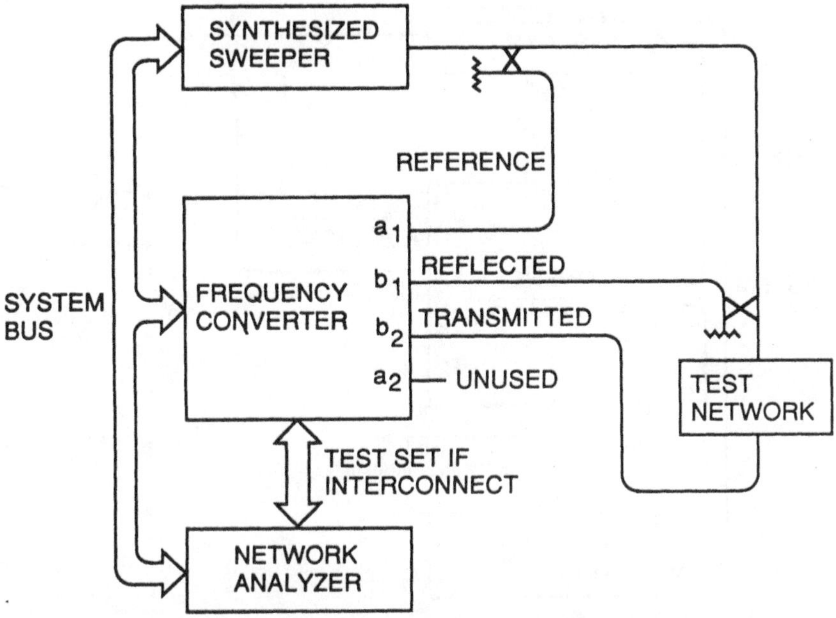


**Figure 5.10** Configuration of a network analyzer for transmission and reflection measurements.

## 5-2.2 Network Analyzer Operation as a Scatterometer

The measurement system of Fig. 5.10 can be reconfigured to allow operation as a scatterometer. In the scatterometer configuration, the "test network" is simply some radar target separated from the measurement system by free space. In this case, the signal from the RF source is coupled to free space with an antenna and travels to the target (i.e., test network) through the air medium instead of a coaxial cable or waveguide. The signal reflected from the target is then received with another antenna and is compared with a sample of the transmitted signal from the RF source, as before. The system can be configured as either a dual-antenna system or a single-antenna system (using a circulator to separate the transmitting and receiving channels) as shown in Fig. 5.11.

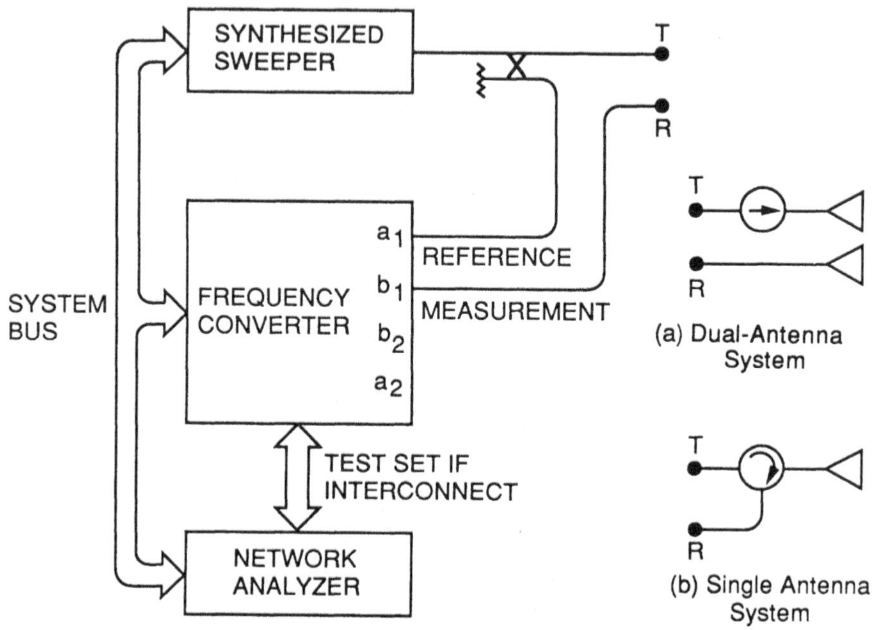


**Figure 5.11** Configuration of a network analyzer for scatterometer measurements; (a) dual-polarized single-antenna system, (b) single-antenna system.

There is a major difference between the transmission-reflection configuration and the scatterometer configuration. When operating as a scatterometer, there is a relatively long time delay between the returned and transmitted signals due to propagation to the target and back. During this delay, the reference signal has changed frequency by an amount $\Delta f$, which corresponds to a phase error in degrees of $\Delta\phi = 360\Delta f\tau$, where $\tau = 2r/c$ is the two-way time delay. If a synthesized source is used, the transmitted signal can be stepped in frequency instead of continuously swept to keep the reference signal at the same frequency as the received signal. However, any phase instability in the source can also produce a phase error. To operate as a coherent system, this phase error must be smaller than a few degrees over the delay time $\tau$. The phase stability over typical scatterometer ranges ($r \leq 20$ m, or $\tau \leq 133$ ns) for synthesized sweepers is negligible, so the phase error can be ignored. For a scatterometer with a continuously swept frequency, the phase error in degrees is related to $\tau$ and the sweep rate $f_s$ (in Hz·s$^{-1}$) by the relation $\Delta\phi = 360 f_s \tau^2$, where $\Delta f = f_s\tau$. For the phase error to be less than one degree, the two-way phase delay $\tau$ must satisfy

$$\tau \leq \sqrt{\frac{1}{360 f_s}} \,. \tag{5.21}$$

A typical sweep rate for the HP 8510 system is $f_s = 20$ MHz·ms$^{-1}$, requiring $\tau \leq 373$ ns (or $r \leq 56$ m). This range is at least twice as large as typical scatterometer ranges, so network analyzers can usually operate effectively as scatterometers by using either synthesized (stepped) frequency or swept frequency sources.

We have so far considered only a single polarization for transmitting and receiving; therefore, the system in Fig. 5.11 represents a traditional (magnitude-only) scatterometer. It must be modified to operate as a polarimetric scatterometer. Because it measures the complete scattering matrix of a given target, a polarimetric scatterometer must be able to transmit and receive a set of orthogonal polarizations. The $v$ and $h$ linear polarizations will be used here for illustrative purposes to remain consistent with the rest of the text. For the single-antenna system, we can add a second polarization channel and apply it to the $b_2$ port of the HP 8511A. For the dual-antenna system, two options are available. An additional channel can be utilized, similar to the single-antenna system, or polarizers can be inserted before the antennas. These different configurations are shown in Fig. 5.12, where the network analyzer is assumed to be configured the same as in Fig. 5.11.

There are many advantages to using a vector network analyzer as the IF processor for a polarimetric scatterometer. The advantages all relate to the many features available with most network analyzers. One of these features is the ease with which the network analyzer can be programmed and controlled by computer over an interface bus. This feature is stressed because it reduces the hardware design to the consideration of only RF components as in Fig. 5.12. The network analyzer supplies all the detection, processing, and data transfer options. To provide a complete description of all these options available with network analyzers is not possible, but we will highlight some that are particularly useful in polarimetric scatterometer applications. One of the most useful processing options available with many network analyzers is time-domain operation. Because the returned signal is measured as a function of frequency, real-time Fourier transform techniques can be used to obtain the time-domain response. In scatterometer applications, the result is a measure of the radar return as a function of range. In addition, all other processing operations are available to both time and frequency domains. The signal, for example, may be displayed in the time domain where the data can be processed as a function of range. Time-gating can then be used to isolate the return due to targets at a particular range. Suppose a measurement of the backscattered power is desired

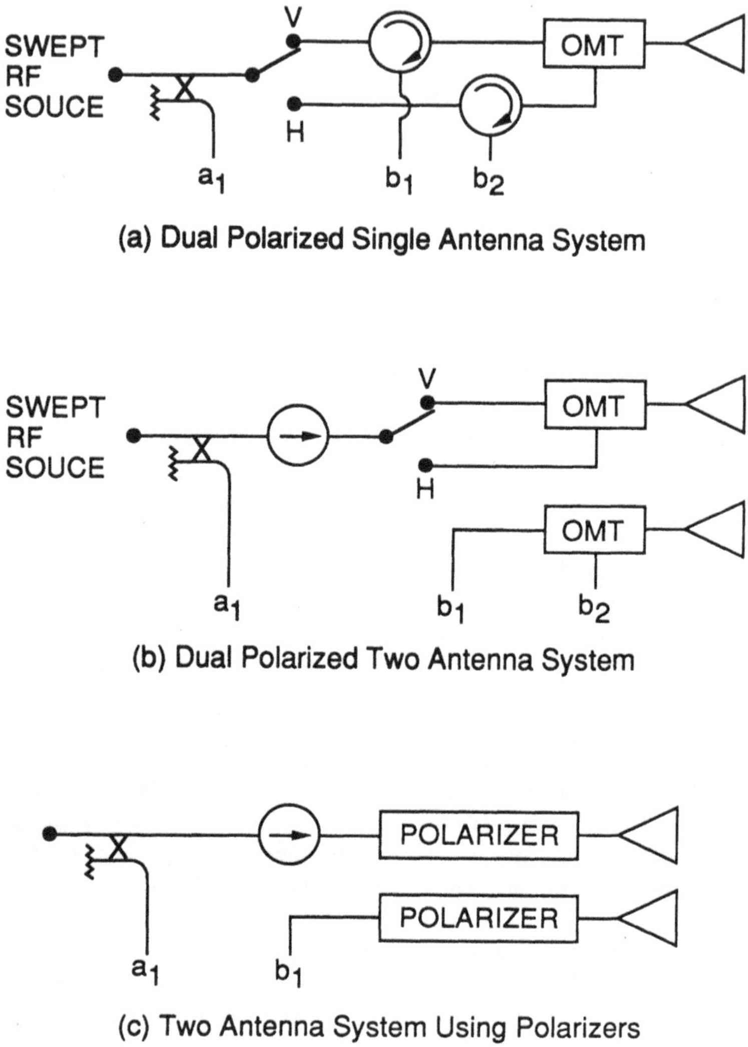


**Figure 5.12** Polarimetric configurations of a network analyzer based scatterometer; (a) dual-polarized single-antenna system, (b) dual-polarized-antenna system, and (c) dual-antenna system using polarizers.

as a function of depth for a volume of scatterers. By using the time-domain option with time-gating, the portion of the returned signal corresponding to a particular depth in the volume can be selected. The position and width of the time gate determines the range and range extent of the small volume within the entire volume of scatterers. The time-gated signal can then be transformed back to the frequency domain to obtain the frequency response of the small volume, as modified by the medium between it and the scatterometer. This process is illustrated in Fig. 5.13.

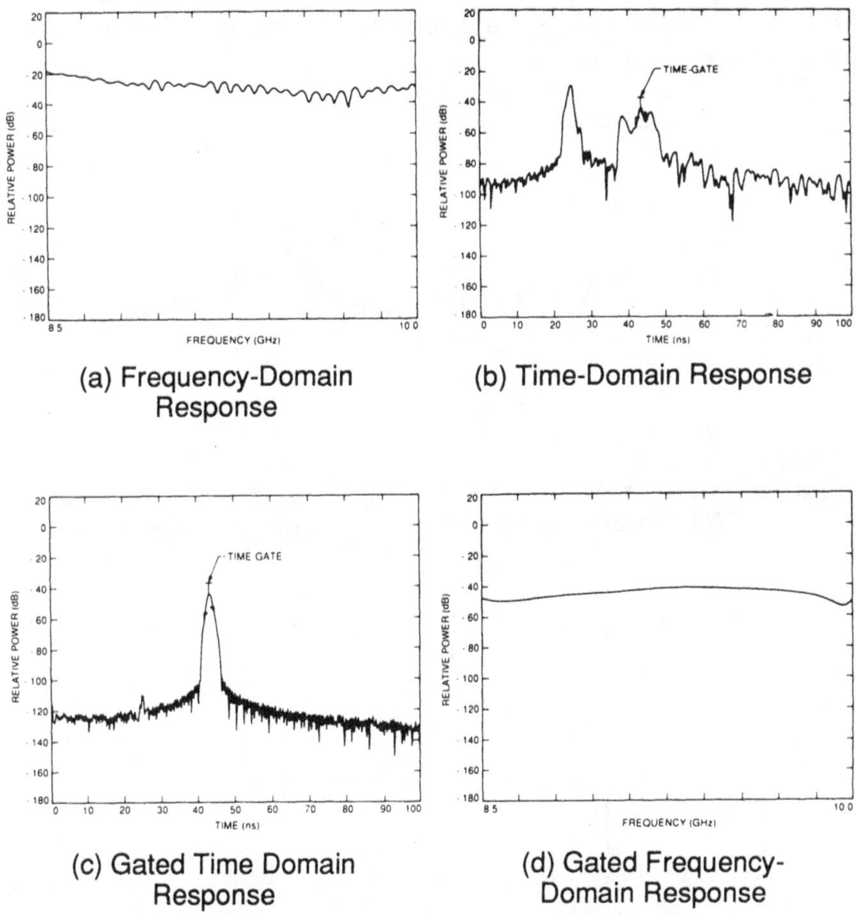


(a) Frequency-Domain Response

(b) Time-Domain Response

(c) Gated Time Domain Response

(d) Gated Frequency-Domain Response

**Figure 5.13** Plots illustrating the use of gating with the time-domain option to measure the frequency response as a function of depth within a random medium; (a) frequency domain response, (b) time domain response, (c) gated time domain response, and (d) gated frequency domain response.

There are other features that can be used to remove nonvarying errors in the return signal. For example, scattering from objects other than the target can contribute to the noise of the measurement. Many network analyzers have data storage and complex math features that can remove these nonvarying errors. First, the response of the test setup without the target is stored and later subtracted from the response when the target is present, resulting in a measurement of the target response alone. The leakage signals caused by multiple reflections that appear at the target range can be removed in the same manner. An example illustrating this procedure is given in Fig. 5.14.

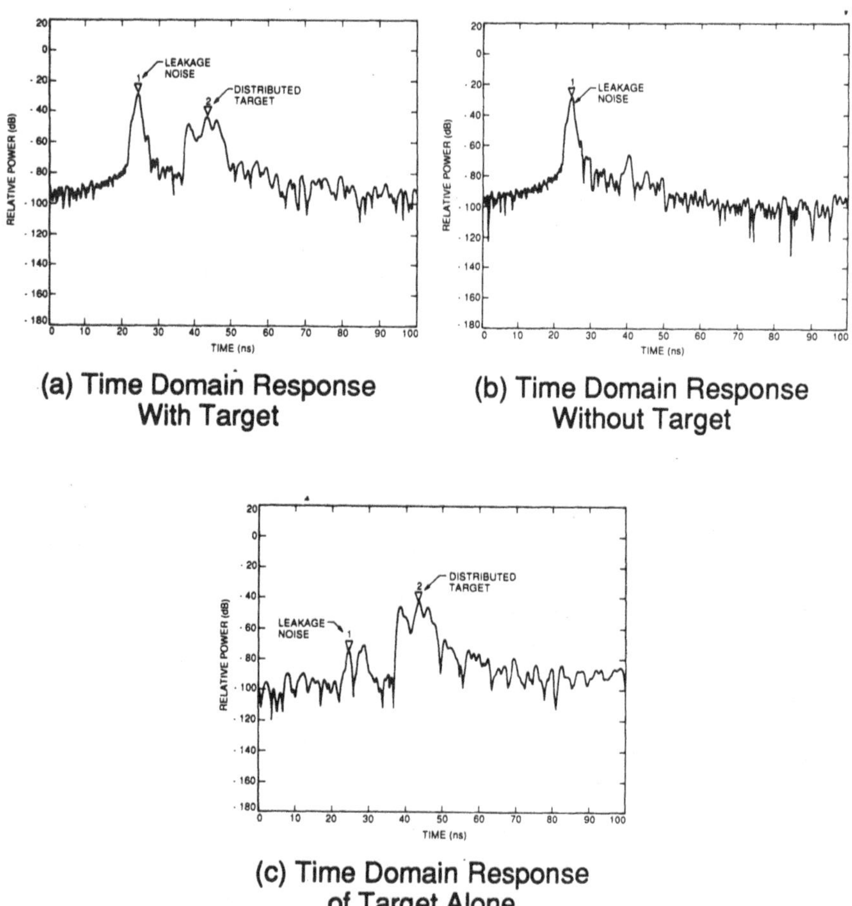


**Figure 5.14** Plots illustrating the use of complex mathematical operations with trace storage to remove leakage signals; (a) time domain response with target, (b) time domain response without target, and (c) time domain response after subtraction of (b) from (a).

Another feature that is useful in reducing random noise contributions is IF averaging. With this feature, a specified number of consecutive data traces are continuously averaged. The constant target signals add coherently, but the random noise adds incoherently, thus improving the SNR. This is similar to integration in traditional radar systems.

## 5-2.3 Microwave Polarimetric Scatterometers

Scatterometers must be able to measure targets with a wide range of radar cross sections. In practice, targets with large radar cross sections are easy to measure, but there is a lower bound on the measurable RCS. This lower bound (minimum detectable target) for a given scatterometer at a fixed range from the target is limited by three major factors. The first factor is the thermal noise level, which is the absolute minimum detectable level, which can be determined by a product of noise figures and system bandwidth. The second limiting factor is the dynamic range of the network analyzer's sampler (or mixer). This problem arises when a single antenna is used or if there is insufficient isolation between the transmitting and receiving antennas. In such cases, part of the transmitted signal returns to the receiver's sampler and sets the minimum detectable signal level. This minimum signal level can be determined by dividing the returned signal level by the dynamic range of the sampler. Finally, the third limiting factor is the effect of multiple reflections that appear at the target range. In a radar system, there usually exists an impedance mismatch between different RF components and between the antenna and free space. A fraction of the transmitted signal corresponding to each mismatch interface returns to the receiver as noise. Multiple reflections between the mismatch interfaces also arrive at the receiver, but with different time delays corresponding to the total path length. The short-range reflections can be distinguished from the target signal in the time domain, but multiple reflections that appear at the target range cannot be separated from the target signal. Therefore, targets with a signal level that is less than the level of the multiple reflections are not detectable.

Because a network analyzer is a coherent device, its bandwidth is very narrow and thermal noise is usually not a limiting factor. However, the dynamic range and multiple reflection problems limit the performance of scatterometers, but their adverse effects can be overcome by improving the isolation between the transmitting and receiving channels of the system.

The two basic schemes for configuring the antenna systems of typical scatterometers were shown in Fig. 5.11 of Section 5-2.2. The single-antenna system shown in Fig. 5.11(a) isolates the transmitting and receiving channels by using a circulator. The isolation is typically on the order of 20 dB, and it is limited by leakage through the circulator and reflection from the mismatch at the antenna. Better isolation is obtained by using the dual-antenna system shown in Fig. 5.11(b). With this configuration, the isolation is limited by the coupling between the antennas, and it is on the order of 50 dB. Although the isolation is superior, there are other physical limitations for the dual-antenna system at microwave frequencies that make a single-antenna system more practical.

To study the characteristics of distributed targets as a function of incidence angle, a scatterometer must have a small angular resolution. This condition is usually achieved by using a small-beamwidth antenna. However, a decrease in beamwidth requires an increase in the size of the antenna aperture. At microwave frequencies, the aperature size becomes too large for dual-antenna systems to be practical. Aside from the obvious increase in weight, the increase in aperature size causes parallax problems due to the large boresight separation between the transmitting and receiving antennas. In this case, the radar is no longer operating in the backscatter mode, and the antenna boresight directions must be adjusted, depending on the target range (see Fig. 5.15).

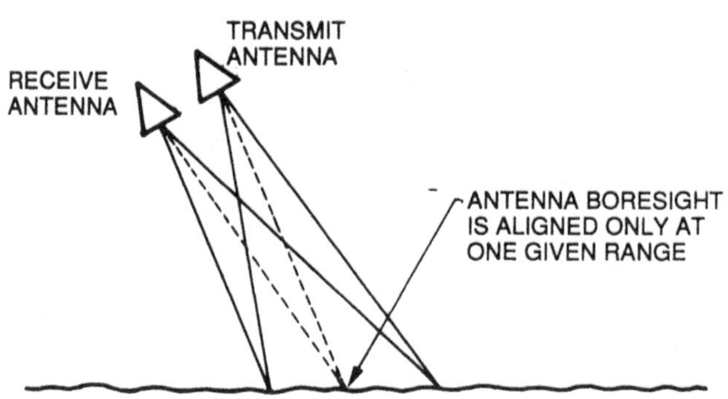


**Figure 5.15** Parallax problem for a dual-antenna system.

As mentioned, one of the disadvantages of single-antenna systems is the problem associated with the dynamic range limitations of network analyzers. For small targets or those that are far from the scatterometer, we may need to add amplifiers in the transmitting or receiving channels. Therefore, the strong short-range signal returns from the antenna mismatch and leakage from the circulator saturate the receiver. To reduce the level of these direct returns, a pulsing technique may be used (Tassoudji *et al.*,

1989; Liepa *et al.*, 1989). In this scheme, the receiver is switched off between transmission and when the signal return from the target arrives at the receiver. Because this switching is done at a much higher rate than the bandwidth of the receiver, the network analyzer does not sense that the incoming signal is pulsed, and it is measured as if it were a CW signal. The pulsing switches are placed in the transmitting and receiving channels as in Fig. 5.16. The isolators and high-pass filters are used to remove switching transients that affect the response of the network analyzer. Note that the pulsing network is not interfaced with the network analyzer, and the pulsing operation is independent of the network analyzer operating sequence. Figures 5.17 and 5.18 show two time and frequency domain responses of a target with and without switching. When pulsing is applied, the short-range returns, which were dominant without pulsing, decrease significantly, and the target return decreases only slightly by an amount dependent on the duty cycle. The overall signal-to-noise improvement in this example is about 10 dB.

### 5-2.4 Millimeter-Wave Polarimetric Scatterometers

The free-space loss associated with the propagation of electromagnetic waves is inversely proportional to the square of the wavelength. If the wavelength is reduced by a factor of 10 (100 GHz instead of 10 GHz), the free-space loss increases by 20 dB. For a millimeter-wave polarimetric scatterometer with the same performance as a microwave system (see Section 5-2.3), we must either increase the transmitted power or reduce the absolute noise level of the system. Because higher power is more difficult to obtain at millimeter-wave frequencies, the most reasonable option is to reduce the noise level. In the microwave system, a single antenna was used because of antenna size limitations, and we found that reflections from the antenna mismatch caused significant noise problems. These noise problems were somewhat improved by adding a pulsing network to remove the major contribution. Much of the noise associated with the single-antenna system can be avoided by using two antennas; therefore, this is the approach adopted with the millimeter-wave system in this section. Because millimeter-wave antennas (with beamwidths comparable to those in the microwave region) are much smaller, dual-antenna systems are not limited by the size of the antennas.

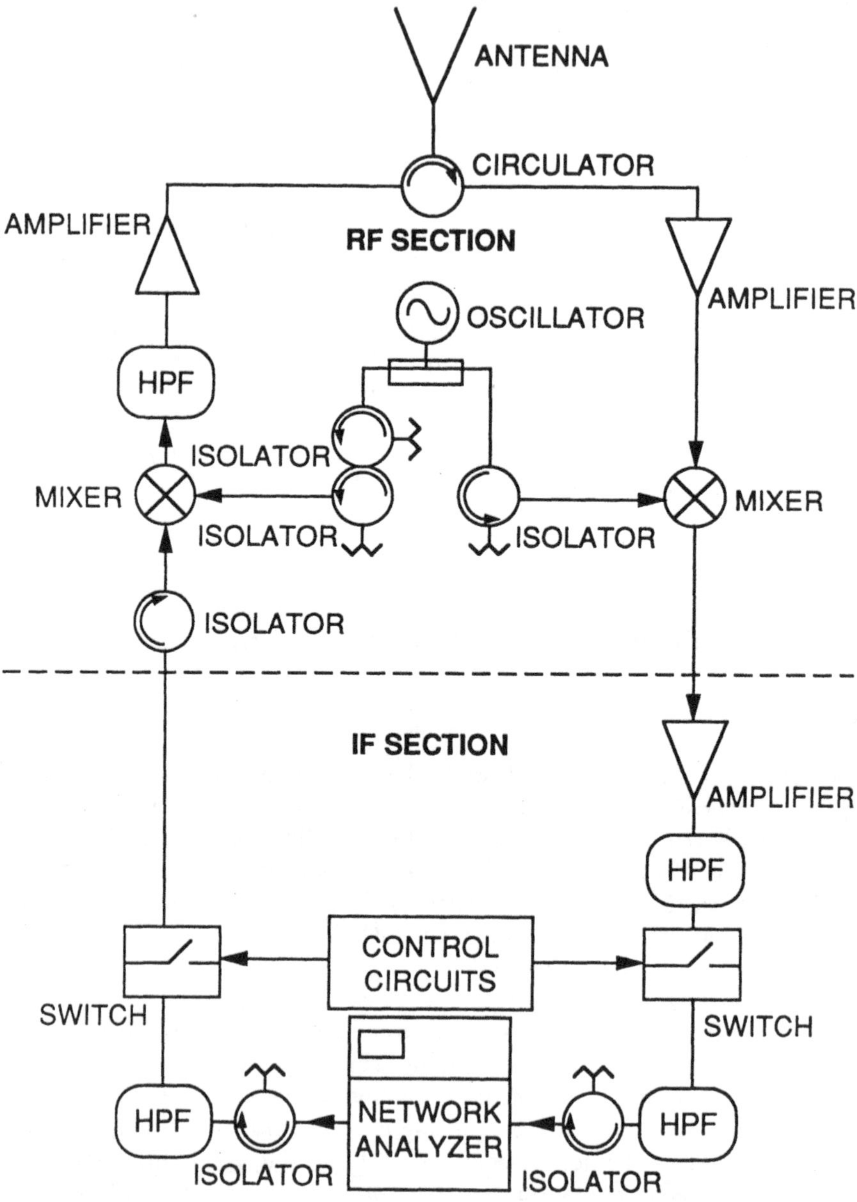


**Figure 5.16** Diagram of scatterometer system with switching.

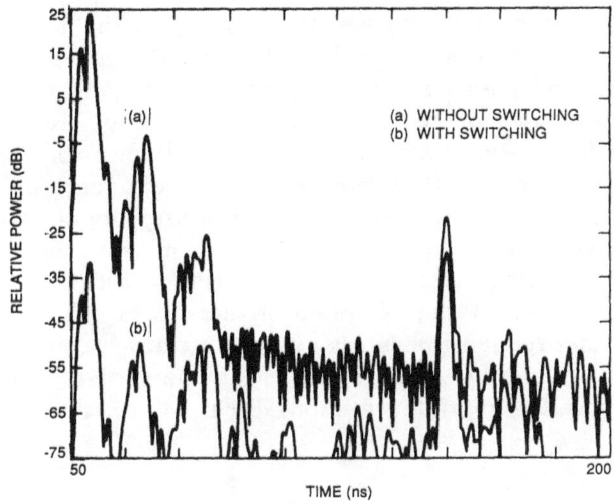


**Figure 5.17** Time-domain response of the system in an anechoic chamber: (a) without switching, (b) with switching.

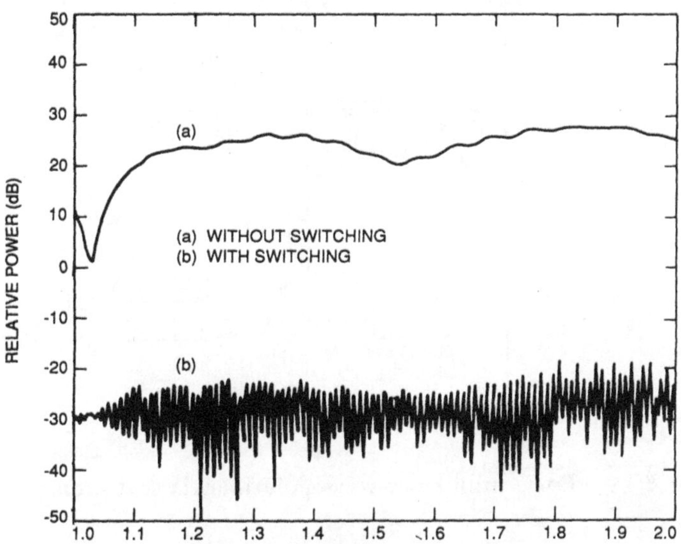


**Figure 5.18** Corresponding frequency-domain trace: (a) without switching, (b) with switching.

The diagram in Fig. 5.19 illustrates a basic polarimetric scatterometer design for millimeter-wave frequencies. To obtain a swept millimeter-wave frequency for transmission, a lower IF (typically in the 1–5 GHz range) is up-converted to the desired RF band by using a single-sideband (SSB) up-converter, which passes only the frequency $f_{RF} = f_{LO} + f_{IF}$. The down-conversion stage used on reception passes only the difference frequency $f_{IF} = f_{RF} - f_{LO}$. Because the up-conversion and down-conversion stages are driven by the same LO (local oscillator), phase coherence is maintained. To reduce the LO power requirements, we can use only one transmitting and one receiving channel (i.e., using only two mixers), and this can be accomplished by using polarizers to provide the polarization agility. The two isolators located at the LO ports of the mixers help to isolate the transmitting and receiving channels by rejecting any leakage components traveling through the LO path. Isolators may be needed throughout the system to reduce the multiple reflections discussed in Section 5-2.3.

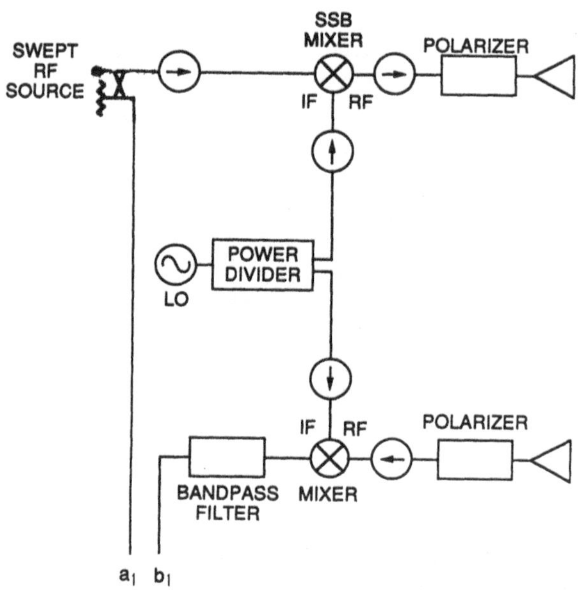


**Figure 5.19** Basic millimeter-wave polarimetric scatterometer.

At millimeter-wave frequencies (above about 120 GHz), the available LO power becomes so low that a single LO may not be able to drive both the up-conversion and down-conversion mixers. However, available LO power can be increased by injection phase-locking two separate sources. Figure 5.20 illustrates one way in which this can be done. All three sources of Fig. 5.20 are designed to operate at the desired LO frequency. The oscillator-circulator combination acts as an unstable amplifier, the output power of which is approximately equal to that of the oscillator. The result is two coherent LO outputs, each with the power of a single oscillator.

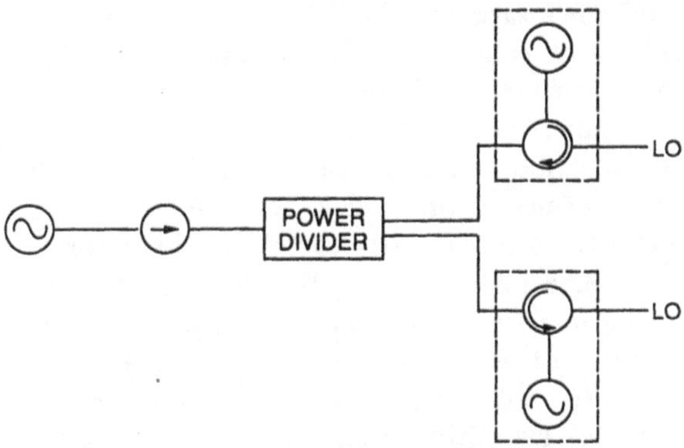


**Figure 5.20** Design used to phase-lock two LO sources.

# 5-3 CALIBRATION OF POLARIMETRIC SCATTEROMETERS

The calibration of polarimetric radar systems has been considered by several investigators in recent years. A technique proposed by Barnes (1986) characterizes the errors introduced by the transmitter and receiver in terms of distortion matrices that alter the measured scattering matrix of the target. This calibration technique requires the measurement of three known targets, some of which have zero elements in their scattering matrices. A technique similar that of Barnes was used by Freeman *et al.*,(1988) where the known target scattering matrices were realized by polarimetric active radar calibrators (PARCs). A technique introduced by Riegger *et al.*,(1987) characterizes the system errors in terms of coupling coefficients between elements of the theoretical scattering matrix and elements of the measured scattering matrix. Riegger's is in essence the same model as used by Barnes, except that Riegger has expanded the matrix product which resulted in twice the number of unknowns. These techniques have been shown to be effective in practical applications, but are limited because the approaches require the measurement of targets with a particular form of scattering matrix that is often difficult to realize.

Two additional polarimetric calibration techniques are of interest because they require only a single nondepolarizing target (such as a sphere or trihedral) to correct for co-polarized channel imbalance and absolute magnitude errors. The cross-polarization coupling (or crosstalk) errors are corrected by using unknown targets. The first technique, proposed by Sarabandi *et al.*, (1990) achieves calibration of the crosstalk errors by measuring any arbitrary depolarizing target. Knowledge of the scattering matrix of the arbitrary target is not required. A similar technique by van Zyl (1990) uses measurements of distributed natural targets to determine the crosstalk errors. The advantage of these techniques is their insensitivity to target positioning, which makes them particularly useful in field calibration.

In this section, we begin by introducing the concept of distortion matrices to model the errors introduced by polarimetric scatterometer systems. With this background, we then show how the technique by Sarabandi *et al.*, (1990) can be used to calibrate polarimetric scatterometers with diagonal distortion matrices. Two calibration targets, namely a metal sphere and any second target with strong cross-polarized RCS, are required to calibrate the scatterometer. The advantage of this method is that (1) it is insensitive to orientation of the calibration targets with respect to the antenna coordinate system and (2) theoretical values for the scattering matrix of the second target are not required. Therefore, the calibration requires only the knowledge of the theoretical scattering matrix for a metal sphere which is well known. The technique is particularly useful for calibrating a radar under field conditions because this method does not require careful alignment of the calibration targets.

For systems where no assumptions can be made about the distortion matrices, we discuss a more recent technique proposed by Whitt *et al.*, (1990). This technique requires the measurement of three known targets to determine the distortion matrices that characterize the effect of the measurement system on the transmitted and received waves. The scattering matrices for the known targets can be of any form, provided that a limited set of constraints are satisfied. The approach involves forming matrix products from the measured scattering matrices to obtain a similarity transformation, where the transformation matrix is the unknown distortion matrix. The relationships between the eigenvalues and eigenvectors of the similar matrices are then used to solve for the distortion matrices. Two algorithms are given on the basis of this technique. The first technique is used for the most general case of nonreciprocal distortion, where the transmitter and receiver distortion matrices are unrelated, and the second technique is used when the transmitter and receiver distortion matrices are the transpose of one another. These techniques are described in Sections 5-3.3 and 5-3.4, respectively.

The final section (Section 5-3.5) compares experimental results obtained for the various calibration techniques discussed.

## 5-3.1 System Distortion Matrices

When using an ideal polarimetric radar, the measured scattering matrix $\overline{\mathcal{M}}$ of a point target would be equal to its theoretical (or actual) scattering matrix $\overline{\mathcal{P}}$. Because this is rarely the case, the errors introduced by the radar system must be determined, and then the process must be inverted to obtain an estimate of the actual scattering matrix.

The calibration problem can be modeled in terms of the generalized schematic representation shown in Fig. 5.21. Figure 5.21(a) represents the condition when the polarization switch of the transmitting antenna is set to transmit a vertically polarized wave of amplitude $E_0$. Because the system is nonideal, the transmitted voltage $\overline{\mathbf{E}}^t$ consists of a vertically polarized component $\overline{T}'_{vv} E_0$ and a horizontally polarized component $\overline{T}'_{hv} E_0$:

$$\overline{\mathbf{E}}^t = E_0 \begin{bmatrix} \overline{T}'_{vv} \\ \overline{T}'_{hv} \end{bmatrix}, \tag{5.22}$$

where $\overline{T}'_{vv}$ represents the amplitude and phase errors in the transmitted signal, and $\overline{T}'_{hv}$ represents the coupling between the $v$- and $h$-channels of the transmitter.

The field scattered back to the radar from a point target with scattering matrix $\overline{\mathcal{P}}$ illuminated by this transmitted field is given by

$$\overline{\mathbf{E}}^s = \frac{1}{r^2}\left[\frac{2\eta_0 P_t G_t}{4\pi}\right]^{1/2} e^{i\phi}\overline{\mathcal{P}} \begin{bmatrix} \overline{T}'_{vv} \\ \overline{T}'_{hv} \end{bmatrix}, \tag{5.23}$$

where $P_t = E_0^2/2\eta_0$ is the transmitted power, $G_t$ is the nominal gain of the transmitting antenna, $\phi = 2k_0 r$ is a phase factor that accounts for propagation between the target and the radar, $r$ is the range to the point target, and

$$\overline{\mathcal{P}} = \begin{bmatrix} \overline{P}_{vv} & \overline{P}_{vh} \\ \overline{P}_{hv} & \overline{P}_{hh} \end{bmatrix}. \tag{5.24}$$

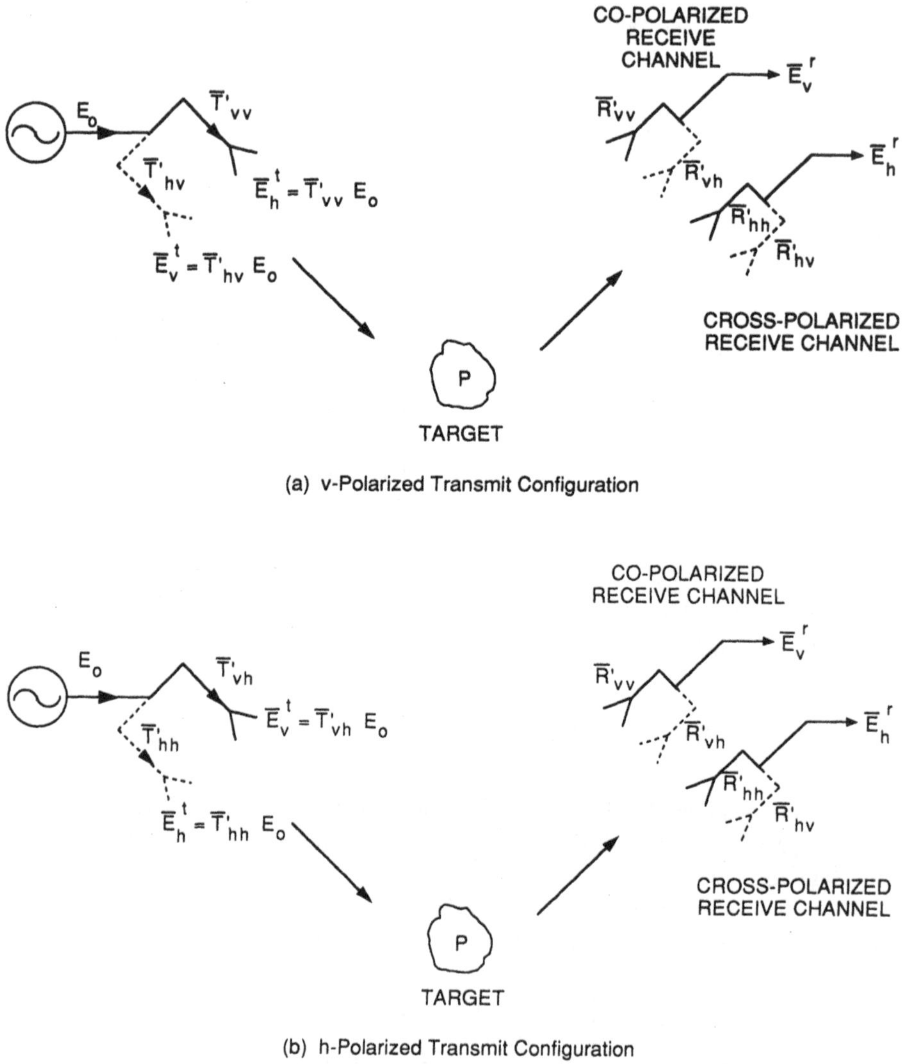


**Figure 5.21** Schematic representation of transmitting and receiving antennas; (a) $v$-polarized transmitting configuration, (b) $h$-polarized transmitting configuration. The dashed lines represent polarization coupling. For an antenna with perfect polarization isolation, the coupling coefficients would be zero.

Again, because the receiving antenna is nonideal, the received voltage is given by

$$\overline{\mathbf{E}}^r = \left[\frac{G_r\lambda^2}{4\pi}\right]^{1/2} \begin{bmatrix} \overline{R}'_{vv} & \overline{R}'_{vh} \\ \overline{R}'_{hv} & \overline{R}'_{hh} \end{bmatrix} \overline{\mathbf{E}}^s, \tag{5.25}$$

where $\overline{R}'_{vv}$ and $\overline{R}'_{hh}$ represent the amplitude and phase errors introduced by the receiver, $\overline{R}'_{vh}$ and $\overline{R}'_{hv}$ represent the cross-polarization coupling terms for the receiver, and $G_r$ is the nominal gain of the receiving antenna. Using (5.23) in (5.25), the received voltage can be written as

$$\overline{\mathbf{E}}^r = e^{i\phi}\frac{K}{r^2} \begin{bmatrix} \overline{R}'_{vv} & \overline{R}'_{vh} \\ \overline{R}'_{hv} & \overline{R}'_{hh} \end{bmatrix} \begin{bmatrix} \overline{P}_{vv} & \overline{P}_{vh} \\ \overline{P}_{hv} & \overline{P}_{hh} \end{bmatrix} \begin{bmatrix} \overline{T}'_{vv} \\ \overline{T}'_{hv} \end{bmatrix}, \tag{5.26}$$

where

$$K = \left[\frac{2\eta_0 P_t G_t G_r \lambda^2}{(4\pi)^2}\right]^{1/2}. \tag{5.27}$$

Figure 5.21(b) represents the condition when the polarization switch of the transmitting antenna is set to transmit horizontal polarization, and an expression similar to (5.26) can be obtained for this case:

$$\overline{\mathbf{E}}^r = e^{i\phi}\frac{K}{r^2} \begin{bmatrix} \overline{R}'_{vv} & \overline{R}'_{vh} \\ \overline{R}'_{hv} & \overline{R}'_{hh} \end{bmatrix} \begin{bmatrix} \overline{P}_{vv} & \overline{P}_{vh} \\ \overline{P}_{hv} & \overline{P}_{hh} \end{bmatrix} \begin{bmatrix} \overline{T}'_{vh} \\ \overline{T}'_{hh} \end{bmatrix}, \tag{5.28}$$

where $\overline{T}'_{hh}$ represents the amplitude and phase errors in the transmitted signal, and $\overline{T}'_{vh}$ represents the cross-polarization coupling.

Equations (5.26) and (5.28) can be combined into the compact form:

$$\overline{\mathbf{E}}^r = \frac{K}{r^2} e^{i\phi} \overline{R}'_{vv} \overline{T}'_{vv} \overline{\boldsymbol{\mathcal{R}}}\,\overline{\boldsymbol{\mathcal{P}}}\,\overline{\boldsymbol{\mathcal{T}}}\,\overline{\mathbf{p}}^t, \tag{5.29}$$

where

$$\overline{\mathbf{p}}^t = \hat{\mathbf{v}} = \begin{bmatrix} 1 \\ 0 \end{bmatrix} \tag{5.30}$$

when the polarization switch of the transmitter is set to transmit vertical polarization, and

$$\overline{\mathbf{p}}^t = \hat{\mathbf{h}} = \begin{bmatrix} 0 \\ 1 \end{bmatrix} \tag{5.31}$$

when it is set to transmit horizontal polarization. The normalized matrices $\overline{\mathcal{R}}$ and $\overline{\mathcal{T}}$ are given by

$$\overline{\mathcal{R}} = \begin{bmatrix} 1 & \overline{R}_{vh} \\ \overline{R}_{hv} & \overline{R}_{hh} \end{bmatrix}, \qquad \overline{\mathcal{T}} = \begin{bmatrix} 1 & \overline{T}_{vh} \\ \overline{T}_{hv} & \overline{T}_{hh} \end{bmatrix}, \tag{5.32}$$

where $\overline{R}_{rt} = \overline{R}'_{rt}/\overline{R}'_{vv}$ and $\overline{T}_{rt} = \overline{T}'_{rt}/\overline{T}'_{vv}$, with $rt \in \{vh, hv, hh\}$.

If the radar is assumed to have distortionless antennas, which implies that $\overline{\mathcal{R}} = \overline{\mathcal{T}} = \overline{\mathcal{I}}$ (where $\overline{\mathcal{I}}$ is the identity matrix) and $\overline{R}'_{vv} = \overline{T}'_{vv} = 1$, the set of four measurements corresponding to polarizations $vv$, $vh$, $hv$, and $hh$ define a "measured" scattering matrix $\overline{\mathcal{N}}$ of the target. The direct scattering from the background and constant leakage terms within the system, represented by the scattering matrix $\overline{\mathcal{B}}$, have been assumed to be subtracted from the raw measurement represented by the scattering matrix $\overline{\mathcal{M}}$; therefore, $\overline{\mathcal{N}} = \overline{\mathcal{M}} - \overline{\mathcal{B}}$. The background scattering matrix $\overline{\mathcal{B}}$ can be directly measured by removing the target and making a standard scattering matrix measurement. Note that a significant contribution to the backscattered field due to interaction between the target and the background may also be present, but because this term is difficult to measure, it will not be considered. The interaction term can usually be ignored if the RCS of the target mount is small and the rest of the background is far from the target. Assuming distortionless antennas, (5.29) becomes

$$\overline{\mathbf{E}}^r = \overline{\mathcal{N}}\overline{\mathbf{p}}^t, \tag{5.33}$$

where the phase reference is at the radar antenna. Upon equating (5.29) and (5.33), we obtain

$$\overline{\mathcal{N}} = c\, e^{i\phi} \overline{\mathcal{R}}\,\overline{\mathcal{P}}\,\overline{\mathcal{T}}, \tag{5.34}$$

where

$$c = \frac{K}{r^2} \overline{R}'_{vv} \overline{T}'_{vv}. \tag{5.35}$$

With some single-antenna systems, the transmitter and receiver distortion matrices are simply the transpose of one another (i.e., the system is reciprocal), resulting in the equation:

$$\overline{\mathcal{N}} = c\, e^{i\phi} \widetilde{\overline{\mathcal{A}}}\,\overline{\mathcal{P}}\,\overline{\mathcal{A}}, \tag{5.36}$$

where the distortion matrix $\overline{\mathcal{A}}$ is given by

$$\overline{\mathcal{A}} = \begin{bmatrix} 1 & \overline{A}_{vh} \\ \overline{A}_{hv} & \overline{A}_{hh} \end{bmatrix}. \tag{5.37}$$

If the matrices $\overline{\mathcal{N}}$, $\overline{\mathcal{T}}$, and $\overline{\mathcal{R}}$ (or $\overline{\mathcal{A}}$ in the reciprocal case) can be determined, the actual scattering matrix $\overline{\mathcal{P}}$ can be obtained from one of the expressions:

$$\overline{\mathcal{P}} = \frac{1}{c} e^{-i\phi} \overline{\mathcal{R}}^{-1} \overline{\mathcal{N}} \overline{\mathcal{T}}^{-1} \tag{5.38}$$

or

$$\overline{\mathcal{P}} = \frac{1}{c} e^{-i\phi} \widetilde{\overline{\mathcal{A}}}^{-1} \overline{\mathcal{N}} \overline{\mathcal{A}}^{-1}, \tag{5.39}$$

which were obtained by rearranging (5.34) and (5.36). This representation assumes that the distortion matrices are invertible, which is generally true for actual radar systems. The propagation phase factor $e^{-i\phi}$ in (5.38) and (5.39) is difficult to measure because the phase center of the target and the target position must be known exactly. In most cases, however, only the relative phase of $\overline{\mathcal{P}}$ is desired.

## 5-3.2 Systems with Diagonal Distortion Matrices

If the scatterometer to be calibrated has good cross-polarization isolation, the distortion matrices are approximately diagonal. A diagonal distortion matrix corresponds to perfect isolation between the $v$- and $h$-ports of the scatterometer; therefore,

$$\overline{T}_{vh} = \overline{T}_{hv} = \overline{R}_{vh} = \overline{R}_{hv} = 0. \tag{5.40}$$

From (5.29), we can show that this condition implies

$$\overline{\mathbf{E}}^r = \frac{K}{r^2} \overline{R}'_{vv} \overline{T}'_{vv} e^{i\phi} \begin{bmatrix} \overline{P}_{vv} & \overline{T}_{hh} \overline{P}_{vh} \\ \overline{R}_{hh} \overline{P}_{hv} & \overline{R}_{hh} \overline{T}_{hh} \overline{P}_{hh} \end{bmatrix} \overline{\mathbf{p}}^t. \tag{5.41}$$

A simplified block diagram of a polarimetric scatterometer with perfect isolation between its antenna ports is shown in Fig. 5.22. In reality, a scatterometer may use a single antenna to provide both transmitting and receiving functions and may also use a polarization switching network capable of exciting either $v$- or $h$-polarized waves in the antenna. If the scatterometer is set to transmit $\overline{\mathbf{p}}^t \in \{\hat{\mathbf{v}}, \hat{\mathbf{h}}\}$ and receive $\overline{\mathbf{p}}^r \in \{\hat{\mathbf{v}}, \hat{\mathbf{h}}\}$ for a point target with scattering matrix $\overline{\mathcal{P}}$, the measured voltage for the receiving-transmitting polarization $rt \in \{vv, vh, hv, hh\}$ is given by

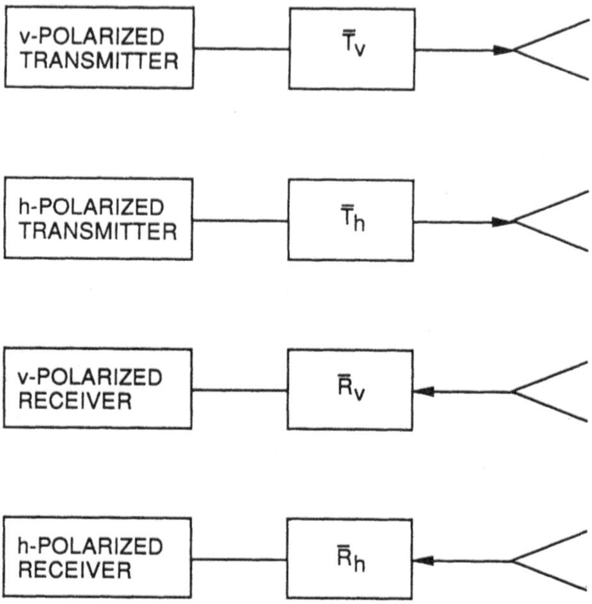


**Figure 5.22** Simplified block diagram of a dual-polarized radar system.

$$\overline{E}_{rt} = \frac{K}{r^2} e^{i2k_o r} \overline{R}_r \overline{T}_t \overline{P}_{rt} \ , \tag{5.42}$$

where the quantities $\overline{R}_r$ and $\overline{T}_t$ are field transfer functions for the receiver and transmitter, respectively, and they are given by

$$\overline{R}_r = \overline{R}'_{rr} \tag{5.43}$$
$$\overline{T}_t = \overline{T}'_{tt} . \tag{5.44}$$

A simple calibration approach involves the use of one reference target of known scattering matrix. Upon measuring $\overline{E}_{rt}$ with $\overline{P}_{rt}$ known, the quantity $K\overline{R}_r\overline{T}_t$ can be determined in amplitude and phase, with the latter being relative to some reference location. In principle, this procedure is straightforward, but the problem arises when we need to select a reference target of known scattering matrix. The metal sphere is the easiest target to align and its scattering matrix can be computed exactly (Ruck *et al.*, 1970). Unfortunately, a sphere can only be used to calibrate the *vv*- and *hh*-channels because $\overline{P}^{sp}_{vh} = \overline{P}^{sp}_{hv} = 0$. Targets that exhibit significant cross-polarized scattering include the dihedral corner reflector, tilted cylinders, and others, but scattering from such targets is inherently sensitive to the orientation of the target relative to the $(\hat{\mathbf{v}}, \hat{\mathbf{h}}, \hat{\mathbf{k}})$ coordinate system. This and other factors, such as edge scattering, may lead to significant errors between the calculated scattering amplitudes and their actual values. The orientation problem may be reduced to an acceptable level when operating in an anechoic chamber under controlled laboratory conditions, but poses a difficulty when calibrating a truck-mounted scatterometer, for example, under field conditions.

To solve this problem, Sarabandi *et al.*, (1990) proposed a calibration technique that uses two reference targets, namely a sphere and any target with strong cross-polarized scattering. As will be shown below, knowledge of the scattering matrix of the second reference target is not necessary to calibrate the radar system.

First, consider a metal sphere of known size at a distance $r_0$ from the radar. The scattering amplitudes of a metal sphere are $\overline{P}^{sp}_{hh} = \overline{P}^{sp}_{vv} = p_0$, and $\overline{P}^{sp}_{hv} = \overline{P}^{sp}_{vh} = 0$. For a given size of sphere, $p_0$ can be computed by using the well-known Mie solution (Ruck *et al.*, 1970). The received voltages for $vv$ and $hh$ polarizations are

$$\overline{E}^0_{vv} = \frac{K}{r_0^2} e^{i2k_or_0} \overline{R}_v \overline{T}_v p_0 \tag{5.45}$$

$$\overline{E}^0_{hh} = \frac{K}{r_0^2} e^{i2k_or_0} \overline{R}_h \overline{T}_h p_0, \tag{5.46}$$

where the subscript and superscript 0 denote quantities associated with the metal sphere. For a test target with *unknown* scattering matrix $\overline{\mathcal{P}}^u$ located at a distance $r_u$ from the radar, the received voltage for the different polarization combinations are

$$\overline{E}^u_{vv} = \frac{K}{r_u^2} e^{i2k_or_u} \overline{R}_v \overline{T}_v \overline{P}^u_{vv} \tag{5.47}$$

$$\overline{E}^u_{hh} = \frac{K}{r_u^2} e^{i2k_or_u} \overline{R}_h \overline{T}_h \overline{P}^u_{hh} \tag{5.48}$$

$$\overline{E}^u_{hv} = \frac{K}{r_u^2} e^{i2k_or_u} \overline{R}_h \overline{T}_v \overline{P}^u_{hv} \tag{5.49}$$

$$\overline{E}^u_{vh} = \frac{K}{r_u^2} e^{i2k_or_u} \overline{R}_v \overline{T}_h \overline{P}^u_{vh}. \tag{5.50}$$

From (5.45) to (5.48), we obtain the following expressions for the unknown co-polarized scattering amplitudes:

$$\overline{P}^u_{vv} = \left(\frac{\overline{E}^u_{vv}}{\overline{E}^0_{vv}}\right) \left(\frac{r_u}{r_0}\right)^2 e^{i2k_o(r_0-r_u)} p_0 \tag{5.51}$$

$$\overline{P}^u_{hh} = \left(\frac{\overline{E}^u_{hh}}{\overline{E}^0_{hh}}\right)\left(\frac{r_u}{r_0}\right)^2 e^{i2k_0(r_0-r_u)}p_0. \tag{5.52}$$

All the quantities on the right-hand side of the above two expressions can be measured directly, except for $p_0$, which can be computed precisely. Small errors in the measurement of the ranges $r_u$ and $r_0$ will have a minor effect on the magnitude of $\overline{P}^u_{hh}$, but may cause a large error in the phase angle of $\overline{P}^u_{hh}$, if the error $\Delta(r_0 - r_u)$ is comparable to the wavelength $\lambda$. However, the phase of $\overline{P}^u_{hh}$ relative to that of $\overline{P}^u_{vv}$ is independent of $(r_0 - r_u)$.

Next, consider any point target that exhibits strong cross-polarized scattering, and then measure the received voltage for $hv$ and $vh$ polarizations:

$$\overline{E}^c_{hv} = \frac{K}{r_c^2}e^{i2k_0r_c}\overline{R}_h\overline{T}_v\overline{P}^c_{hv} \tag{5.53}$$

$$\overline{E}^c_{vh} = \frac{K}{r_c^2}e^{i2k_0r_c}\overline{R}_v\overline{T}_h\overline{P}^c_{vh}, \tag{5.54}$$

where the subscript and superscript $c$ refer to the cross-polarization calibration target. The reciprocity theorem states that, in the backscattering direction, the cross-polarized scattering amplitudes are always equal. Hence,

$$\overline{P}^c_{hv} = \overline{P}^c_{vh}, \tag{5.55}$$

and, consequently,

$$K_1 = \frac{\overline{E}^c_{hv}}{\overline{E}^c_{vh}} = \frac{\overline{R}_h\overline{T}_v}{\overline{R}_v\overline{T}_h} \tag{5.56}$$

$$K_2 = \overline{E}^0_{vv}\overline{E}^0_{hh} = \frac{K^2}{r_0^4}e^{i4k_0r_0}\overline{R}_v\overline{T}_v\overline{R}_h\overline{T}_h p_0^2. \tag{5.57}$$

Using these expressions with (5.49) and (5.50), we obtain the following relations for the cross-polarized scattering amplitudes of the test target:

$$\overline{P}^u_{hv} = \frac{\overline{E}^u_{hv}}{\sqrt{K_1K_2}}\left(\frac{r_u}{r_0}\right)^2 e^{-i2k(r_0-r_u)}p_0 \tag{5.58}$$

$$\overline{P}^u_{vh} = \sqrt{\frac{K_1}{K_2}}\,\overline{E}^u_{vh}\left(\frac{r_u}{r_0}\right)^2 e^{-i2k(r_0-r_u)}p_0. \tag{5.59}$$

Equations (5.51), (5.52), (5.58), and (5.59) provide expressions for the four scattering amplitudes of the unknown target. Note that knowledge of the scattering amplitude of the second calibration target is not required.

If the radar antennas individually do not have good polarization isolation between their $v$- and $h$-ports, we need to characterize each antenna by a polarization distortion matrix that accounts for the coupling between the two ports, and to use at least two, preferably three, targets with known scattering matrices to calibrate the radar completely. This case is the subject of the next section.

### 5-3.3 Nonreciprocal Systems with Full Distortion Matrices

Consider measurements of the form in (5.34) conducted for three different targets with known scattering matrices. Using subscripts to denote the corresponding target and measurement scattering matrices, we obtain three matrix equations;

$$\overline{\mathcal{N}}_j = c\, e^{i\phi_j} \overline{\mathcal{R}}\, \overline{\mathcal{P}}_j \overline{\mathcal{T}}, \quad \text{with} \quad j = 1, 2, 3 \tag{5.60}$$

where $\overline{\mathcal{N}}_j$ and $\overline{\mathcal{P}}_j$ are known matrices, but $\overline{\mathcal{R}}$ and $\overline{\mathcal{T}}$ are unknown. Notice that a subscript is also used on the absolute phase to account for the positioning of the targets. The phase centers of the three targets will generally be located at different distances from the radar.

The following derivation requires that the target scattering matrix $\overline{\mathcal{P}}_j$ be invertible for at least one of the targets. Without loss of generality, we assume that this requirement is satisfied with the first target. Premultiplying both $\overline{\mathcal{N}}_2$ and $\overline{\mathcal{N}}_3$ by $\overline{\mathcal{N}}_1^{-1}$ and denoting the products as $\mathcal{N}_T$ and $\underline{\mathcal{N}}_T$, we obtain the similarity transformations:

$$\begin{aligned} \mathcal{N}_T &= e^{i(\phi_2-\phi_1)} \overline{\mathcal{T}}^{-1} \mathcal{P}_T \overline{\mathcal{T}} &(5.61)\\ \underline{\mathcal{N}}_T &= e^{i(\phi_3-\phi_1)} \overline{\mathcal{T}}^{-1} \underline{\mathcal{P}}_T \overline{\mathcal{T}}, &(5.62)\end{aligned}$$

where $\mathcal{N}_T = \overline{\mathcal{N}}_1^{-1}\overline{\mathcal{N}}_2$, $\underline{\mathcal{N}}_T = \overline{\mathcal{N}}_1^{-1}\overline{\mathcal{N}}_3$, $\mathcal{P}_T = \overline{\mathcal{P}}_1^{-1}\overline{\mathcal{P}}_2$, and $\underline{\mathcal{P}}_T = \overline{\mathcal{P}}_1^{-1}\overline{\mathcal{P}}_3$.

We now consider an important property relating the eigenvalues and eigenvectors of similar matrices (Gere and Weaver, 1983, pp. 165–166). The eigenvalues and eigenvectors of the matrices $\mathcal{N}_T$ and $\mathcal{P}_T$ in (5.61) satisfy the relations:

$$\begin{aligned} \mathcal{P}_T \mathcal{X}_T &= \mathcal{X}_T \Lambda'_T &(5.63)\\ \mathcal{N}_T \mathcal{Y}_T &= \mathcal{Y}_T \Lambda_T, &(5.64)\end{aligned}$$

where $\Lambda'_T$ and $\Lambda_T$ are the diagonal matrices composed of the eigenvalues of $\mathcal{P}_T$ and $\mathcal{N}_T$, respectively. The corresponding eigenvectors of $\mathcal{P}_T$ and $\mathcal{N}_T$

form the columns of $\mathcal{X}_T$ and $\mathcal{Y}_T$, respectively. The eigenvalue matrices are related by the expression (Gere and Weaver, 1983, pp. 165–166):

$$\mathbf{\Lambda}_T' = \mathbf{\Lambda}_T \; e^{i(\phi_1 - \phi_2)}. \tag{5.65}$$

Notice that the propagation phase difference, $\phi_1 - \phi_2$, between the phase centers of any two known targets can be determined by using (5.65), although this fact is not used in the present development. Equations (5.63) to (5.65) state that the eigenvalues of similar matrices are equal. Furthermore, the eigenvectors of $\mathcal{P}_T$ and $\mathcal{N}_T$ are related by the expression:

$$\mathcal{Y}_T = \overline{\mathcal{T}}^{-1} \mathcal{X}_T. \tag{5.66}$$

However, (5.66) does not uniquely specify $\overline{\mathcal{T}}$ because the eigenvectors comprising $\mathcal{X}_T$ and $\mathcal{Y}_T$ have arbitrary scale factors. Upon independently solving the eigenvalue problems in (5.63) and (5.64) for the matrices $\mathcal{X}_T$ and $\mathcal{Y}_T$ (arbitrarily choosing the scale factors), the matrix $\overline{\mathcal{T}}$ is uniquely specified by

$$\mathcal{Y}_T = \overline{\mathcal{T}}^{-1} \mathcal{X}_T \mathcal{C} \tag{5.67}$$

or

$$\overline{\mathcal{T}} = \mathcal{X}_T \mathcal{C} \mathcal{Y}_T^{-1}, \tag{5.68}$$

where $\mathcal{C}$ is the diagonal matrix with elements $c_1 \neq 0$ and $c_2 \neq 0$ on the diagonal.

If the eigenvalues of $\mathcal{P}_T$ are distinct, then the corresponding eigenvectors are linearly independent. Therefore, the matrix $\mathcal{X}_T$ has rank two, which means that it is nonsingular and invertible (Gere and Weaver, 1983, p. 149). From (5.67), $\mathcal{Y}_T$ is invertible if and only if $\mathcal{X}_T$ is invertible. We can easily show that distinct eigenvalues are obtained when

$$4\Delta(\mathcal{P}_T) \neq [\mathrm{tr}\,(\mathcal{P}_T)]^2. \tag{5.69}$$

The notations $\Delta(\cdot)$ and $\mathrm{tr}\,(\cdot)$ denote the determinant and the trace, respectively.

In the same way, the eigenvalues and eigenvectors of the matrices $\underline{\mathcal{N}}_T$ and $\underline{\mathcal{P}}_T$ in (5.62) satisfy the relations:

$$\underline{\mathcal{P}}_T \underline{\mathcal{X}}_T = \underline{\mathcal{X}}_T \underline{\mathbf{\Lambda}}_T' \tag{5.70}$$

$$\underline{\mathcal{N}}_T \underline{\mathcal{Y}}_T = \underline{\mathcal{Y}}_T \underline{\mathbf{\Lambda}}_T, \tag{5.71}$$

where $\underline{\Lambda}'_T$ and $\underline{\Lambda}_T$ are composed of the eigenvalues, and $\underline{\mathcal{X}}_T$ and $\underline{\mathcal{Y}}_T$ are matrices with columns given by the corresponding eigenvectors. The eigenvalues of $\underline{\mathcal{P}}_T$ and $\underline{\mathcal{N}}_T$ are related by the expression:

$$\underline{\Lambda}'_T = \underline{\Lambda}_T \, e^{i(\phi_1 - \phi_3)}. \tag{5.72}$$

From these results, we obtain another matrix equation for $\overline{\mathcal{T}}$:

$$\overline{\mathcal{T}} = \underline{\mathcal{X}}_T \underline{\mathcal{C}} \, \underline{\mathcal{Y}}_T^{-1}, \tag{5.73}$$

where $\underline{\mathcal{C}}$ is the diagonal matrix with elements $\underline{c}_1 \neq 0$ and $\underline{c}_2 \neq 0$ on the diagonal. Here, we have assumed that

$$4\Delta(\underline{\mathcal{P}}_T) \neq [\mathrm{tr}\,(\underline{\mathcal{P}}_T)]^2 . \tag{5.74}$$

Equating (5.68) and (5.73), we obtain

$$\mathcal{X}_T \mathcal{C} \mathcal{Y}_T^{-1} = \underline{\mathcal{X}}_T \underline{\mathcal{C}} \, \underline{\mathcal{Y}}_T^{-1}, \tag{5.75}$$

and, after rearranging, (5.75) becomes

$$\mathcal{C} \mathcal{Y}_T^{-1} \underline{\mathcal{Y}}_T = \mathcal{X}_T^{-1} \underline{\mathcal{X}}_T \underline{\mathcal{C}}. \tag{5.76}$$

Expanding (5.76) and writing it in terms of four scalar equations, we have

$$c_1 \Delta(\mathcal{X}_T)(y_{22}\underline{y}_{11} - y_{12}\underline{y}_{21}) = \underline{c}_1 \Delta(\mathcal{Y}_T)(x_{22}\underline{x}_{11} - x_{12}\underline{x}_{21}) \tag{5.77}$$

$$c_1 \Delta(\mathcal{X}_T)(y_{22}\underline{y}_{12} - y_{12}\underline{y}_{22}) = \underline{c}_2 \Delta(\mathcal{Y}_T)(x_{22}\underline{x}_{12} - x_{12}\underline{x}_{22}) \tag{5.78}$$

$$c_2 \Delta(\mathcal{X}_T)(y_{11}\underline{y}_{21} - y_{21}\underline{y}_{11}) = \underline{c}_1 \Delta(\mathcal{Y}_T)(x_{11}\underline{x}_{21} - x_{21}\underline{x}_{11}) \tag{5.79}$$

$$c_2 \Delta(\mathcal{X}_T)(y_{11}\underline{y}_{22} - y_{21}\underline{y}_{12}) = \underline{c}_2 \Delta(\mathcal{Y}_T)(x_{11}\underline{x}_{22} - x_{21}\underline{x}_{12}), \tag{5.80}$$

where $x_{mn}$, $y_{mn}$, $\underline{x}_{mn}$, and $\underline{y}_{mn}$ are the elements of the matrices $\mathcal{X}_T$, $\mathcal{Y}_T$, $\underline{\mathcal{X}}_T$, and $\underline{\mathcal{Y}}_T$, respectively. Assuming that (5.77) to (5.80) are all nonzero, we can obtain two expressions for $c_2/c_1$ and two for $\underline{c}_2/\underline{c}_1$:

$$\frac{c_2}{c_1} = \frac{(x_{11}\underline{x}_{21} - x_{21}\underline{x}_{11})(y_{22}\underline{y}_{11} - y_{12}\underline{y}_{21})}{(x_{22}\underline{x}_{11} - x_{12}\underline{x}_{21})(y_{11}\underline{y}_{21} - y_{21}\underline{y}_{11})} \tag{5.81}$$

$$\frac{c_2}{c_1} = \frac{(x_{11}\underline{x}_{22} - x_{21}\underline{x}_{12})(y_{22}\underline{y}_{12} - y_{12}\underline{y}_{22})}{(x_{22}\underline{x}_{12} - x_{12}\underline{x}_{22})(y_{11}\underline{y}_{22} - y_{21}\underline{y}_{12})} \tag{5.82}$$

$$\frac{\underline{c}_2}{\underline{c}_1} = \frac{(x_{22}\underline{x}_{11} - x_{12}\underline{x}_{21})(y_{22}\underline{y}_{12} - y_{12}\underline{y}_{22})}{(x_{22}\underline{x}_{12} - x_{12}\underline{x}_{22})(y_{22}\underline{y}_{11} - y_{12}\underline{y}_{21})} \tag{5.83}$$

$$\frac{\underline{c}_2}{\underline{c}_1} = \frac{(x_{11}\underline{x}_{21} - x_{21}\underline{x}_{11})(y_{11}\underline{y}_{22} - y_{21}\underline{y}_{12})}{(x_{11}\underline{x}_{22} - x_{21}\underline{x}_{12})(y_{11}\underline{y}_{21} - y_{21}\underline{y}_{11})}. \tag{5.84}$$

If only one equation is zero, then either (5.81) or (5.82) for $c_2/c_1$ will be indeterminate. The same applies for either (5.83) or (5.84) for $\underline{c}_2/\underline{c}_1$. However, the remaining two expressions can be used to determine both $c_2/c_1$ and $\underline{c}_2/\underline{c}_1$.

Based on the constraints in (5.69) and (5.74), we know that $\mathcal{X}_T^{-1}\underline{\mathcal{X}}_T$ is invertible and has a nonzero determinant. Hence, if two equations are zero, they must be either the diagonal elements or the off-diagonal elements of $\mathcal{X}_T^{-1}\underline{\mathcal{X}}_T$. In this situation, (5.81) to (5.84) are all indeterminate, and a solution for $\overline{\mathcal{T}}$ cannot be found. We must therefore require that

$$\underline{\mathcal{X}}_T \neq \mathcal{X}_T\mathcal{U} \quad \text{and} \quad \underline{\mathcal{X}}_T \neq \mathcal{X}_T\mathcal{V}, \tag{5.85}$$

where

$$\mathcal{U} = \begin{bmatrix} u_1 & 0 \\ 0 & u_2 \end{bmatrix}, \qquad \mathcal{V} = \begin{bmatrix} 0 & v_1 \\ v_2 & 0 \end{bmatrix}. \tag{5.86}$$

This constraint states that $\underline{\mathcal{P}}_T$ and $\mathcal{P}_T$ can have no more than one common eigenvector.

The transmitted distortion matrix $\overline{\mathcal{T}}$ can now be determined from either (5.68) or (5.73). Without loss of generality, we will show the result using (5.68). The first element of $\overline{\mathcal{T}}$ must be unity, and from (5.68) it is given by

$$\frac{1}{\Delta(\mathcal{Y}_T)}(c_1 x_{11} y_{22} - c_2 x_{12} y_{21}) = 1. \tag{5.87}$$

This expression can be used to obtain $c_1$ and $c_2$ in terms of the ratio $c_2/c_1$;

$$c_1 = \Delta(\mathcal{Y}_T)\left(x_{11}y_{22} - \frac{c_2}{c_1}x_{12}y_{21}\right)^{-1} \tag{5.88}$$

$$c_2 = \Delta(\mathcal{Y}_T)\left(\frac{c_1}{c_2}x_{11}y_{22} - x_{12}y_{21}\right)^{-1}. \tag{5.89}$$

The matrix $\overline{\mathcal{R}}$ can also be determined using a similar procedure. Post-multiplying both $\overline{\mathcal{N}}_2$ and $\overline{\mathcal{N}}_3$ by $\overline{\mathcal{N}}_1^{-1}$, we obtain

$$\mathcal{N}_R = e^{i(\phi_2-\phi_1)}\overline{\mathcal{R}}\mathcal{P}_R\overline{\mathcal{R}}^{-1} \tag{5.90}$$

$$\underline{\mathcal{N}}_R = e^{i(\phi_3-\phi_1)}\overline{\mathcal{R}}\underline{\mathcal{P}}_R\overline{\mathcal{R}}^{-1}, \tag{5.91}$$

where $\mathcal{N}_R = \overline{\mathcal{N}}_2\overline{\mathcal{N}}_1^{-1}$, $\underline{\mathcal{N}}_R = \overline{\mathcal{N}}_3\overline{\mathcal{N}}_1^{-1}$, $\mathcal{P}_R = \overline{\mathcal{P}}_2\overline{\mathcal{P}}_1^{-1}$, and $\underline{\mathcal{P}}_R = \overline{\mathcal{P}}_3\overline{\mathcal{P}}_1^{-1}$. Because (5.90) and (5.91) are again similarity transformations, we can uniquely specify $\overline{\mathcal{R}}$ by the relations:

$$\overline{\mathcal{R}} = \mathcal{X}_R\mathcal{D}\mathcal{Y}_R^{-1} \tag{5.92}$$

$$\overline{\mathcal{R}} = \underline{\mathcal{X}}_R\underline{\mathcal{D}}\underline{\mathcal{Y}}_R^{-1}, \tag{5.93}$$

where the eigenvector matrices $\mathcal{X}_R$, $\mathcal{Y}_R$, $\underline{\mathcal{X}}_R$, and $\underline{\mathcal{Y}}_R$ satisfy the relations:

$$\mathcal{N}_R\mathcal{X}_R = \mathcal{X}_R\Lambda'_R \quad (5.94)$$

$$\overline{\mathcal{P}}_R\mathcal{Y}_R = \mathcal{Y}_R\Lambda_R \quad (5.95)$$

$$\underline{\mathcal{N}}_R\underline{\mathcal{X}}_R = \underline{\mathcal{X}}_R\underline{\Lambda}'_R \quad (5.96)$$

$$\underline{\mathcal{P}}_R\underline{\mathcal{Y}}_R = \underline{\mathcal{Y}}_R\underline{\Lambda}_R, \quad (5.97)$$

and the eigenvalue matrices are related by the expressions:

$$\Lambda'_R = \Lambda_R\, e^{i(\phi_2-\phi_1)} \quad (5.98)$$

$$\underline{\Lambda}'_R = \underline{\Lambda}_R\, e^{i(\phi_3-\phi_1)}. \quad (5.99)$$

The matrices $\mathcal{D}$ and $\underline{\mathcal{D}}$ are analogous to $\mathcal{C}$ and $\underline{\mathcal{C}}$; they are diagonal with elements ($d_1 \neq 0$, $d_2 \neq 0$) and ($\underline{d}_1 \neq 0$, $\underline{d}_2 \neq 0$), respectively. Equations (5.92) and (5.93) are of the same form as (5.68) and (5.73); therefore, (5.81) to (5.89) can be used to find expressions for $d_1$ and $d_2$ by replacing $c$ with $d$ and letting $x_{mn}$, $y_{mn}$, $\underline{x}_{mn}$, and $\underline{y}_{mn}$ denote the elements of the matrices $\mathcal{X}_R$, $\mathcal{Y}_R$, $\underline{\mathcal{X}}_R$, and $\underline{\mathcal{Y}}_R$, respectively.

Because $\mathcal{P}_R = \mathcal{P}_T^T$ and $\underline{\mathcal{P}}_R = \underline{\mathcal{P}}_T^T$ due to reciprocity, the constraints on $\mathcal{P}_T$ and $\underline{\mathcal{P}}_T$ are sufficient to guarantee a solution for $\overline{\mathcal{R}}$ as well as $\overline{\mathcal{T}}$. The proof uses the relationship between the eigenvalues and eigenvectors of a matrix and its transpose (Gere and Weaver, 1983, p. 164).

With the distortion matrices known, the absolute magnitude can be obtained by substituting back into one of the original measurements. Equating the elements of the matrices on both sides of (5.60) and taking the magnitude, we obtain for the $mn$th element:

$$|c| = \frac{|(\mathcal{N}_j)_{mn}|}{|(\overline{\mathcal{R}}\,\overline{\mathcal{P}}_j\overline{\mathcal{T}})_{mn}|}. \quad (5.100)$$

The best estimate of $|c|$ will be obtained by choosing the target for which the theoretical matrix $\overline{\mathcal{P}}_j$ is most accurate.

Using (5.100), the scattering matrix $\overline{\mathcal{P}}^u$ for an unknown point target can be written in terms of the measured scattering matrix $\overline{\mathcal{N}}^u$:

$$\overline{\mathcal{P}}^u = e^{-i\phi'}\frac{1}{|c|}\overline{\mathcal{R}}^{-1}\overline{\mathcal{N}}^u\overline{\mathcal{T}}^{-1}, \quad (5.101)$$

where $\phi'$ is the unknown absolute phase given by

$$\phi' = \phi + \tan^{-1}\left(\frac{\mathrm{Im}\{c\}}{\mathrm{Re}\{c\}}\right). \quad (5.102)$$

### 5-3.4 Reciprocal Systems with Full Distortion Matrices

In general, single-antenna radar systems, like dual-antenna systems, have different distortion matrices for transmitting and receiving. Although the single antenna affects the transmitted and received waves in a similar manner, the remaining portions of the transmitting and receiving paths through the system affect them differently. Therefore, (5.34) should be used to describe the measured scattering matrix for a general single-antenna system, and the technique of Section 5-3.3 should be used in calibration.

In some cases, the antenna assembly is the major contributor to the distortion errors, and the contributions due to the different transmitting and receiving paths is negligible. With such systems, care must be taken to make the transmitting and receiving paths practically identical, and then only antenna effects need to be considered. The measured scattering matrix can then be described with (5.36), and a slightly different technique can be used in calibration. The major advantage of the technique is that only two known targets need to be measured to calibrate fully the measurement system. The technique is described in the following development.

Consider measurements of the form in (5.36) conducted for two different targets with known scattering matrices. Using subscripted notation similar to that of (5.60), the measured scattering matrix of the $j$th target is

$$\overline{\mathcal{N}}_j = c\,e^{i\phi_j}\widetilde{\overline{\mathcal{A}}}\,\overline{\mathcal{P}}_j\overline{\mathcal{A}}; \quad \text{with} \quad j = 1, 2, \tag{5.103}$$

where $\overline{\mathcal{P}}_j$ is the known scattering matrix of the target. The unknown distortion matrix is $\overline{\mathcal{A}}$. The calibration technique to be described requires a more restrictive condition on the form of the known scattering matrices than does the general technique of Section 5-3.3. With the present technique, both calibration targets must have invertible scattering matrices, whereas only one was required with the general technique.

Forming the products $\mathcal{N} = \overline{\mathcal{N}}_2^{-1}\overline{\mathcal{N}}_1$ and $\underline{\mathcal{N}} = \widetilde{\overline{\mathcal{N}}}_1^{-1}\widetilde{\overline{\mathcal{N}}}_2$, we obtain the two similarity transformations:

$$\mathcal{N} = e^{i(\phi_1-\phi_2)}\overline{\mathcal{A}}^{-1}\mathcal{P}\,\overline{\mathcal{A}} \tag{5.104}$$

$$\underline{\mathcal{N}} = e^{i(\phi_2-\phi_1)}\overline{\mathcal{A}}^{-1}\underline{\mathcal{P}}\,\overline{\mathcal{A}}, \tag{5.105}$$

where $\mathcal{P} = \overline{\mathcal{P}}_2^{-1}\overline{\mathcal{P}}_1$ and $\underline{\mathcal{P}} = \widetilde{\overline{\mathcal{P}}}_1^{-1}\widetilde{\overline{\mathcal{P}}}_2$. This method can be applied to the specialized single-antenna system only because the transmitter and receiver distortion matrices are related by a transpose. In a manner similar to that in Section 5-3.3, we can uniquely specify $\overline{\mathcal{A}}$ by the relations:

$$\overline{\mathcal{A}} = \mathcal{X}\,\mathcal{G}\,\mathcal{Y}^{-1} \tag{5.106}$$

$$\overline{\mathcal{A}} = \underline{\mathcal{X}}\,\underline{\mathcal{G}}\,\underline{\mathcal{Y}}^{-1}, \tag{5.107}$$

where the matrices $\mathcal{G}$ and $\underline{\mathcal{G}}$ are diagonal with elements ($g_1 \neq 0$, $g_2 \neq 0$) and ($\underline{g}_1 \neq 0$, $\underline{g}_2 \neq 0$), respectively. The eigenvector matrices $\mathcal{X}$, $\mathcal{Y}$, $\underline{\mathcal{X}}$, and $\underline{\mathcal{Y}}$ satisfy the expressions:

$$\mathcal{P}\,\mathcal{X} = \mathcal{X}\,\Lambda' \tag{5.108}$$

$$\mathcal{N}\,\mathcal{Y} = \mathcal{Y}\,\Lambda \tag{5.109}$$

$$\underline{\mathcal{P}}\,\underline{\mathcal{X}} = \underline{\mathcal{X}}\,\underline{\Lambda}' \tag{5.110}$$

$$\underline{\mathcal{N}}\,\underline{\mathcal{Y}} = \underline{\mathcal{Y}}\,\underline{\Lambda}, \tag{5.111}$$

where the eigenvalue matrices $\Lambda'$, $\Lambda$, $\underline{\Lambda}'$, and $\underline{\Lambda}$ are related by

$$\Lambda' = \Lambda\, e^{i(\phi_1 - \phi_2)} \tag{5.112}$$

$$\underline{\Lambda}' = \underline{\Lambda}\, e^{i(\phi_2 - \phi_1)}. \tag{5.113}$$

Equating (5.106) and (5.107) and then rearranging, we obtain

$$\mathcal{G}\mathcal{Y}^{-1}\underline{\mathcal{Y}} = \mathcal{X}^{-1}\underline{\mathcal{X}}\,\underline{\mathcal{G}}. \tag{5.114}$$

Equation (5.114) is of the same form as (5.76), so $g_1$ and $g_2$ are given by (5.88) and (5.89) with $g$ replacing $c$. Here, the elements $x_{mn}$, $y_{mn}$, $\underline{x}_{mn}$, and $\underline{y}_{mn}$ denote the elements of the matrices $\mathcal{X}$, $\mathcal{Y}$, $\underline{\mathcal{X}}$, and $\underline{\mathcal{Y}}$, respectively. As in Section 5-3.3, we require the following constraints:

$$4\Delta(\mathcal{P}) \neq [\mathrm{tr}\,(\mathcal{P})]^2 \tag{5.115}$$

$$4\Delta(\underline{\mathcal{P}}) \neq [\mathrm{tr}\,(\underline{\mathcal{P}})]^2 \tag{5.116}$$

$$\underline{\mathcal{X}} \neq \mathcal{X}\mathcal{U} \quad \text{and} \quad \underline{\mathcal{X}} \neq \mathcal{X}\mathcal{V}, \tag{5.117}$$

where $\mathcal{U}$ and $\mathcal{V}$ are of the form in (5.86).

By substituting the matrix $\overline{\mathcal{A}}$ resulting from (5.106) into (5.103) and equating the elements on both sides, we obtain for the $mn$th element:

$$|c| = \frac{|(\mathcal{N}_j)_{mn}|}{|(\widetilde{\overline{\mathcal{A}}}\,\overline{\mathcal{P}}_j\,\overline{\mathcal{A}})_{mn}|}. \tag{5.118}$$

As in the general technique, the best estimate of $|c|$ will be obtained by using the known target for which the theoretical matrix $\overline{\mathcal{P}}_j$ is most accurate.

We now have the effect of the distortion matrices determined to within an unknown phase factor. The scattering matrix $\overline{\mathcal{P}}^u$ for some unknown point target can be written as

$$\overline{\mathcal{P}}^u = e^{-i\phi'} \frac{1}{|c|} \widetilde{\overline{\mathcal{A}}}^{-1} \overline{\mathcal{N}}^u \overline{\mathcal{A}}^{-1}, \tag{5.119}$$

where $\phi'$ is the unknown absolute phase given by

$$\phi' = \phi + \tan^{-1}\left(\frac{\mathrm{Im}\{c\}}{\mathrm{Re}\{c\}}\right). \tag{5.120}$$

### 5-3.5 Comparison with Measured Data

In Sections 5-3.2 and 5-3.3, we introduced two calibration techniques. The technique described in Section 5-3.2 was based on the assumption that the radar system has perfect isolation between the $v$- and $h$-polarized channels. We shall refer to this technique as the *isolated antenna calibration technique* (IACT). The technique described in Section 5-3.3 was a general technique which required measurements of three targets with arbitrary but known scattering matrices. We shall refer to this technique as the *generalized calibration technique* (GCT).

The performance of these two calibration techniques has been compared using the same measurement system and calibration targets (Whitt *et al.*, 1990). In each case, the radar was calibrated and then used to measure the scattering matrix of a test target over the 9–10 GHz operating frequency range of the system. The measured scattering matrix of the test target was then compared to the theoretical scattering matrix. Both the GCT and the IACT used a conducting sphere and a conducting cylinder oriented at 45° relative to vertical for calibration targets. The diameter of the sphere was 15 cm, and the length and diameter of the cylinder were 27.1 cm and 0.79 cm, respectively. The third calibration target required by the GCT was realized by orienting the cylinder horizontally. Two test targets were considered, a conducting sphere and a conducting cylinder oriented vertically. The diameter of the sphere was 20.3 cm, and the length and diameter of the cylinder were 27.1 cm and 0.79 cm, respectively. Table 5.1 provides a summary of the calibration and test targets used for each technique.

The measured results for the conducting sphere are given in Fig. 5.23. Figures 5.23(a) and (b) show spectral plots of the co-polarized RCS $\sigma_{vv}$ and $\sigma_{hh}$, respectively. Each figure contains a plot of the RCS computed by using the standard Mie solution and the measured RCS after calibration by using

**Table 5.1** Calibration and Test Targets Used

| Targets | GCT | IACT |
|---|---|---|
| Cal Target 1 | 15-cm diameter metal sphere | 15-cm diameter metal sphere |
| Cal Target 2 | 45° cylinder, l=27.1 cm d=0.79 cm | 45° cylinder, l=27.1 cm d=0.79 cm |
| Cal Target 3 | horizontal cylinder, l=27.1 cm d=0.79 cm | NONE |
| Test Target 1 | 20.3-cm diameter metal sphere | 20.3-cm diameter metal sphere |
| Test Target 2 | vertical cylinder, l=27.1 cm, d=0.79 cm | vertical cylinder, l=27.1 cm, d=0.79 cm |

either the IACT or the GCT. We observe that the IACT and GCT results are identical and within 0.3 dB of the theoretical value. Figure 5.23(c) shows the "measured" cross-polarized RCS $\sigma_{vh}$. According to theory, a conducting sphere does not depolarize; therefore, $\sigma_{vh}$ should be zero (or $-\infty$ on a dB scale). For a given calibration technique, the measured value of $\sigma_{vh}$, relative to $\sigma_{vv}$ and $\sigma_{hh}$, represents the *effective polarization isolation* provided by that technique. From the data in Fig. 5.23(a–c), the effective polarization isolation is found to be $\approx$ 30 dB better for the GCT than for the IACT. The effective polarization isolation is $\approx$ 50 dB for the GCT and $\approx$ 20 dB for the IACT, which was approximately the polarization isolation of the antenna. According to Fig. 5.23(d), the measured phase of $\overline{P}_{hh}$ relative to that of $\overline{P}_{vv}$, namely $(\phi_{hh} - \phi_{vv})$, is within 2° of the theoretical value for both techniques.

The second test target was a vertical cylinder, and the measured results for $\sigma_{vv}$ and $\sigma_{hh}$ shown in Fig. 5.24(a) and (b) lead to the same conclusions as obtained with the sphere. Figure 5.24(c) presents the vertical cylinder results for $(\phi_{hh} - \phi_{vv})$.

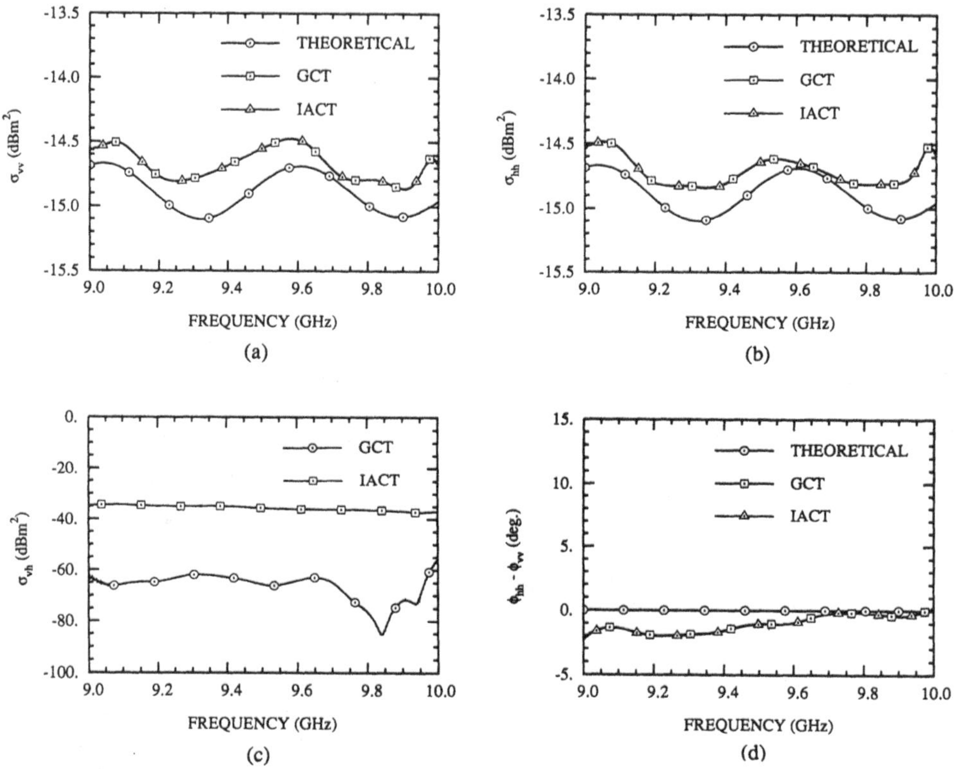


**Figure 5.23** Plots of the theoretical and measured spectra for the metal sphere with diameter $d = 20.3$ cm: (a) $\sigma_{vv}$, (b) $\sigma_{hh}$, (c) $\sigma_{vh}$, and (d) $\phi_{hh} - \phi_{vv}$.

Overall, both the IACT and GCT provide excellent results, with amplitude and phase errors of less than 0.3 dB and 3°, respectively. The IACT, however, does not significantly improve the polarization isolation of the radar, while the GCT provides an effective polarization isolation of $\approx$ 30 dB better than that of the uncalibrated radar. The effective polarization isolation using the GCT is on the order of 50 dB. Except when measuring targets with cross-polarized scattering amplitudes ($\overline{P}_{vh}$ and $\overline{P}_{hv}$) that are much larger than their co-polarized scattering amplitudes, the IACT provides results that are essentially identical with those observed for the GCT. The effective polarization isolation obtained with the IACT, however, is no better than for the uncalibrated system.

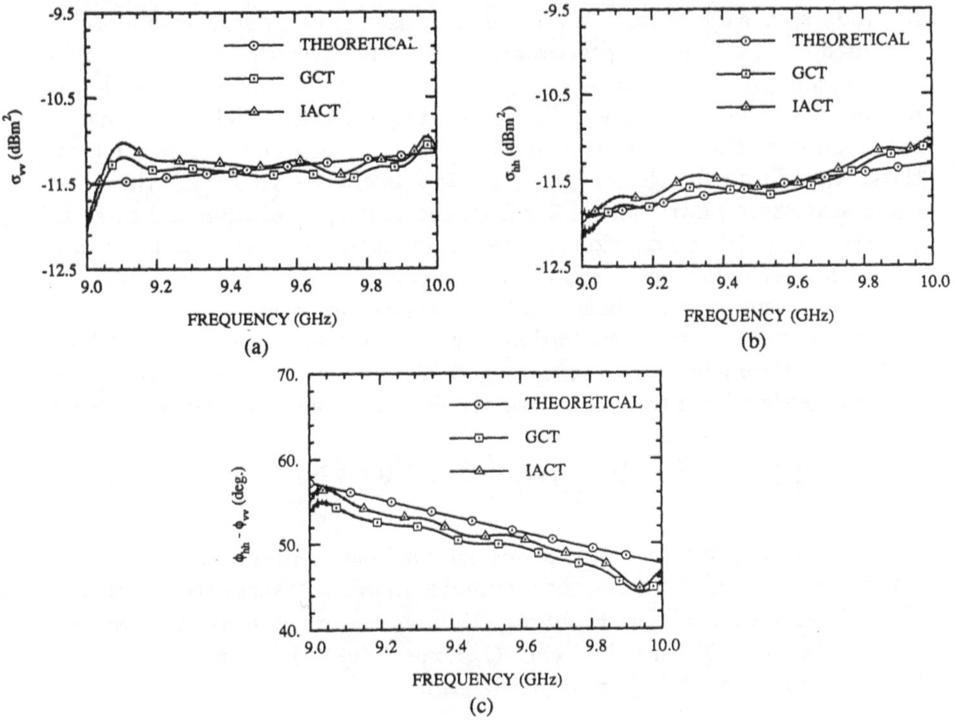


**Figure 5.24** Plots of the theoretical and measured spectra for the metal vertical cylinder with length $\ell = 27.1$cm and diameter $d = 0.79$ cm: (a) $\sigma_{vv}$, (b) $\sigma_{hh}$, and (c) $\phi_{hh} - \phi_{vv}$.

# 5-4 SYNTHESIZED POLARIZATION RESPONSE OF DISTRIBUTED TARGETS

The concept of polarization synthesis was introduced in Chapter 2, where it was discussed for point targets and distributed targets as measured by imaging radars. In these two cases, the far-field illumination over the extent of the target can be considered as a uniform plane wave. If the angle subtended by the point target is small relative to the antenna beamwidth, the illumination can be assumed uniform over the entire target. For imaging radars, a wide-beamwidth antenna is used, and spatial resolution is obtained with additional processing. Therefore, over the size of one resolution cell, the illumination is again relatively uniform. Because this is not the case when measuring distributed targets with a field polarimetric scatterometer, the polarization synthesis techniques differ somewhat from those introduced in Chapter 2. The statistics associated with the measured scattering matrix also differ from the uniformly illuminated case. In this section, both polarization synthesis and scattering matrix statistics are considered for polarimetric scatterometer measurements, and examples are given for scattering from both random surfaces and volumes.

Consider a random distribution of scatterers illuminated by a polarimetric scatterometer as shown in Fig. 5.25. The electric field incident on the $j$th particle for a transmitter power $P_t$ and polarization $\overline{\mathbf{p}}^t$ is given by

$$\overline{\mathbf{E}}_j^i = \sqrt{\frac{2\eta_0 P_t G_t}{4\pi r_j^2}}\, \overline{T}'_{vv}(\mathbf{r}_j)\, e^{ik_o r_j}\, e^{i\overline{\mathcal{K}}(\mathbf{r}_j) r_j}\, \overline{\mathcal{T}}(\mathbf{r}_j)\overline{\mathbf{p}}^t, \tag{5.121}$$

where $\mathbf{r}_j$ is the position vector of the $j$th particle relative to the transmitting antenna, $G_t$ is the boresight nominal gain of the transmitting antenna, $\overline{T}'_{vv}(\mathbf{r}_j)$ is the transmitter amplitude gain pattern for $v$ polarization normalized to the boresight nominal gain $G_t$, $\exp(i\overline{\mathcal{K}}(\mathbf{r}_j)r_j)$ is a field extinction term, which accounts for propagation, and

$$\overline{\mathcal{T}}(\mathbf{r}_j) = \begin{bmatrix} 1 & \overline{T}_{vh}(\mathbf{r}_j) \\ \overline{T}_{hv}(\mathbf{r}_j) & \overline{T}_{hh}(\mathbf{r}_j) \end{bmatrix} \tag{5.122}$$

is the transmitter distortion matrix defined relative to the $v$-polarized gain.

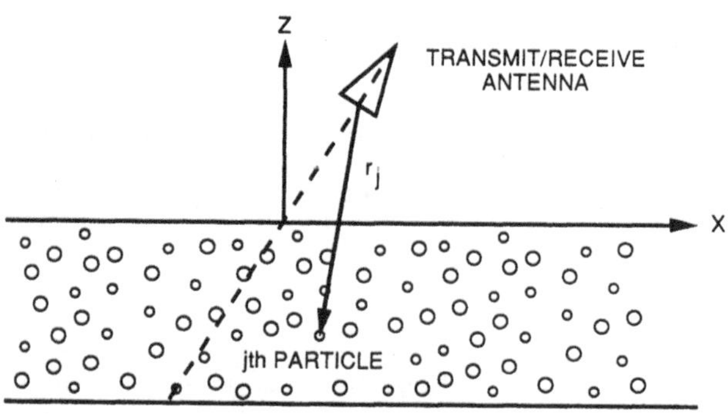


**Figure 5.25** Random distribution of particles illuminated by a scatterometer.

The matrix $\overline{\mathcal{K}}(\mathbf{r}_j)$ is a $2 \times 2$ complex matrix, the elements of which are functions of the position vector $\mathbf{r}_j$:

$$\overline{\mathcal{K}}(\mathbf{r}_j) = \begin{bmatrix} \overline{K}_{vv}(\mathbf{r}_j) & \overline{K}_{vh}(\mathbf{r}_j) \\ \overline{K}_{hv}(\mathbf{r}_j) & \overline{K}_{hh}(\mathbf{r}_j) \end{bmatrix}. \tag{5.123}$$

The notation $\exp(i\overline{\mathcal{K}}(\mathbf{r}_j)r_j)$ represents the matrix defined by

$$e^{i\overline{\mathcal{K}}(\mathbf{r}_j)r_j} = \overline{\mathcal{P}}\,\overline{\mathcal{D}}\,\overline{\mathcal{Q}} \tag{5.124}$$

where

$$\overline{\mathcal{P}} = \begin{bmatrix} \overline{p}_{11} & \overline{p}_{12} \\ \overline{p}_{21} & \overline{p}_{22} \end{bmatrix} \quad \overline{\mathcal{Q}} = \begin{bmatrix} \overline{q}_{11} & \overline{q}_{12} \\ \overline{q}_{21} & \overline{q}_{22} \end{bmatrix} \tag{5.125}$$

$$\overline{\mathcal{D}} = \begin{bmatrix} e^{i\overline{K}_1(\mathbf{r}_j)r_j} & 0 \\ 0 & e^{i\overline{K}_2(\mathbf{r}_j)r_j} \end{bmatrix}. \tag{5.126}$$

The matrices $\overline{\mathcal{P}}$ and $\overline{\mathcal{Q}}$ are composed of the normalized right-hand and left-hand eigenvectors of $\overline{\mathcal{K}}(\mathbf{r}_j)$, and $\overline{K}_n(\mathbf{r}_j)$ (with $n \in \{1, 2\}$) are the eigenvalues of $\overline{\mathcal{K}}(\mathbf{r}_j)$. Mathematically, this is stated as

$$\overline{\mathcal{K}}(\mathbf{r}_j) \cdot \hat{\mathbf{p}}_n = \overline{K}_n(\mathbf{r}_j)\hat{\mathbf{p}}_n \tag{5.127}$$

$$\hat{\mathbf{q}}_n \cdot \overline{\mathcal{K}}(\mathbf{r}_j) = \overline{K}_n(\mathbf{r}_j)\hat{\mathbf{q}}_n, \tag{5.128}$$

where the normalized right-hand and left-hand eigenvectors, respectively, are

$$\hat{\mathbf{p}}_n = \begin{bmatrix} \overline{p}_{1n} \\ \overline{p}_{n1} \end{bmatrix}, \qquad \hat{\mathbf{q}}_n = \begin{bmatrix} \overline{q}_{n1} \\ \overline{q}_{n2} \end{bmatrix}. \tag{5.129}$$

The scattered field at the receive antenna from the $j^{th}$ particle is

$$\overline{\mathbf{E}}_j^s = \sqrt{\frac{2\eta_0 P_t G_t}{4\pi r_j^4}}\, \overline{T}'_{vv}(\mathbf{r}_j)\, e^{i2k_o r_j}\, \overline{\mathcal{S}}_j \overline{\mathcal{T}}(\mathbf{r}_j)\overline{\mathbf{p}}^t, \tag{5.130}$$

where the scattering matrix for the $j$th particle is $\overline{\mathcal{S}}'(\mathbf{r}_j)$, and the matrix

$$\overline{\mathcal{S}}_j = e^{i\overline{\mathcal{K}}(-\mathbf{r}_j)r_j}\, \overline{\mathcal{S}}'_j\, e^{i\overline{\mathcal{K}}(\mathbf{r}_j)r_j} \tag{5.131}$$

represents the particle scattering matrix modified by the intervening medium.

Assuming single-particle scattering, the total scattered field is the coherent sum of the scattered fields from the $N$ particles. Accounting for the receiver characteristics in a manner similar to the transmitter, the voltage measured by the scatterometer for $rt$ polarization is therefore given by

$$\overline{V}_{rt} = c \sum_{j=1}^{N} g_r(\mathbf{r}_j) g_t(\mathbf{r}_j) \frac{r_0^2}{r_j^2}\, e^{i2k_0 r_j} \overline{\mathbf{p}}^r \cdot \overline{\mathcal{R}}(\mathbf{r}_j) \overline{\mathcal{S}}_j \overline{\mathcal{T}}(\mathbf{r}_j) \overline{\mathbf{p}}^t, \tag{5.132}$$

where $\overline{\mathbf{p}}^r$ is the receiver polarization, and the receiver distortion matrix is defined by

$$\overline{\mathcal{R}}(\mathbf{r}_j) = \begin{bmatrix} 1 & \overline{R}_{vh}(\mathbf{r}_j) \\ \overline{R}_{hv}(\mathbf{r}_j) & \overline{R}_{hh}(\mathbf{r}_j) \end{bmatrix}. \tag{5.133}$$

Similar to (5.35), the constant $c$ is defined as

$$c = \overline{R}'_{vv}(\mathbf{r}_0) \overline{T}'_{vv}(\mathbf{r}_0) \left[ \frac{2\eta_0 P_t G_t G_r \lambda^2}{(4\pi)^2 r_0^4} \right]^{1/2}, \tag{5.134}$$

where $r_0$ is the average boresight range, and

$$\overline{R}'_{vv}(\mathbf{r}_j) = \overline{R}'_{vv}(\mathbf{r}_0) g_r(\mathbf{r}_j) \tag{5.135}$$

$$\overline{T}'_{vv}(\mathbf{r}_j) = \overline{T}'_{vv}(\mathbf{r}_0) g_t(\mathbf{r}_j). \tag{5.136}$$

The quantities $\overline{T}'_{vv}(\mathbf{r}_0)$ and $\overline{R}'_{vv}(\mathbf{r}_0)$ are the boresight gain for transmitting and receiving, respectively; therefore, $g_t(\mathbf{r}_j) g_r(\mathbf{r}_j)$ is the normalized gain pattern for $vv$ polarization such that

$$g_r(\mathbf{r}_0) g_t(\mathbf{r}_0) = 1. \tag{5.137}$$

Equation (5.132) also represents the voltage received when illuminating a random surface. If the random surface is divided into $N$ independent facets or scattering centers as in Fig. 5.26, $\overline{\mathcal{S}}_j$ (for $j = 1$ to $N$) is the scattering matrix for the $j$th facet. Because scattering occurs only at the surface, the extinction factors used for volume scattering are not present.

In general, the transmitter and receiver distortion matrices are functions of the position vector $\mathbf{r}_j$, and they can be represented in terms of the distortion matrix elements for boresight by

$$\overline{\mathcal{T}}(\mathbf{r}_j) = \begin{bmatrix} 1 & g^t_{vh}(\mathbf{r}_j) \overline{T}_{vh}(\mathbf{r}_0) \\ g^t_{hv}(\mathbf{r}_j) \overline{T}_{hv}(\mathbf{r}_0) & g^t_{hh}(\mathbf{r}_j) \overline{T}_{hh}(\mathbf{r}_0) \end{bmatrix} \tag{5.138}$$

$$\overline{\mathcal{R}}(\mathbf{r}_j) = \begin{bmatrix} 1 & g^r_{vh}(\mathbf{r}_j) \overline{R}_{vh}(\mathbf{r}_0) \\ g^r_{hv}(\mathbf{r}_j) \overline{R}_{hv}(\mathbf{r}_0) & g^r_{hh}(\mathbf{r}_j) \overline{R}_{hh}(\mathbf{r}_0) \end{bmatrix}. \tag{5.139}$$

The real functions $g^t_{rt}(\mathbf{r}_j)$ and $g^r_{rt}(\mathbf{r}_j)$ with $rt \in \{vh, hv, hh\}$ are the co-polarized and cross-polarized gain patterns for the transmitter and receiver, normalized with respect to the gain pattern for $vv$ polarization. When $r = t$, the function represents the co-polarized pattern, and when $r \neq t$, it represents the cross-polarized pattern. It has been assumed that the relative phase between the distortion matrix elements is constant over the antenna beam, which is a reasonable approximation over the main lobe. If the normalized gain patterns and the distortion matrices at boresight are known, the distortion introduced by the transmitter and receiver is fully characterized. However, to use the information to separate the contributions due to antenna characteristics from those due to the random medium is difficult.

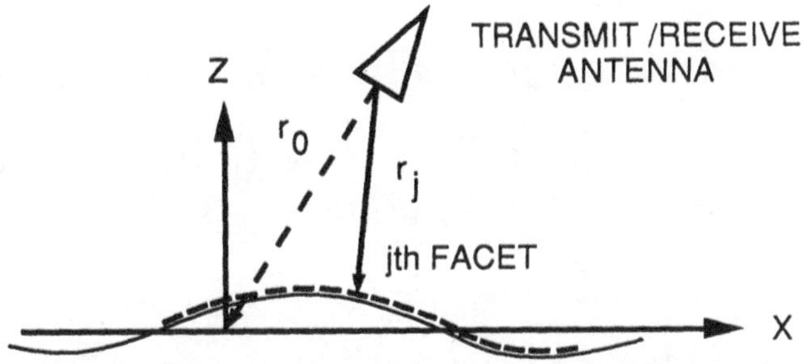


**Figure 5.26** Random surface illuminated by a scatterometer.

For many antennas, the normalized gain patterns for all combinations $rt \in \{vh, hv, hh\}$ are approximately equal to that of $vv$ over the main lobe; therefore, $g^t_{rt}(\mathbf{r}_j) \approx g^r_{rt}(\mathbf{r}_j) \approx 1$, and

$$\overline{\mathcal{T}}(\mathbf{r}_j) \approx \overline{\mathcal{T}}(\mathbf{r}_0) \tag{5.140}$$

$$\overline{\mathcal{R}}(\mathbf{r}_j) \approx \overline{\mathcal{R}}(\mathbf{r}_0). \tag{5.141}$$

Using this assumption and calibrating the scatterometer by using the techniques discussed in Section 5-3, the measured scattering matrix is

$$\overline{\mathcal{S}} = \sum_{j=1}^{N} g_r(\mathbf{r}_j) g_t(\mathbf{r}_j) \frac{r_0^2}{r_j^2} \, e^{i2k_0 r_j} \overline{\mathcal{S}}_j . \tag{5.142}$$

Because of the presence of the normalized pattern $g_r(\mathbf{r}_j)g_t(\mathbf{r}_j)$, the quantity $\overline{\mathcal{S}}$ is not independent of the scatterometer. With additional approximations, however, the target properties can be determined. The formulation for this process is somewhat different for volume scattering and surface scattering, and so the next two sections discuss these separately.

### 5-4.1 Surface Scattering

To compute the scattering properties of distributed targets for an arbitrary receiving-transmitting polarization combination, polarization synthesis techniques similar to those described in Chapter 2 are used. Because the scattering matrix measured for a single spatial sample of a distributed target is a function of scatterometer parameters, the techniques must be modified. As with the techniques in Chapter 2, however, the correlation of the scattering matrix elements in (5.142) must be computed. Assuming independence of the individual scatterers, the correlation is given by

$$\langle \overline{S}_{pq}\overline{S}^*_{mn}\rangle = \sum_{j=1}^{N} g_r^2(\mathbf{r}_j)g_t^2(\mathbf{r}_j)\frac{r_0^4}{r_j^4}\langle \overline{S}_{jpq}\overline{S}^*_{jmn}\rangle. \tag{5.143}$$

Since the correlation on the left of (5.143) must be related to the properties of the random surface, the quantity $\langle \overline{S}^0_{pq}(\overline{S}^0_{mn})^*\rangle$ is defined such that

$$\langle \overline{S}^0_{pq}(\overline{S}^0_{mn})^*\rangle = \frac{1}{\Delta A_j}\langle \overline{S}_{jpq}\overline{S}^*_{jmn}\rangle, \tag{5.144}$$

where $\Delta A_j$ is the horizontal area of the $j$th scatterer (or facet). When $pq = mn$, the above definition is equivalent to the definition

$$\sigma^0_{pq} = \frac{1}{\Delta A_j}\langle \sigma_{jpq}\rangle, \tag{5.145}$$

where $\sigma^0_{pq}$ is the differential scattering coefficient for $pq$ polarization, and the quantity $\langle \sigma_{jpq}\rangle$ is the average RCS. Because of this analogy, the quantity $\langle \overline{S}^0_{jpq}(\overline{S}^0_{jmn})^*\rangle$ will be called the differential correlation. Using (5.144) in (5.143) and converting the summation to an integral over the illumination, the correlation of the measured scattering matrix elements becomes

$$\langle \overline{S}_{pq}\overline{S}^*_{mn}\rangle = \int g_r^2(\mathbf{r})g_t^2(\mathbf{r})\frac{r_0^4}{r^4}\langle \overline{S}^0_{pq}(\overline{S}^0_{mn})^*\rangle \, \mathrm{d}A, \tag{5.146}$$

where $\mathrm{d}A$ is the differential area in the plane of the mean surface. If the beamwidth is sufficiently small, the differential correlation is approximately constant over the illumination, and

$$\langle \overline{S}_{pq}\overline{S}^*_{mn}\rangle \approx \langle \overline{S}^0_{pq}(\overline{S}^0_{mn})^*\rangle I_s(r_0,\theta_0), \tag{5.147}$$

where $\theta_0$ is the boresight incidence angle, and $I_s(r_0,\theta_0)$ is the surface illumination integral defined by

$$I_s(r_0,\theta_0) = \int g_r^2(\mathbf{r})g_t^2(\mathbf{r})\frac{r_0^4}{r^4}\, \mathrm{d}A. \tag{5.148}$$

If the normalized gain functions $g_r(\mathbf{r})$ and $g_t(\mathbf{r})$ are constant over the illuminated area $A$, then $I_s(r_0, \theta_0) \approx A$. Under these conditions, (2.75) to (2.78) of Section 2-3 are used to describe $\sigma^0$ as a function of polarization. However, for general scatterometer systems, these equations must be replaced by

$$\sigma^0_{rt}(\psi_r, \chi_r, \psi_t, \chi_t) = \frac{4\pi}{I_s(r_0, \theta_0)} \overline{\mathbf{A}}^r \cdot \langle \overline{\mathcal{M}} \rangle \overline{\mathbf{A}}^t \tag{5.149}$$

$$\sigma^0_{rt}(\psi_r, \chi_r, \psi_t, \chi_t) = \frac{4\pi}{I_s(r_0, \theta_0)} \overline{\mathbf{A}}^r_m \cdot \langle \overline{\mathcal{M}}_m \rangle \overline{\mathbf{A}}^t_m \tag{5.150}$$

$$\sigma^0_{rt}(\psi_r, \chi_r, \psi_t, \chi_t) = \frac{4\pi}{I_s(r_0, \theta_0)} \overline{\mathbf{B}}^* \cdot \langle \overline{\mathcal{C}} \rangle \overline{\mathbf{B}} \tag{5.151}$$

$$\sigma^0_{rt}(\psi_r, \chi_r, \psi_t, \chi_t) = \frac{4\pi}{I_s(r_0, \theta_0)} \langle | \overline{\mathbf{p}}^r \cdot \overline{\mathcal{S}} \overline{\mathbf{p}}^t |^2 \rangle, \tag{5.152}$$

where $\overline{\mathcal{M}}$, $\overline{\mathcal{M}}_m$, and $\overline{\mathcal{C}}$ are given in terms of the measured scattering matrix $\overline{\mathcal{S}}$ by (2.62), (2.65), and (2.73).

The relationship in (5.147) motivates the definition of a differential scattering matrix:

$$\overline{\mathcal{S}}^0 = \frac{1}{\sqrt{I_s(r_0, \theta_0)}} \overline{\mathcal{S}} \tag{5.153}$$

for the illuminated area of the distributed target. Here, $\overline{\mathcal{S}}$ is the scattering matrix measured by the scatterometer, and the division of $\overline{\mathcal{S}}$ by $\sqrt{I_s(r_0, \theta_0)}$ appoximately removes the dependence on system parameters, namely the nonuniform illumination across the antenna beam. Therefore, the scattering matrix $\overline{\mathcal{S}}^0$ represents a system-independent property of the target.

Related to the differential scattering matrix $\overline{\mathcal{S}}^0$, quantities similar to $\overline{\mathcal{M}}$, $\overline{\mathcal{M}}_m$, and $\overline{\mathcal{C}}$ can be defined by using the elements of $\overline{\mathcal{S}}^0$ instead of $\overline{\mathcal{S}}$. Using a superscript 0 to distinguish these differential matrices from those above, we write

$$\overline{\mathcal{M}}^0 = \widetilde{\mathcal{R}}^{-1} \overline{\mathcal{W}}^0 \mathcal{R}^{-1} \tag{5.154}$$

$$\overline{\mathcal{M}}^0_m = \widetilde{\mathcal{V}}^{-1} \overline{\mathcal{W}}^0 \mathcal{V}^{-1} \tag{5.155}$$

$$\overline{\mathcal{C}}^0 = \overline{\mathbf{X}}^0 (\widetilde{\overline{\mathbf{X}}}^0)^*, \tag{5.156}$$

where

$$\overline{\mathcal{W}}^0 = \begin{bmatrix} \overline{S}^0_{vv}(\overline{S}^0_{vv})^* & \overline{S}^0_{vh}(\overline{S}^0_{vh})^* & \overline{S}^0_{vh}(\overline{S}^0_{vv})^* & \overline{S}^0_{vv}(\overline{S}^0_{vh})^* \\ \overline{S}^0_{hv}(\overline{S}^0_{hv})^* & \overline{S}^0_{hh}(S^0_{hh})^* & \overline{S}^0_{hh}(\overline{S}^0_{hv})^* & \overline{S}^0_{hv}(\overline{S}^0_{hh})^* \\ \overline{S}^0_{hv}(\overline{S}^0_{vv})^* & \overline{S}^0_{hh}(S^0_{vh})^* & \overline{S}^0_{hh}(\overline{S}^0_{vv})^* & \overline{S}^0_{hv}(\overline{S}^0_{vh})^* \\ \overline{S}^0_{vv}(\overline{S}^0_{hv})^* & \overline{S}^0_{vh}(\overline{S}^0_{hh})^* & \overline{S}^0_{vh}(\overline{S}^0_{hv})^* & \overline{S}^0_{vv}(\overline{S}^0_{hh})^* \end{bmatrix} \tag{5.157}$$

$$\overline{\mathbf{X}}^0 = \begin{bmatrix} \overline{S}^0_{vv} \\ \overline{S}^0_{vh} \\ \overline{S}^0_{hv} \\ \overline{S}^0_{hh} \end{bmatrix}. \tag{5.158}$$

The matrices $\mathcal{R}$ and $\mathcal{V}$ are the same as those defined in Chapter 2.

Equations (5.149) to (5.152) can now be written in terms of the differential matrices $\overline{\mathcal{M}}^0$, $\overline{\mathcal{M}}^0_m$, $\overline{\mathcal{C}}^0$, and $\overline{\mathcal{S}}^0$ as

$$\sigma^0_{rt}(\psi_r, \chi_r, \psi_t, \chi_t) = 4\pi \overline{\mathbf{A}}^r \cdot \langle \overline{\mathcal{M}}^0 \rangle \overline{\mathbf{A}}^t \tag{5.159}$$

$$\sigma^0_{rt}(\psi_r, \chi_r, \psi_t, \chi_t) = 4\pi \overline{\mathbf{A}}^r_m \cdot \langle \overline{\mathcal{M}}^0_m \rangle \overline{\mathbf{A}}^t_m \tag{5.160}$$

$$\sigma^0_{rt}(\psi_r, \chi_r, \psi_t, \chi_t) = 4\pi \overline{\mathbf{B}}^* \cdot \langle \overline{\mathcal{C}}^0 \rangle \overline{\mathbf{B}} \tag{5.161}$$

$$\sigma^0_{rt}(\psi_r, \chi_r, \psi_t, \chi_t) = 4\pi \langle | \overline{\mathbf{p}}^r \cdot \overline{\mathcal{S}}^0 \overline{\mathbf{p}}^t |^2 \rangle. \tag{5.162}$$

## 5-4.2 Volume Scattering

The major difference with polarimetric scatterometer measurements of volumes and surfaces is the depth of penetration into the medium. For surfaces of lossy media such as wet soil, the scattering originates at the surface, and volume effects due to penetration within the medium are usually ignored. For volume scattering media such as vegetation and snow cover, the penetration may be significant and scattering within the volume may become the dominant backscatter contribution. With vegetation, for example, the interface between the air and the vegetation is usually assumed to be diffuse or nonscattering; that is, all the energy incident on the interface is transmitted into the vegetation medium. For some media, however, both the surface and volume scattering contributions must be considered. In the present discussion, surface scattering effects at the interface between free space and a volume of random particles will be ignored.

Ideally, a polarimetric scatterometer has a ranging capability such that the returns from the volume can be separated in range. The network analyzer-based polarimetric scatterometers described in Section 5-2 have this capability. If ranging is used, the scattering matrix measured by the scatterometer can be divided into the sum of scattering matrices, each representing the contribution from a different range "bin." The angular resolution of the illuminated volume is determined by the antenna beamwidths, and the range resolution is determined by the ranging capability as shown in Fig. 5.27. Suppose we consider only one of these scattering matrices due to contributions from the range bin located at range $r_l$ and with range extent $\Delta r_l$. The scattering matrix due to the $l$th range bin has the same form as (5.142):

$$\overline{\mathcal{S}}^l = \sum_{j=1}^{N^l} g_r(\mathbf{r}_j) g_t(\mathbf{r}_j) \frac{r_0^2}{r_j^2}\, e^{i2k_o r_j} \overline{\mathcal{S}}_j^l, \tag{5.163}$$

where the superscript $l$ has been added to denote the $lth$ range bin.

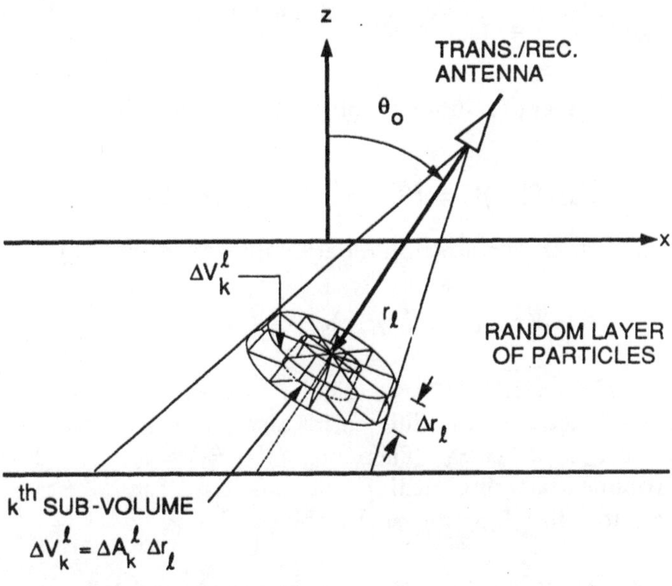


**Figure 5.27** Random distribution of particles divided into subvolumes over which the illumination is constant.

The correlation of the measured scattering matrix elements can now be written as

$$\langle \overline{S}^l_{pq} (\overline{S}^l_{mn})^* \rangle = \sum_{j=1}^{N} g_r^2(\mathbf{r}_j) g_t^2(\mathbf{r}_j) \frac{r_0^4}{r_j^4} \langle \overline{S}^l_{jpq} (\overline{S}^l_{jmn})^* \rangle . \tag{5.164}$$

In this case, the properties of the medium are denoted by a volume differential quantity analogous to (5.144):

$$\langle \overline{S}^{lv}_{pq} (\overline{S}^{lv}_{mn})^* \rangle = \frac{1}{\Delta A^l_j \Delta r_l} \langle \overline{S}^l_{jpq} (\overline{S}^l_{jmn})^* \rangle , \tag{5.165}$$

which is equivalent to

$$\sigma^{lv}_{pq} = \frac{1}{\Delta A^l_j \Delta r_l} \langle \sigma^l_{jpq} \rangle \tag{5.166}$$

when $pq = mn$. Here, $\sigma^{lv}_{pq}$ is the volume differential scattering coefficient for the $l$th range bin (modified by the intervening medium), and is analogous to $\sigma^0_{pq}$ for the random surface. Converting the summation to an integral, the correlation in (5.164) becomes

$$\langle \overline{S}^l_{pq} (\overline{S}^l_{mn})^* \rangle = \int g_r^2(\mathbf{r}) g_t^2(\mathbf{r}) \frac{r_0^4}{r^4} \langle \overline{S}^{lv}_{pq} (\overline{S}^{lv}_{mn})^* \rangle \, \mathrm{d}A_\perp \Delta r_l . \tag{5.167}$$

If $\langle \overline{S}^{lv}_{pq} (\overline{S}^{lv}_{mn})^* \rangle$ is approximately constant over the illumination, (5.167) becomes

$$\langle \overline{S}^l_{pq} (\overline{S}^l_{mn})^* \rangle = \langle \overline{S}^{lv}_{pq} (\overline{S}^{lv}_{mn})^* \rangle I_v(r_l, \theta_0) \Delta r_l , \tag{5.168}$$

where $I_v(r_l, \theta_0)$ is the volume illumination integral defined by

$$I_v(r_l, \theta_0) = \int g_r^2(\mathbf{r}) g_t^2(\mathbf{r}) \frac{r_l^4}{r^4} \, \mathrm{d}A_\perp . \tag{5.169}$$

The differential area $\mathrm{d}A_\perp$ is perpendicular to the antenna boresight direction, and it is related to the differential area on an equivalent surface by the relation $\mathrm{d}A_\perp = \mathrm{d}A \cos\theta_0$. Therefore, $I_v(r_l, \theta_0) \approx I_s(r_l, \theta_0) \cos\theta_0$.

For a volume scattering medium, the quantity of interest is $\sigma^{lv}_{rt}$, which can be computed as a function of the polarization $rt$ by using

$$\sigma^{lv}_{rt}(\psi_r, \chi_r, \psi_t, \chi_t) = \frac{4\pi}{\Delta r_l I_v(r_l, \theta_0)} \overline{\mathbf{A}}^r \cdot \langle \overline{\mathcal{M}} \rangle \overline{\mathbf{A}}^t \tag{5.170}$$

$$\sigma^{lv}_{rt}(\psi_r, \chi_r, \psi_t, \chi_t) = \frac{4\pi}{\Delta r_l I_v(r_l, \theta_0)} \overline{\mathbf{A}}^r_m \cdot \langle \overline{\mathcal{M}}_m \rangle \overline{\mathbf{A}}^t_m \tag{5.171}$$

$$\sigma_{rt}^{lv}(\psi_r,\chi_r,\psi_t,\chi_t) = \frac{4\pi}{\Delta r_l I_v(r_l,\theta_0)}\overline{\mathbf{B}}^*\cdot\langle\overline{\mathcal{C}}\rangle\overline{\mathbf{B}} \tag{5.172}$$

$$\sigma_{rt}^{lv}(\psi_r,\chi_r,\psi_t,\chi_t) = \frac{4\pi}{\Delta r_l I_v(r_l,\theta_0)}\langle|\,\overline{\mathbf{p}}^r\cdot\overline{\mathcal{S}}\,\overline{\mathbf{p}}^t\,|^2\rangle. \tag{5.173}$$

In this case, (5.168) motivates the definition of a volume differential scattering matrix $\overline{\mathcal{S}}^{lv}$ for the $l$th range bin (as modified by the intervening medium):

$$\overline{\mathcal{S}}^{lv} = \frac{1}{\sqrt{I_v(r_l,\theta_0)\Delta r_l}}\overline{\mathcal{S}}. \tag{5.174}$$

In addition, volume differential matrices $\overline{\mathcal{M}}^{lv}$, $\overline{\mathcal{M}}_m^{lv}$, and $\overline{\mathcal{C}}^{lv}$ are defined similar to those for surfaces in (5.154) to (5.156), with the superscript $lv$ instead of 0:

$$\overline{\mathcal{M}}^{lv} = \widetilde{\mathcal{R}}^{-1}\overline{\mathcal{W}}^{lv}\mathcal{R}^{-1} \tag{5.175}$$

$$\overline{\mathcal{M}}_m^{lv} = \widetilde{\mathcal{V}}^{-1}\overline{\mathcal{M}}^{lv}\mathcal{V}^{-1} \tag{5.176}$$

$$\overline{\mathcal{C}}^{lv} = \overline{\mathbf{X}}^{lv}(\widetilde{\overline{\mathbf{X}}}^{lv})^*, \tag{5.177}$$

where

$$\overline{\mathcal{W}}^{lv} = \begin{bmatrix} \overline{S}_{vv}^{lv}(\overline{S}_{vv}^{lv})^* & \overline{S}_{vh}^{lv}(\overline{S}_{vh}^{lv})^* & \overline{S}_{vh}^{lv}(\overline{S}_{vv}^{lv})^* & \overline{S}_{vv}^{lv}(\overline{S}_{vh}^{lv})^* \\ \overline{S}_{hv}^{lv}(\overline{S}_{hv}^{lv})^* & \overline{S}_{hh}^{lv}(\overline{S}_{hh}^{lv})^* & \overline{S}_{hh}^{lv}(\overline{S}_{hv}^{lv})^* & \overline{S}_{hv}^{lv}(\overline{S}_{hh}^{lv})^* \\ \overline{S}_{hv}^{lv}(\overline{S}_{vv}^{lv})^* & \overline{S}_{hh}^{lv}(\overline{S}_{vh}^{lv})^* & \overline{S}_{hh}^{lv}(\overline{S}_{vv}^{lv})^* & \overline{S}_{hv}^{lv}(\overline{S}_{vh}^{lv})^* \\ \overline{S}_{vv}^{lv}(\overline{S}_{hv}^{lv})^* & \overline{S}_{vh}^{lv}(\overline{S}_{hh}^{lv})^* & \overline{S}_{vh}^{lv}(\overline{S}_{hv}^{lv})^* & \overline{S}_{vv}^{lv}(\overline{S}_{hh}^{lv})^* \end{bmatrix} \tag{5.178}$$

$$\overline{\mathbf{X}}^{lv} = \begin{bmatrix} \overline{S}_{vv}^{lv} \\ \overline{S}_{vh}^{lv} \\ \overline{S}_{hv}^{lv} \\ \overline{S}_{hh}^{lv} \end{bmatrix}. \tag{5.179}$$

Hence, (5.170) to (5.173) are given in terms of volume differential matrices by the expressions:

$$\sigma_{rt}^{lv}(\psi_r,\chi_r,\psi_t,\chi_t) = 4\pi\overline{\mathbf{A}}^r\cdot\langle\overline{\mathcal{M}}^{lv}\rangle\overline{\mathbf{A}}^t \tag{5.180}$$

$$\sigma_{rt}^{lv}(\psi_r,\chi_r,\psi_t,\chi_t) = 4\pi\overline{\mathbf{A}}_m^r\cdot\langle\overline{\mathcal{M}}_m^{lv}\rangle\overline{\mathbf{A}}_m^t \tag{5.181}$$

$$\sigma_{rt}^{lv}(\psi_r,\chi_r,\psi_t,\chi_t) = 4\pi\overline{\mathbf{B}}^*\cdot\langle\overline{\mathcal{C}}^{lv}\rangle\overline{\mathbf{B}} \tag{5.182}$$

$$\sigma_{rt}^{lv}(\psi_r,\chi_r,\psi_t,\chi_t) = 4\pi\langle|\,\overline{\mathbf{p}}^r\cdot\overline{\mathcal{S}}^{lv}\overline{\mathbf{p}}^t\,|^2\rangle. \tag{5.183}$$

Imaging radars usually do not have the ranging capability available with scatterometers; therefore, volume scattering media (like vegetation) are generally treated in the same manner as a random surface. To compare the scatterometer results with those obtained by using an imaging radar, the relationship between the two methods of analysis must be determined. The total scattering matrix for a given spatial location is obtained by

$$\overline{S} = \sum_{l=1}^{n_v} \overline{S}^l, \tag{5.184}$$

where $n_v$ is the number of range bins used by the scatterometer. From this expression, the correlation $\langle \overline{S}_{pq} \overline{S}^*_{mn} \rangle$ is given by

$$\langle \overline{S}_{pq} \overline{S}^*_{mn} \rangle = \sum_{l=1}^{n_v} \langle \overline{S}^l_{pq} (\overline{S}^l_{mn})^* \rangle, \tag{5.185}$$

where independence of the $n_v$ range bins has been assumed. Using (5.168) in (5.185) and the definition in (5.147), the surface equivalent differential correlation $\langle \overline{S}^0_{pq} (\overline{S}^0_{mn})^* \rangle$ for the volume medium is given by

$$\langle \overline{S}^0_{pq} (\overline{S}^0_{mn})^* \rangle = \frac{1}{I_s(r_0, \theta_0)} \sum_{l=1}^{n_v} \langle \overline{S}^{lv}_{pq} (\overline{S}^{lv}_{mn})^* \rangle I_v(r_l, \theta_0) \Delta r_l. \tag{5.186}$$

## 5-4.3 Measurement Precision

In addition to the average property of the random medium, the statistics associated with the variation (or fading) can be obtained. Again, the presence of the gain functions $g_t(\mathbf{r}_j)$ and $g_r(\mathbf{r}_j)$ modify the measured statistics, but the effects can be approximately separated from the actual target statistics. Consider the scattering matrix described by (5.142) for surface scattering, and divide the illuminated area into subareas over which the illumination is constant (see Fig. 5.28). The scattering matrix can now be written as

$$\overline{S} = \sum_{k=1}^{n_s} g_t(\mathbf{r}_k) g_r(\mathbf{r}_k) \frac{r_0^2}{r_k^2} \left( \frac{1}{\Delta A_k} \sum_{j=1}^{N_s} e^{i2k_0 r_j} \overline{S}^k_j \right) \Delta A_k, \tag{5.187}$$

where $n_s$ is the number of subareas, $N_s$ is the number of facets (or scatterers) in each subarea, and the superscript $k$ has been added to the scattering

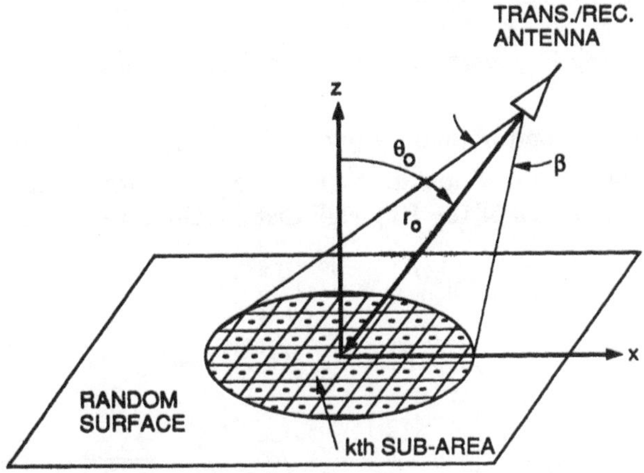


**Figure 5.28** Random surface divided into subareas over which the illumination is constant.

matrix $\overline{\mathcal{S}}_j$ to denote a facet in the $k$th subarea. The quantity within paranthesis in (5.187) is a random matrix that shall be denoted as $\overline{\mathcal{S}}^0_k$. If the product beamwidth of the antennas is small such that the target is not varying across the illumination, the $\overline{\mathcal{S}}^0_k$ (for $k = 1$ to $n_s$) values are independent and identically distributed. Using this notation in (5.187), the measured scattering matrix $\overline{\mathcal{S}}$ is given by

$$\overline{\mathcal{S}} = \sum_{k=1}^{n_s} g_t(\mathbf{r}_k) g_r(\mathbf{r}_k) \frac{r_0^2}{r_k^2} \overline{\mathcal{S}}^0_k \Delta A_k. \tag{5.188}$$

Assuming that each of the subareas is composed of a large number of point scatterers (none of which dominates the total scattering), the real and imaginary parts of the elements of $\overline{\mathcal{S}}^0_k$ can be considered as independent, identically distributed Gaussian random variables with zero mean and standard deviation $s^0_{rt}$ (this quantity should not be confused with the scattering matrix element for $rt$ polarization, $\overline{S}^0_{rt}$). From the properties for sums of scalar multiples of Gaussian random variables, the variance of the real and imaginary parts of the element $\overline{S}_{rt}$ is

$$s^2_{rt} = (s^0_{rt})^2 \sum_{k=1}^{n_s} g^2_t(\mathbf{r}_k) g^2_r(\mathbf{r}_k) \frac{r_0^4}{r_k^4} \Delta A_k, \tag{5.189}$$

and the mean remains zero. Replacing the summation by an integral over the illumination, (5.189) becomes

$$s_{rt}^2 = (s_{rt}^0)^2 \int g_t^2(\mathbf{r}) g_r^2(\mathbf{r}) \frac{r_0^4}{r^4}\, dA = (s_{rt}^0)^2 I_s(r_0, \theta_0). \tag{5.190}$$

Because the real and imaginary parts of both $\overline{S}_{rt}$ and $\overline{S}_{rt}^0$ are Gaussian random variables with zero mean, $|\overline{S}_{rt}|$ and $|\overline{S}_{rt}^0|$ are Rayleigh distributed. Based on the properties of the Rayleigh distribution, we can show that

$$\langle|\overline{S}_{rt}^0|\rangle = \sqrt{\frac{\pi}{2}} s_{rt}^0 \tag{5.191}$$

$$\langle|\overline{S}_{rt}^0|^2\rangle = 2(s_{rt}^0)^2, \tag{5.192}$$

and

$$\langle|\overline{S}_{rt}|\rangle = \sqrt{\frac{\pi}{2}} s_{rt} \approx \sqrt{\frac{\pi}{2}} s_{rt}^0 \sqrt{I_s(r_0, \theta_0)} \tag{5.193}$$

$$\langle|\overline{S}_{rt}|^2\rangle = 2 s_{rt}^2 \approx 2(s_{rt}^0)^2 I_s(r_0, \theta_0). \tag{5.194}$$

The standard deviation-to-mean ratios for $|\overline{S}_{rt}^0|$ and $|\overline{S}_{rt}|$, however, are the same. Thus,

$$\left(\frac{s}{\mu}\right)_{|\overline{S}_{rt}^0|} = \frac{\sqrt{2(s_{rt}^0)^2 - \frac{\pi}{2}(s_{rt}^0)^2}}{\sqrt{\frac{\pi}{2}} s_{rt}^0} = 0.5227, \tag{5.195}$$

and

$$\left(\frac{s}{\mu}\right)_{|\overline{S}_{rt}|} = \frac{\sqrt{2 s_{rt}^2 - \frac{\pi}{2} s_{rt}^2}}{\sqrt{\frac{\pi}{2}} s_{rt}} = 0.5227. \tag{5.196}$$

In the same way, the distributions for both $|\ \overline{S}_{rt}\ |^2$ and $|\ \overline{S}_{rt}^0\ |^2$ are exponentially distributed, and we can show that

$$\left(\frac{s}{\mu}\right)_{|\overline{S}_{rt}^0|^2} = \left(\frac{s}{\mu}\right)_{|\overline{S}_{rt}|^2} = 1. \tag{5.197}$$

To consider the statistics for volume scattering media, the scattering matrix of the $l$th range bin is divided into contributions from subvolumes ($\Delta V_k^l = \Delta A_k^l \Delta r_l$) over which the gain functions $g_t(\mathbf{r}_j)$ and $g_r(\mathbf{r}_j)$ are constant (see Fig. 5.27):

$$\overline{\mathcal{S}}^l = \sum_{k=1}^{n_v^l} g_t(\mathbf{r}_k) g_r(\mathbf{r}_k) \frac{r_0^2}{r_k^2} \left( \frac{1}{\Delta A_k^l \Delta r_l} \sum_{j=1}^{N_v^l} e^{i2k_o r_j} \overline{\mathcal{S}}_j^{lk} \right) \Delta A_k^l \Delta r_l, \tag{5.198}$$

where $n_v^l$ is the number of subvolumes, and $N_v^l$ is the number of particles within each subvolume. Again, the quantity within parentheses in (5.198) is a random matrix to be denoted as $\overline{\boldsymbol{S}}_k^{lv}$, and the measured scattering matrix is written as

$$\overline{\boldsymbol{S}}^l = \sum_{k=1}^{n_v^l} g_t(\mathbf{r}_k) g_r(\mathbf{r}_k) \frac{r_0^2}{r_k^2} \overline{\boldsymbol{S}}_k^{lv} \Delta A_k^l \Delta r_l. \tag{5.199}$$

If the standard deviation of the real and imaginary parts of $\overline{S}_{krt}^{lv}$ is denoted by $s_{rt}^{lv}$, and that of $\overline{S}_{rt}^l$ by $s_{rt}^l$, the relationship between the variances is

$$(s_{rt}^l)^2 = (s_{rt}^{lv})^2 \sum_{k=1}^{n_v^l} g_t^2(\mathbf{r}_k) g_r^2(\mathbf{r}_k) \frac{r_0^4}{r_k^4} \Delta A_k^l \Delta r_l. \tag{5.200}$$

In integral form, this becomes

$$(s_{rt}^l)^2 = (s_{rt}^{lv})^2 \int g_t^2(\mathbf{r}) g_r^2(\mathbf{r}) \frac{r_0^4}{r^4} \mathrm{d}A_\perp \Delta r_l = (s_{rt}^{lv})^2 I_v(r_l, \theta_0) \Delta r_l. \tag{5.201}$$

Similar to $|\overline{S}_{rt}|$ and $|\overline{S}_{rt}^0|$ ($|\overline{S}_{rt}|^2$ and $|\overline{S}_{rt}^0|^2$) for surfaces, $|\overline{S}_{rt}^l|$ and $|\overline{S}_{rt}^{lv}|$ ($|\overline{S}_{rt}^l|^2$ and $|\overline{S}_{rt}^{lv}|^2$) for the $l$th range bin are also Rayleigh (exponentially) distributed; therefore, relations similar to (5.191) to (5.197) are obtained. These relations are

$$\langle |\overline{S}_{rt}^{lv}| \rangle = \sqrt{\frac{\pi}{2}} s_{rt}^{lv} \tag{5.202}$$

$$\langle |\overline{S}_{rt}^{lv}|^2 \rangle = 2(s_{rt}^{lv})^2 \tag{5.203}$$

$$\langle |\overline{S}_{rt}^{l}| \rangle = \sqrt{\frac{\pi}{2}} s_{rt}^{l} \approx \sqrt{\frac{\pi}{2}} s_{rt}^{lv} \sqrt{I_s(r_0, \theta_0)} \tag{5.204}$$

$$\langle |\overline{S}_{rt}^{l}|^2 \rangle = 2(s_{rt}^l)^2 \approx 2(s_{rt}^{lv})^2 I_s(r_0, \theta_0) \tag{5.205}$$

$$\left(\frac{s}{\mu}\right)_{|\overline{S}_{rt}^{lv}|} = \left(\frac{s}{\mu}\right)_{|\overline{S}_{rt}^{l}|} = 0.5227 \tag{5.206}$$

$$\left(\frac{s}{\mu}\right)_{|\overline{S}_{rt}^{lv}|^2} = \left(\frac{s}{\mu}\right)_{|\overline{S}_{rt}^{l}|} = 1. \tag{5.207}$$

### 5-4.4 Examples of Polarimetric Measurements of Surfaces and Volumes

The previous three sections discussed polarization synthesis techniques used with polarimetric scatterometer data and the associated measurement precision. Here, we present the data obtained in several polarimetric scatterometer experiments conducted for surfaces and volumes at both microwave and millimeter-wave frequencies.

The first set of data was obtained at 35 GHz (Whitt and Ulaby, 1988) for three surfaces consisting of rocks, visually rough sand, and visually smooth sand. The polarimetric scatterometer used for the measurements was the MMP (millimeter-wave polarimeter), a network analyzer-based polarimetric scatterometer developed at the University of Michigan (Ulaby *et al.*, 1988a). The incidence angle relative to normal incidence was 64°. By measuring the differential scattering matrix $\overline{\mathcal{S}}^0$ for 100 different spatial samples, three sets of information were obtained: (1) distributions for $\sigma^0_{rt}$ with $rt \in \{vv, hv, hh\}$, (2) distributions for the relative phase of the scattering matrix elements $(\phi_{rt} - \phi_{hh})$ with $rt \in \{vv, hv\}$, and (3) polarization responses.

The distributions for $\sigma^0_{rt}$ and $(\phi_{rt} - \phi_{hh})$ for the rock surface are shown in Figs. 5.29 and 5.30, respectively. The standard deviation-to-mean ratio of $\sigma^0_{rt}$ is approximately 1, which is to be expected from Rayleigh fading statistics. The $\sigma^0_{rt}$ results obtained for the smooth sand surface and the rough sand surface are significantly lower in level than those for the rocks; however, the shapes of the distributions are very similar. Table 5.2 gives the mean values of $\sigma^0_{rt}$ for all three surfaces. For relative phase, the cross-polarized phase distribution $(\phi_{hv} - \phi_{hh})$ is approximately uniform between $-180°$ and $180°$ while the co-polarized phase distribution $(\phi_{vv} - \phi_{hh})$ is approximately Gaussian with a mean value of 8.48° and a standard deviation of 65.5°. For all three surfaces, the distributions for $(\phi_{hv} - \phi_{hh})$ are the same. However, the distributions for $(\phi_{vv} - \phi_{hh})$ differ with respect to their mean values, but the shapes are approximately the same. The means and standard deviations of $(\phi_{rt} - \phi_{hh})$ for all three surfaces are given in Table 5.3. For the rough sand surface, the mean of the distribution for $(\phi_{vv} - \phi_{hh})$ shown in Fig. 5.31 is $-30.31°$, which is a significant shift from 0° based on the standard error $s/\sqrt{N} \approx 7°$.

The polarization responses of the three surfaces are shown in Fig. 5.32. These were obtained using the techniques given in Section 5-4.1. For the rock surface, the co-polarized response has a maximum at the linear polarizations ($\chi = 0°$) with $\sigma^0_{vv}$ only 0.4 dB higher than $\sigma^0_{hh}$, as expected from the mean values in Table 5.2.

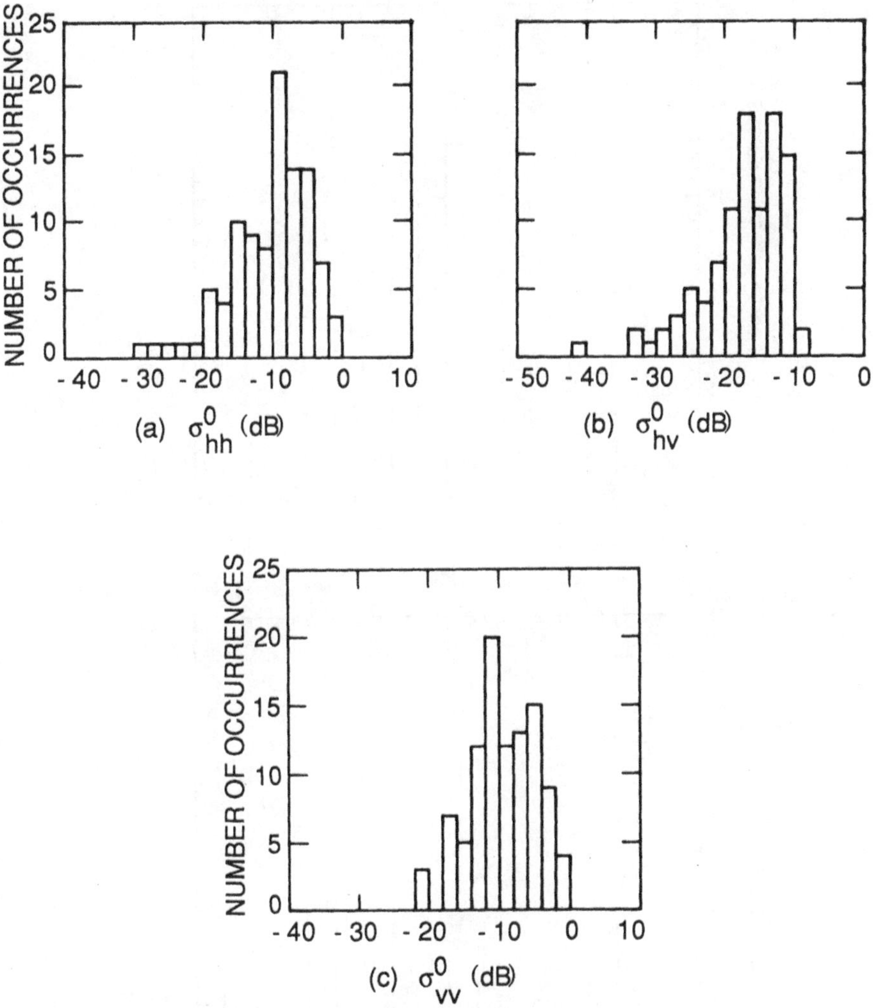


**Figure 5.29** Measured distributions of $\sigma^\circ$ for the rock surface: (a) $\sigma^\circ_{hh}$, (b) $\sigma^\circ_{hv}$, and (c) $\sigma^\circ_{vv}$.

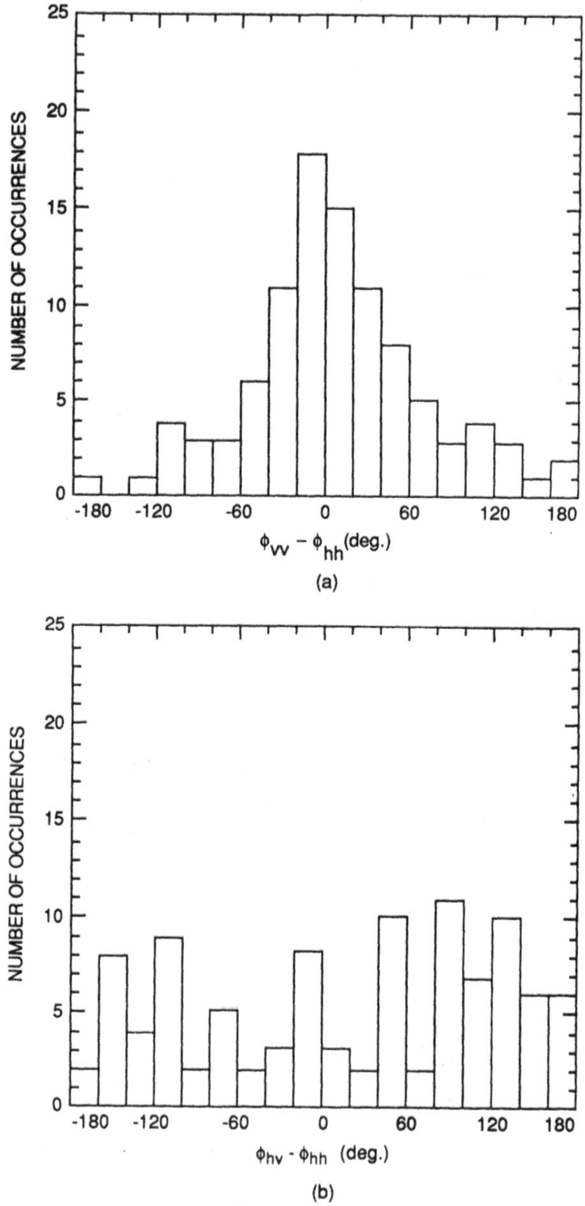


**Figure 5.30** Measured distributions of relative phase for the rock surface: (a) $\phi_{vv} - \phi_{hh}$ and (b) $\phi_{hv} - \phi_{hh}$.

**Table 5.2** Mean Values of $\sigma^0_{rt}$ in dB for the Three Surfaces Measured

| Surface | $\sigma^0_{hh}$ | $\sigma^0_{hv}$ | $\sigma^0_{vv}$ |
|---|---|---|---|
| Rocks | −7.5 | −14.8 | −7.1 |
| Smooth Sand | −15.2 | −22.2 | −12.6 |
| Rough Sand | −13.1 | −20.4 | −12.1 |

**Table 5.3** Mean Values and Standard Deviations of $\phi_{rt} - \phi_{hh}$ in Degrees for the Three Surfaces Measured

| Surface | | Mean | Std. Dev. |
|---|---|---|---|
| Rocks | $vv$ | 8.48 | 65.5 |
| | $hv$ | 18.72 | 106.2 |
| Smooth Sand | $vv$ | −11.13 | 69.1 |
| | $hv$ | 16.31 | 108.7 |
| Rough Sand | $vv$ | −30.31 | 62.6 |
| | $hv$ | 8.15 | 110.4 |

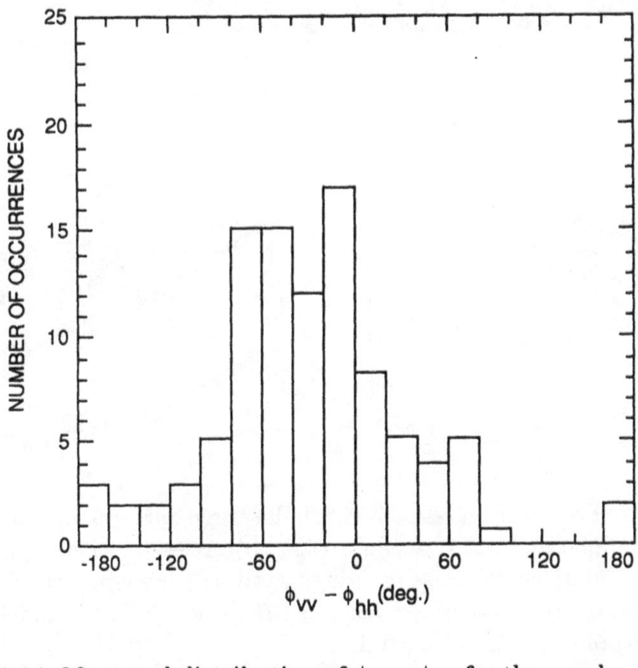


**Figure 5.31** Measured distribution of $\phi_{vv} - \phi_{hh}$ for the rough sand surface.

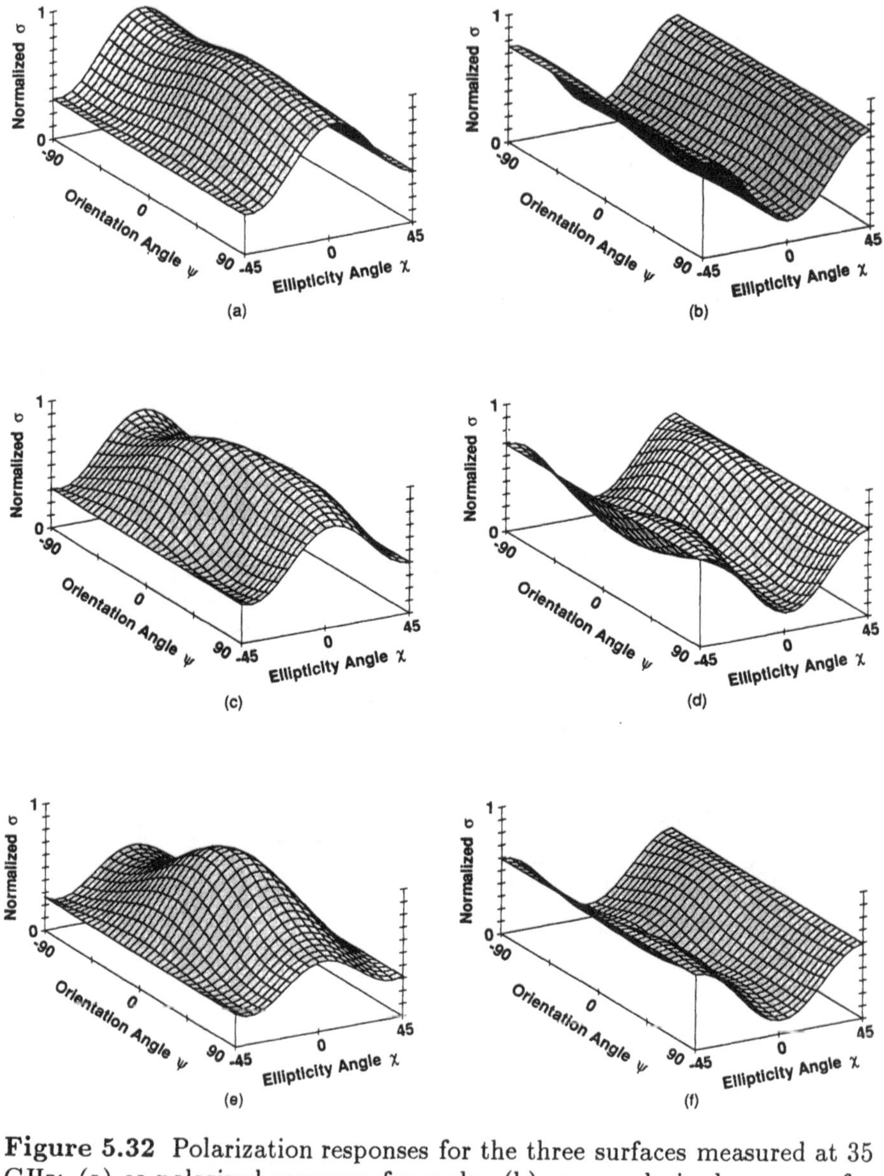


**Figure 5.32** Polarization responses for the three surfaces measured at 35 GHz: (a) co-polarized response for rocks, (b) cross-polarized response for rocks, (c) co-polarized response for rough sand, (d) cross-polarized response for rough sand, (e) co-polarized response for smooth sand, and (f) cross-polarized response for smooth sand.

The minimum in the co-polarized response occurs at the circular polarizations ($\chi = \pm 45°$). The cross-polarized response has a minimum at the linear polarizations and a maximum at circular. This type of polarization response is characteristic of specular reflection from a smooth dielectric surface at normal incidence (van Zyl *et al.*, 1987). not smooth, but the scattering is dominated by specular reflection from smooth facets where the local surface normal is oriented toward the radar.

The smooth sand surface and the rough sand surface have polarization responses similar to the rock surface. For both sand surfaces, the normalized co-polarized $\sigma^0$ is maximum at $vv$ polarization. Because the polarization response is an average power representation of the target, the difference between $vv$ and $hh$ is equal to the difference in the mean values as given in Table 5.2. In contrast to the rock surface, the minima (maxima) for the co-polarized (cross-polarized) response are at elliptical polarizations instead of circular.

The polarization responses for all three surfaces show the presence of an unpolarized component of the scattered field in the form of a pedestal (nonzero minimum level), which is approximately the same for all surfaces. The pedestal can arise from various effects, including multiple scattering which is probably a significant mechanism at 35 GHz for all three surfaces measured.

For comparison with the surface measurements at 35 GHz, Fig. 5.33 gives the measured polarization response for a grass surface at L-band with incidence angles of 30° and 50°. For this experiment, 50 independent measurements of $\overline{\mathcal{S}}^0$ were used to generate the polarization response. These plots are very similar to those in Fig. 5.32, with the co-polarized maxima (cross-polarized minima) at linear polarizations and the cross-polarized maxima (co-polarized minima) at circular polarization. However, the relative phase distributions in Fig. 5.34 vary somewhat from the 35 GHz results. At an incidence angle of 30°, the relative phase $(\phi_{vv} - \phi_{hh})$ is again approximately Gaussian with a mean near 0°, whereas $(\phi_{hv} - \phi_{hh})$ is closer to a uniform distribution. However, at an incidence angle of 50°, both the $(\phi_{vv} - \phi_{hh})$ and $(\phi_{hv} - \phi_{hh})$ distributions are approximately uniform. Therefore, the polarization response is only one piece of information available from polarimetric measurements. The phase statistics, if they can be obtained, can provide additional information about the target.

We have shown examples of measured polarimetric data for random surfaces, and now we will give examples for volume scattering media. The measured L-band polarization response for three different types of trees (walnut, oak, and pine) are given in Fig. 5.35. The response for the walnut trees was obtained using 30 independent spatial samples, and the responses for both the oak and pine trees were obtained using 50 independent spatial samples. The incidence angle was 55° for the walnut trees and 50° for the oak and pine trees. The character of the three polarization responses is very different. The oak response shows a co-polarized maximum at $vv$ polarization, whereas that for the pine is at $hh$ polarization. The character of the walnut response shows equal $vv$ and $hh$ returns. Also, the size of the pedestal, representing the unpolarized component, decreases from the oak (largest pedestal) to the walnut (smallest pedestal). The cross-polarized responses also differ in the ratio of maximum-to-minimum cross-polarized returns. The walnut response has a much higher circular cross-polarized return than linear cross-polarized return, whereas the response for both the oak and the pine has less variation over the range of ellipticity and orientation angles. The relative phase distributions for the three types of trees are given in Fig. 5.36. The distributions of $(\phi_{vv} - \phi_{hh})$ and $(\phi_{hv} - \phi_{hh})$ for both the oak and pine trees are approximately uniform. For the walnut trees, the distribution for $(\phi_{vv} - \phi_{hh})$ is approximately Gaussian, but $(\phi_{hv} - \phi_{hh})$ is uniform.

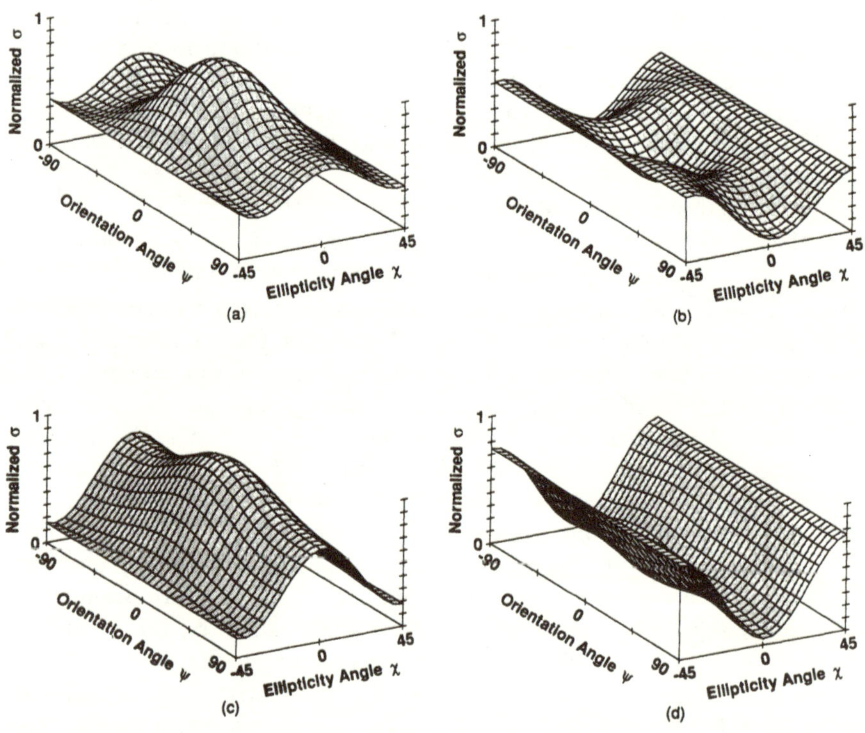


**Figure 5.33** Polarization responses for a grass surface measured at L-band for two incidence angles: (a) co-polarized response for $\theta = 50°$, (b) cross-polarized response for $\theta = 50°$, (c) co-polarized response for $\theta = 30°$, (d) cross-polarized response for $\theta = 30°$.

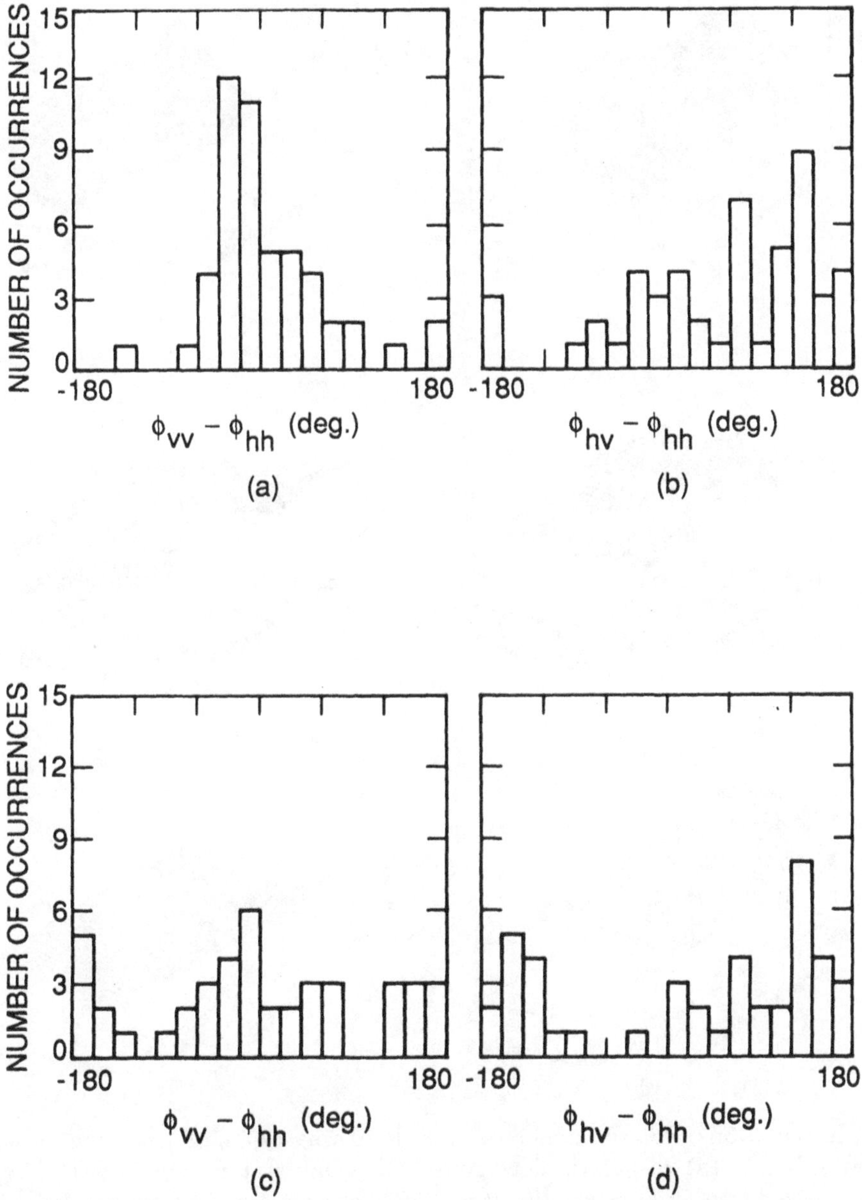


**Figure 5.34** Relative phase distributions for a grass surface measured at L-band for two incidence angles: (a) $\phi_{vv} - \phi_{hh}$ for $\theta = 30°$, (b) $\phi_{hv} - \phi_{hh}$ for $\theta = 30°$, (c) $\phi_{vv} - \phi_{hh}$ for $\theta = 50°$, (d) $\phi_{hv} - \phi_{hh}$ for $\theta = 50°$.

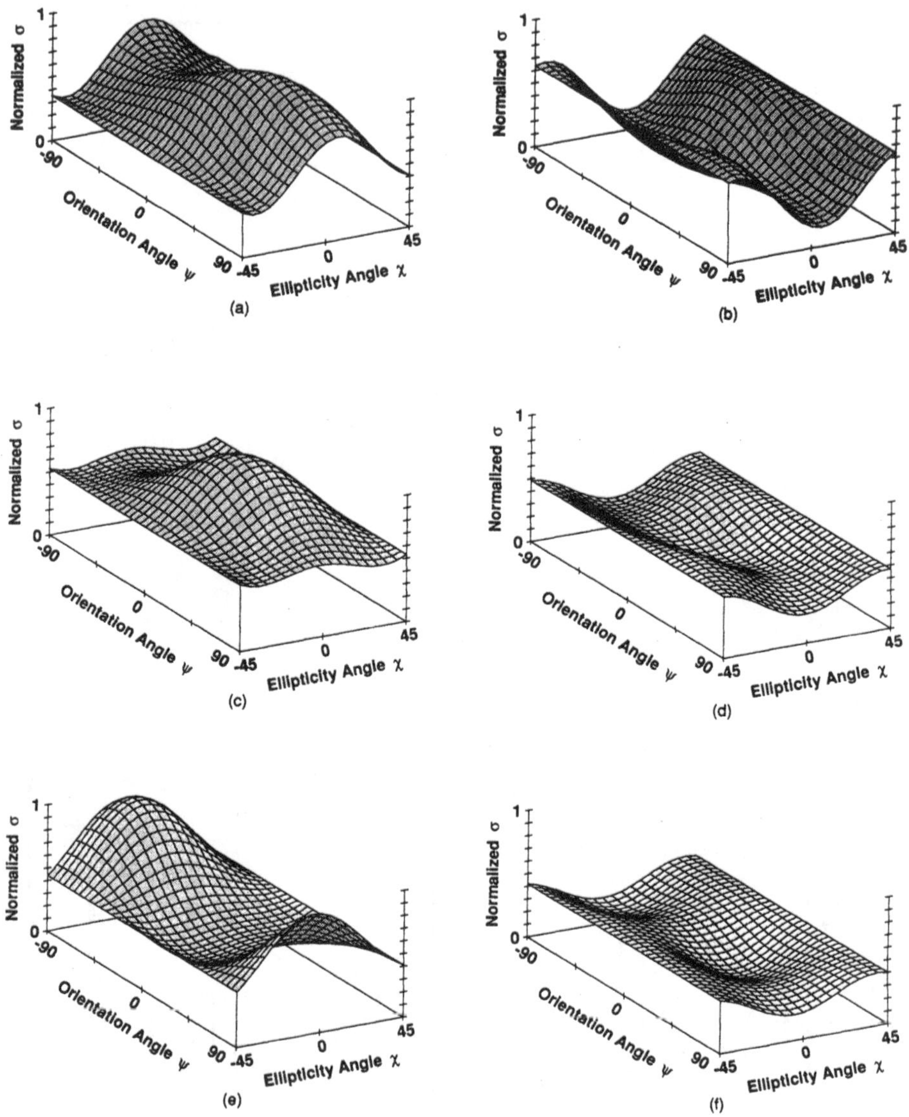


**Figure 5.35** Polarization responses for three types of trees measured at L-band: (a) co-polarized response for walnuts at $\theta = 55°$, (b) cross-polarized response for walnuts at $\theta = 55°$, (c) co-polarized response for oak at $\theta = 50°$, (d) cross-polarized response for oak at $\theta = 50°$, (e) co-polarized response for red pine at $\theta = 50°$, (f) cross-polarized response for red pine at $\theta = 50°$.

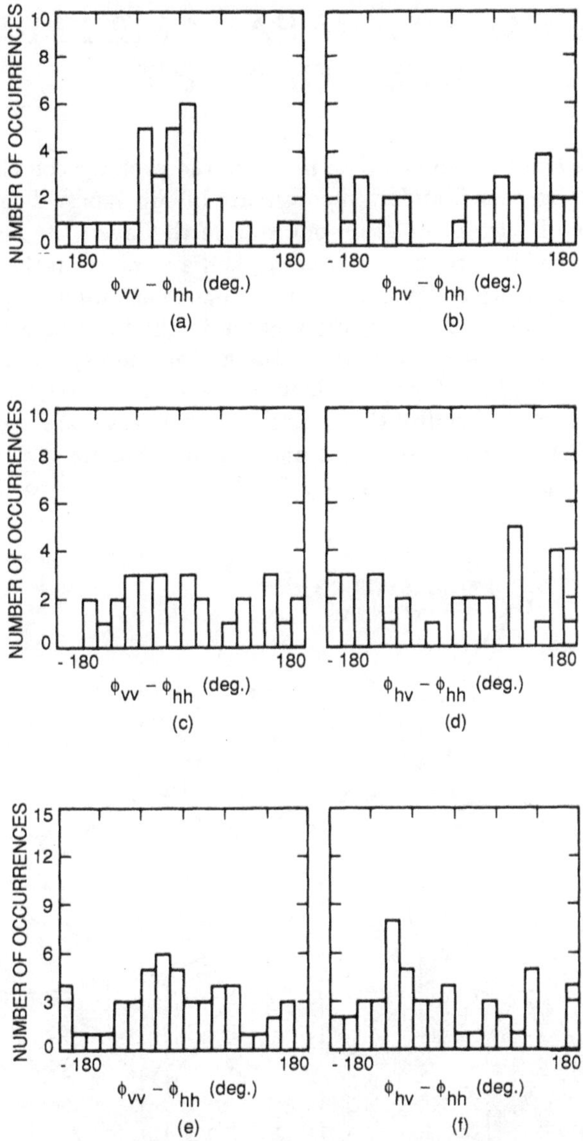


**Figure 5.36** Relative phase distributions for three types of trees measured at L-band: (a) $\phi_{vv} - \phi_{hh}$ for walnuts at $\theta = 55°$, (b) $\phi_{hv} - \phi_{hh}$ for walnuts at $\theta = 55°$, (c) $\phi_{vv} - \phi_{hh}$ for oak at $\theta = 50°$, (d) $\phi_{hv} - \phi_{hh}$ for oak at $\theta = 50°$, (e) $\phi_{vv} - \phi_{hh}$ for red pine at $\theta = 50°$, (f) $\phi_{hv} - \phi_{hh}$ for red pine at $\theta = 50°$.

# 5-5 MEASUREMENT OF THE PROPAGATION PARAMETERS OF A FOREST CANOPY

The approaches traditionally used to measure the propagation characteristics of an inhomogeneous random medium are of two types. The first type involves the use of a transmitter on one side of the layer, the propagation characteristics of which are to be measured, and a receiver on the other side as in Fig. 5.37. Because the medium is inhomogeneous, we need to measure the loss and phase delay through many paths in the layer to establish the mean values and associated statistics of the attenuation coefficient $\alpha_p$ and phase constant $\beta_p$. The subscript $p$ denotes the polarization configuration under consideration. Usually, the quantities of interest are $\alpha_h$, $\alpha_v$, and $\Delta\beta = \beta_v - \beta_h$, where the subscripts $h$ and $v$ denote horizontal and vertical polarizations, respectively, and $\hat{\mathbf{h}}$ is defined to be parallel to the earth's surface.

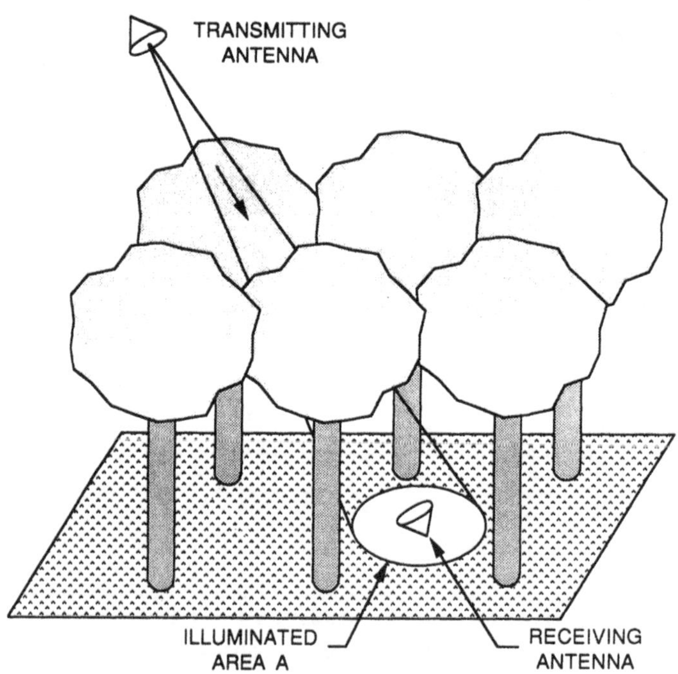


**Figure 5.37** Sketch showing a receiving antenna placed underneath the trees for measuring the propagation properties of a canopy.

For a forest canopy, the receiver-transmitter technique is somewhat cumbersome to use, particularly at oblique incidence, because of problems associated with proper pointing and positioning of the antennas.

In the second approach, which is a modified version of the first, the transmitter and receiver are collocated (radar mode) on one side of the layer as in Fig. 5.38, and the quantity measured by the receiver is the signal backscattered from a strong target of known RCS placed on the other side of the layer. Because the received signal contains backscattering contributions from both the reference target located behind (or below) the layer and from the layer itself, the precision associated with the measured two-way attenuation of the layer is governed by the amplitude ratio $|\overline{\mathbf{E}}_r|/|\overline{\mathbf{E}}_b|$. The quantities $\overline{\mathbf{E}}_r$ and $\overline{\mathbf{E}}_b$ are the fields backscattered from the reference target and the "background" layer, respectively.

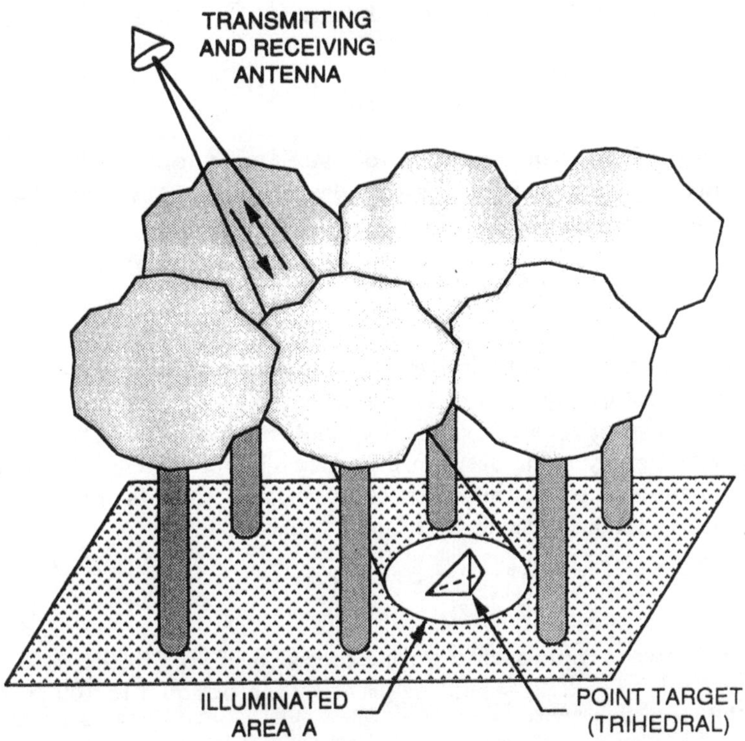


**Figure 5.38** Sketch showing a point target placed underneath the trees for measuring the attenuated backscatter using a downward-looking radar.

The total electric field of the backscattered signal is

$$\overline{\mathbf{E}} = \overline{\mathbf{E}}_r + \overline{\mathbf{E}}_b = |\overline{\mathbf{E}}_{r0}|\, \mathrm{e}^{-\tau}\, \mathrm{e}^{-i\phi_r} + |\overline{\mathbf{E}}_b|\, \mathrm{e}^{-i\phi_b}, \tag{5.208}$$

where $|\overline{\mathbf{E}}_{r0}|$ is the magnitude of the field backscattered from the reference target in the absence of the intervening layer, $\phi_r$ is the phase associated with propagation to the reference target and back, $|\overline{\mathbf{E}}_b|$ and $\phi_b$ are the magnitude and phase of the field backscattered from the layer, and

$$\tau = \sec\theta \int_0^h 2\alpha(z)\, \mathrm{d}z \tag{5.209}$$

is the slant-path optical thickness of the forest canopy. The quantities $\alpha(z)$, $h$, and $\theta$ are the attenuation coefficient at height $z$, the total vertical thickness of the layer, and the incidence angle, respectively. A traditional radar measures $|\overline{\mathbf{E}}|^2$ and, through calibration, $|\overline{\mathbf{E}}|^2$ is converted to units of RCS, $\sigma$. Thus,

$$\sigma = K|\overline{\mathbf{E}}|^2, \tag{5.210}$$

and similar relations may be used to relate $|\overline{\mathbf{E}}_{r0}|^2$ to $\sigma_r$ and $|\overline{\mathbf{E}}_b|^2$ to $\sigma_b$. Upon multiplying (5.208) by its complex conjugate and simplifying, we obtain the expression:

$$\sigma = \sigma_r\, \mathrm{e}^{-2\tau} + \sigma_b + 2\, \mathrm{e}^{-\tau} \sqrt{\sigma_r \sigma_b} \cos\phi, \tag{5.211}$$

where $\phi = \phi_b - \phi_r$.

The quantity we seek to determine is the two-way loss factor $\mathrm{e}^{(-2\tau)}$. The radar measures $\sigma$, the RCS of the illuminated scene, and can be used to measure $\sigma_r$ and $\sigma_b$ separately, but it cannot measure $\phi$ directly. Moreover, $\phi$ may assume any value between 0 and $2\pi$. Hence, the usual solution to this problem is to select the reference target such that its cross section $\sigma_r$ is sufficiently large to ensure that

$$\frac{\sigma_r}{\sigma_b}\, \mathrm{e}^{-2\tau} \gg 1. \tag{5.212}$$

If the objective is to measure $\tau$ with a precision of $\pm 1$ dB, the ratio on the left-hand side of (5.212) should be greater than or equal to 100 (Ulaby *et al.*, 1986a, pp. 768–770).

As an example, assume that a narrow-beam 1.6 GHz radar antenna is used to illuminate a flat metal plate oriented normal to the illumination and placed underneath a tree canopy. A flat metal plate is chosen for this example because, among passive reflectors, it provides the highest radar cross section for a given physical cross section $A_r$. The scattering properties of the tree canopy are described in terms of the backscattering coefficient $\sigma_b^0 = \sigma_b/A$, where $\sigma_b$ is the RCS, and $A$ is the area illuminated by the antenna beam. Using typical values for the illuminated area and the backscattering coefficient, $A = 5$ m$^2$ and $\sigma_b^0 = 0.1$ m$^2$m$^{-2}$, the RCS is $\sigma_b = \sigma_b^0 A = 0.5$ m$^2$. Assuming a 10 dB one-way loss due to the layer, the two-way loss term $\exp(-2\tau) = 0.01$. If the requirement $(\sigma_r/\sigma_b)\, e^{(-2\tau)} = 100$ is used with the values given above, $\sigma_r$ must be 5000 m$^2$. From the relationship:

$$\sigma_r = \frac{4\pi A_r^2}{\lambda^2}, \tag{5.213}$$

$A_r$ must be 3.7 m$^2$ at 1.6 GHz. The half-power beamwidth of a square or circular metal plate with such an area is about 2.5°, which causes difficulty in pointing the reflector properly with respect to the radar antenna. In essence, the pointing accuracy must be better than about 0.7° for the error associated with the estimated value of $\tau$ to be less than 1 dB. A wider-beam trihedral corner reflector may be used instead, but in that case the length $L$ of its aperture edges will have to be 3.6 m each for $\sigma_r$ to be 5000 m$^2$; for a trihedral corner reflector, $\sigma_r = \pi L^4/3\lambda^2$ (Ulaby *et al.*, 1986a, p. 776). Clearly, this is a large trihedral to use in a forest canopy. Moreover, if a measurement precision of better than ±1 dB is desired, the trihedral will need to be even larger and heavier.

Instead of this "brute force" approach, the polarimetric calibration techniques introduced in Section 5-3 can be used to measure the extinction and propagation characteristics of random media. The remaining subsections illustrate the methods used, and experimental results are given for a simplified case.

### 5-5.1 Polarimetric Radar Model for Trees with a Point Target Underneath

For a distributed target such as a forest canopy, the backscattering behavior is characterized by the differential scattering matrix $\overline{\boldsymbol{T}}^0$ and a corresponding scattering matrix $\overline{\boldsymbol{T}}$, given by

$$\overline{\boldsymbol{T}} = \sqrt{A}\,\overline{\boldsymbol{T}}^0 = \begin{bmatrix} \overline{T}_{vv} & \overline{T}_{vh} \\ \overline{T}_{hv} & \overline{T}_{hh} \end{bmatrix}, \tag{5.214}$$

where the elements $\overline{T}_{rt}$ for $rt \in \{vv, vh, hv, hh\}$ are complex, and

$$\sigma_{rtT}^0 = \frac{4\pi}{A}\langle |\,\overline{T}_{rt}\,|^2\rangle. \tag{5.215}$$

If measurements of the canopy were obtained by using a polarimetric scatterometer, the illuminated area $A$ in (5.214) and (5.215) should be replaced by the illumination integral $I_s(r_0, \theta_0)$.

When a point target is present underneath a vegetation canopy as in Fig. 5.38, the backscattered signal is composed of contributions due to

1. Backscatter from the canopy;
2. Backscatter from the trihedral attenuated by the canopy (first-order term);
3. Second and higher order multiple scatter terms involving the canopy and the point target.

These contributions (up to second order) are illustrated in Fig. 5.39. The multiple-scatter terms above second order involve at least two bistatic scattering interactions within the canopy, and current models and experimental results indicate these to be negligible at microwave frequencies below C-band (Ulaby *et al.*, 1988b). The remaining second-order scattering terms involve one bistatic scattering interaction within the canopy and one for the target. To obtain a first-order model for the scattering matrix of the canopy with the point target present, this second-order term is also assumed to be negligible. The following argument supports this assumption.

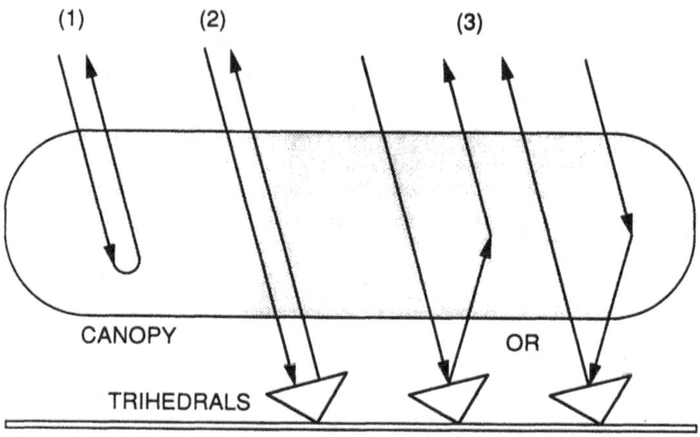


**Figure 5.39** Sketch illustrating the scattering mechanisms up to second order for a trihedral under a tree canopy.

Consider a trihedral placed underneath a tree canopy. The co-polarized ($vv$ and $hh$) bistatic scattering cross sections are much smaller than the co-polarized backscattering cross sections. Therefore, the dominant co-polarized scattering term involving the trihedral is the attenuated backscatter (term 2), and the second and higher order multiple scatter (term 3) can be neglected for co-polarization. However, the cross-polarized (*vh* and *hv*) backscattering cross section for the trihedral is zero, whereas the cross-polarized bistatic cross section is nonzero. This means that, in order to neglect term 3 for *vh* and *hv*, any cross-polarized scattering due to bistatic interactions with the trihedral must be much smaller than the cross-polarized backscatter from the canopy alone. Because this condition is satisfied for many canopies, term 3 will be neglected for cross-polarization as well. This assumption can be justified for a given canopy by measuring the backscatter with and without a trihedral present underneath the canopy. If the cross-polarized backscatter term does not change between the two situations, the assumption is valid. Similar arguments can be used to support the first-order model assumption for other point targets placed underneath the canopy.

Based on the preceding assumptions, the effect of the canopy can be modeled (up to first order) by an equation similar to (5.34) of Section 5-3. Arbitrarily choosing the phase reference to correspond with the upper boundary of the canopy, the measured scattering matrix for one spatial sample of the canopy with a point target present is given by

$$\overline{\mathcal{M}} = \overline{\mathcal{T}} + c\, e^{i\phi} \overline{\mathcal{U}}\, \overline{\mathcal{P}}\, \overline{\mathcal{D}}, \tag{5.216}$$

where $\overline{\mathcal{T}}$ is the scattering matrix of the canopy without the point target, $\overline{\mathcal{P}}$ is the scattering matrix of the point target, and $\overline{\mathcal{U}}$ and $\overline{\mathcal{D}}$ are normalized propagation or "distortion" matrices representing the effect of the canopy on the upward and downward propagating waves. The constant $c$ is a complex normalization factor as in the calibration problem, and here the phase $\phi$ represents the two-way phase shift due to free-space propagation over the distance traveled from the upper interface of the canopy to the point target and back. Because the form of (5.216) is the same as (5.34) for the calibration problem, the propagation matrices $\overline{\mathcal{U}}$ and $\overline{\mathcal{D}}$ can be determined by using the calibration procedures of Section 5-3. By measuring these matrices for different spatial samples within the canopy, the associated statistics can be determined.

As an example, we consider a recent experiment (Ulaby *et al.*, 1988c; Ulaby *et al.*, 1990) to determine the propagation matrices for a stand of red pines. The technique used in the experiment was more direct than any of the calibration procedures given in Section 5-3 because the canopy distortion matrices were assumed to be reciprocal and diagonal. To maintain matrix relations until the very end, a somewhat different derivation than that given in the original paper (Ulaby *et al.*, 1990) will be used. In this case, the expression for the measured scattering matrix can be written as

$$\overline{\mathcal{M}} = \overline{\mathcal{T}} + \overline{\mathcal{L}}\,\overline{\mathcal{P}}\,\overline{\mathcal{L}}, \tag{5.217}$$

where the matrix $\overline{\mathcal{L}}$ is the unnormalized propagation matrix, given by

$$\overline{\mathcal{L}} = \begin{bmatrix} \overline{L}_v\ e^{i\phi_{vL}} & 0 \\ 0 & \overline{L}_h\ e^{i\phi_{hL}} \end{bmatrix}, \tag{5.218}$$

and the elements of $\overline{\mathcal{L}}$ are related to the canopy attenuation and phase constants ($\alpha_p$ and $\beta_p$ for polarization $p \in \{v, h\}$) by the expressions:

$$\overline{L}_p = \exp\left(-\int_0^r \alpha_p\ dr\right) \tag{5.219}$$

$$\phi_{pL} = \int_0^r \beta_p\ dr. \tag{5.220}$$

Here, $r$ is the total path length through the canopy to the point target.

Due to the simple form of (5.217), only a single trihedral needs to be measured to determine $\overline{\mathcal{L}}$. For a trihedral, the scattering matrix is given by

$$\overline{\mathcal{P}} = \begin{bmatrix} \overline{P}_{vv} & 0 \\ 0 & \overline{P}_{hh} \end{bmatrix}. \tag{5.221}$$

Because both $\overline{\mathcal{L}}$ and $\overline{\mathcal{P}}$ are diagonal, the product $\overline{\mathcal{L}}\,\overline{\mathcal{P}}\,\overline{\mathcal{L}} = \overline{\mathcal{L}}^2\overline{\mathcal{P}}$, and the expression for the measured scattering matrix becomes

$$\overline{\mathcal{M}} = \overline{\mathcal{T}} + \overline{\mathcal{L}}^2\overline{\mathcal{P}}. \tag{5.222}$$

From this equation, the loss $\overline{L}_p$ can be determined, as can the relative phase $2(\phi_{vL} - \phi_{hL})$ to within an integer multiple of $2\pi$. Defining the complex variable $C_p$ (for $p \in \{v, h\}$) such that

$$C_p = \frac{\overline{M}_{pp} - \overline{T}_{pp}}{\overline{P}_{pp}}, \tag{5.223}$$

the loss and relative phase are then given by

$$\overline{L}_p = |\,C_p\,| \tag{5.224}$$

$$2(\phi_{vL} - \phi_{hL}) = \tan^{-1}\left\{\frac{\mathrm{Im}(C_v)}{\mathrm{Re}(C_v)}\right\} - \tan^{-1}\left\{\frac{\mathrm{Im}(C_h)}{\mathrm{Re}(C_h)}\right\} + 2n\pi, \tag{5.225}$$

where $n$ is an unknown integer.

The results obtained for 26 independent spatial samples of the red pine canopy are given in Fig. 5.40. Figure 5.40(a) and (b) shows the distributions for $\overline{L}_h^2$ and $\overline{L}_v^2$, the one-way power loss factors, in dB. The mean values were $-9.31$ dB for $h$-polarization and $-9.16$ dB for $v$-polarization. Figure 5.40(c) shows the distribution of $2(\phi_{vL} - \phi_{hL})$. The mean value of $2(\phi_{vL} - \phi_{hL})$ was 4.23°. These results — namely that the mean values of $\overline{L}_v$ and $\overline{L}_h$ are approximately equal and the mean value of the propagation phase difference is approximately zero — suggest that the pine canopy is approximately an isotropic medium.

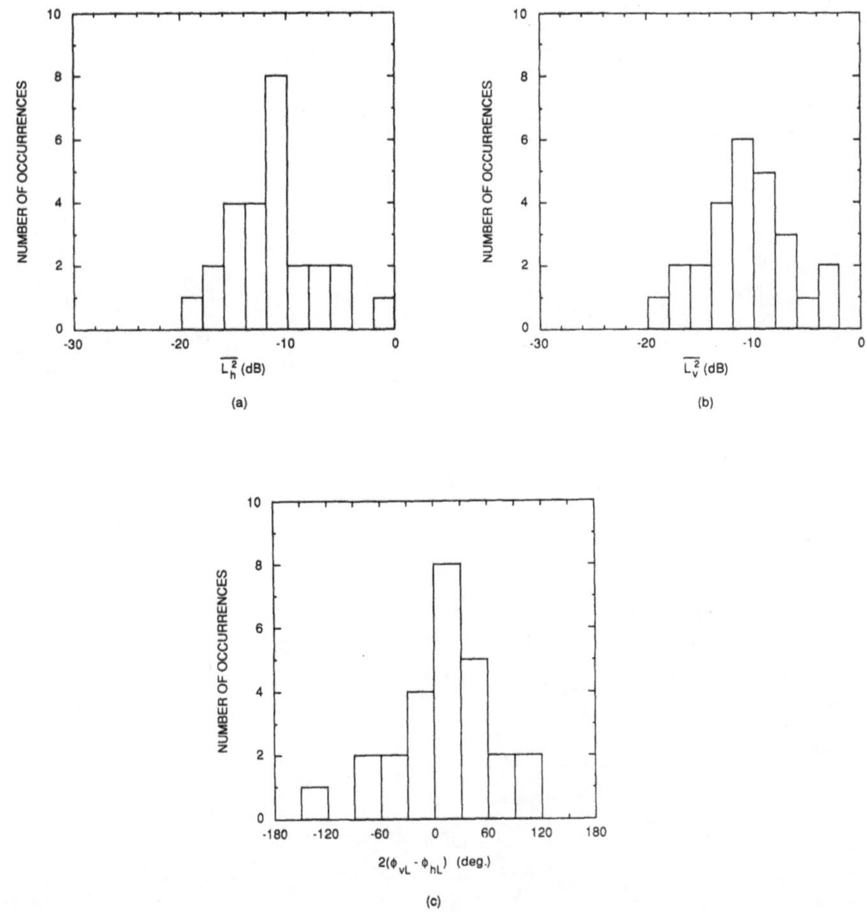


**Figure 5.40** Measured distributions of (a) $\overline{L}_h^2$, (b) $\overline{L}_v^2$, and (c) $2(\phi_{vl} - \phi_{hl})$.

The precision associated with the values of the canopy loss factor measured by using the polarimetric technique is estimated to be on the order of $\pm 0.3$ dB. The primary source of error is orientation-related variations in $P_0$ (the scattering amplitude of the trihedral reflector) arising from deployment errors. The trihedral reflector used in the experiment described in the preceding section had an aperture of $A_t = \sqrt{3}L^2/4 = 0.3$ m$^2$. In comparison, with the amplitude-only measurement technique, $A_t$ would need to have been about 17.7 m$^2$ to achieve the above accuracy of $\pm 0.3$ dB, even if we were to ignore altogether the deployment-related errors.

# References

Barnes, R. M., "Polarimetric Calibration Using In-scene Reflectors," Rep. TT-65, MIT Lincoln Laboratory, Lexington, MA, September 1986.

Blanchard, A. J., and J. Rochier, "Bistatic Radar Cross Section Measurement Facility," *Proceedings of IGARSS '87 Symposium*, Ann Arbor, MI, Vol. I, pp. 545–548, May 1987.

Farnett, E.C., T.B. Howard, and G.H. Stevens, Pulse-Compression Radar, Ch. 20 in *Radar Handbook*, M.I. Skolnik, ed., McGraw-Hill, pp. 20-26, 1970.

Freeman, A., C. Werner, and Yuhshen Shen, "Calibration of Multipolarisation Imaging Radar," *Proceedings of IGARSS '88 Symposium*, Edinburgh, Scotland, Vol. I, pp. 335–339, September 1988.

Gere, J.M., and W. Weaver, Jr., *Matrix Algebra for Engineers*, Second Edition, Brooks/Cole Publishing Company, Monterey, CA, 1983.

Liepa, V.V., K. Sarabandi, and M.A. Tassoudji, "A Pulsed Network Analyzer Based Scatterometer," *Proceedings of IGARSS '89 Symposium*, Vancouver, British Columbia, pp. 1826–1828, July 10–14, 1989.

Riegger, S., W. Wiesbeck, and A. J. Sieber, "On the Origin of Cross-polarization in Remote Sensing," *Proceedings of IGARSS '87 Symposium*, Ann Arbor, MI, Vol. I, pp. 577–580, May 1987.

Ruck, G.T., D. E. Barrick, W. D. Stuart, and C. K. Krichbaum, *Radar Cross Section Handbook, Vol. I*, New York, Plenum Press, 1970.

Sarabandi, K., F.T. Ulaby, and M.A. Tassoudji, "Calibration of Polarimetric Radar Systems With Good Polarization Isolation," *IEEE Trans. on Geoscience and Remote Sensing*, Vol. GE-28, 1990.

Saunders, W.K., "CW and FM Radar," *Radar Handbook*, Ch.16 M.I. Skolnik, ed., McGraw-Hill, New York, pp. 16-21 to 16-37, 1970.

Skolnik, M.I., *Introduction to Radar Systems*, McGraw-Hill, New York, 1980.

Tassoudji, M.A., K. Sarabandi, and F.T. Ulaby, "Design Consideration And Implementation Of The LCX Polarimetric Scatterometer (POLARSCAT)," Rep. No. 022486-T-2, Radiation Laboratory, The University of Michigan, Ann Arbor, June 1989.

Ulaby, F.T., R.K. Moore, and A.K. Fung, *Microwave Remote Sensing: Active and Passive*, Volume II, Artech House, Norwood, MA, 1986a.

Ulaby, F.T., R.K. Moore, and A.K. Fung, *Microwave Remote Sensing: Active and Passive*, Volume III, Artech House, Norwood, MA, 1986b.

Ulaby, F.T., T.F. Haddock, J. East, and M.W. Whitt, "A Millimeter-Wave Network Analyzer Based Scatterometer," *IEEE Trans. on Geoscience and Remote Sensing*, Vol. GE-26, pp. 75–81, January 1988a.

Ulaby, F.T., K. Sarabandi, K. McDonald, M.W. Whitt, and M.C. Dobson, "Michigan Microwave Canopy Scattering Model (MIMICS)," Rep. 022486-T-1, Radiation Laboratory, The University of Michigan, Ann Arbor, July, 1988b.

Ulaby, F.T., M.W. Whitt, and M.C. Dobson, "Radar Polarimetric Observations of a Tree Canopy," *Proceedings of IGARSS '88 Symposium*, Edinburgh, Scotland, Vol. II, pp. 1005–1008, September 1988c.

Ulaby, F.T., M.W. Whitt, and M.C. Dobson, "Measuring the Propagation Properties of a Forest Canopy Using a Polarimetric Scatterometer," *IEEE Trans. on Antennas and Propagation*, Vol. AP-38, February, 1990.

Ulaby, F.T., and M.C. Dobson, *Handbook of Radar Scattering Statistics for Terrain*, Artech House, Norwood, MA, 1989.

van Zyl, J.J., H.A. Zebker, and C. Elachi, "Imaging Radar Polarization Signatures: Theory and Observation," *Radio Science*, Vol. 22, No. 4, pp. 529–543, July-August 1987.

van Zyl, J.J., "Calibration of Polarimetric Images Using Only Image Parameters and Trihedral Corner Reflector Responses," *IEEE Trans. on Geoscience and Remote Sensing*, Vol. GE-28, 1990.

Whitt, M.W., and F.T. Ulaby, "Millimeter-Wave Polarimetric Measurements of Artificial and Natural Targets," *IEEE Trans. on Geoscience and Remote Sensing*, Vol. GE-26, pp. 562–573, September 1988.

Whitt, M.W., F.T. Ulaby, P. Polatin, and V. Liepa, "A General Polarimetric Calibration Technique," *IEEE Trans. on Antennas and Propagation*, Vol. AP-38, 1990.

# Chapter 6

# Polarimetric Radar System Design

**Howard A. Zebker, Jakob J. van Zyl, and Charles Elachi**

## 6-1 POLARIMETRIC AND CONVENTIONAL RADAR SYSTEMS

### 6-1.1 Introduction

Conventional imaging radars operate with a single, fixed-polarization antenna for both transmission and reception of radio frequency signals. In these systems, a single scattering coefficient is measured for a specific transmitting and receiving polarization combination for many thousands of points in a scene. A result of this single-channel approach is that only one component of the scattered wave, a vector quantity, is measured, and any additional information about the surface contained in the polarization properties of the reflected signals is lost. To ensure that all the information in the scattered wave is retained, its polarization must be measured through a vector measurement process. A device which measures the polarization properties of radiation is called a *polarimeter*, and an imaging radar which permits measurement of the full polarization response of every resolution element in an image is hence called an *imaging radar polarimeter*. Imaging polarimetry permits measurement of the amplitude and relative phase of all polarization configurations of the transmitting and receiving antennas for all independent resolution elements in a scene.

Polarimetric radar systems differ from single-polarization radar systems in that they are sensitive to the polarization states of both the transmitted and received radar signals. While early implementations of radar polarimeters utilized conventional techniques with variable physical antenna polarization (see, for example, Hagfors's work (1967) on lunar scattering), recent developments in technology, most notably the availability of digital data recording and general purpose computers for data reduction, permit the implementation of full imaging radar polarimeters that measure the complete scattering matrix of an object (see Chapter 2). Polarimetry, for the purposes of this chapter, refers to instrumental techniques for the determination of the complex backscatter coefficients of radar echoes for any and all transmitting and receiving polarizations.

In this chapter we will present design approaches and considerations that are particular to polarimetric imaging radar systems, as opposed to general radar design practice. We will present material that is also applicable to single-channel radar systems so that notation, terms, and concepts are adequately defined. Treatment of standard radar design parameters has been presented in many general texts, and we do not intend to duplicate that material here. Because most of the hardware design trade-offs required for optimal operation of polarimetric systems are little different from those encountered in conventional radar design, we will not dwell on them here, and refer the interested reader to any of the general radar textbooks (for example, Brookner (1988) or Barton (1988)). One exception, which we will review, is the calculation of power received by a radar system, the "radar equation," and associated design control tables. This section, while not significantly different from corresponding treatment in the general texts, is required for understanding the relation of the measured signals to components of the radar system.

However, where design concerns are of particular interest to polarimetry, we will discuss the issues in greater detail. Specifically, data processing, compression, and calibration are intimately related to polarimeter design and performance; thus, we will emphasize those aspects in the material that follows. We will detail algorithms now in use and present results that illustrate current technology.

Several airborne polarimeter systems are currently in use or under development by NASA, the Canada Centre for Remote Sensing, and the Environmental Research Institute of Michigan (ERIM). These instruments are described in the appropriate section below. In addition, NASA is in the process of developing spaceborne polarimetric radar systems. The Shuttle

Imaging Radar C (SIR-C) instrument represents the first application of this technology to a spaceborne sensor. NASA also plans to include a synthetic aperture imaging polarimeter on the Earth Observing System (Eos)[1] mission, where the radar will be designed to permit long-term monitoring of the earth's environment from space.

### 6-1.2 Note on Coordinate Systems

In this chapter we will be utilizing the *backscatter alignment coordinate* system (BSA); thus, the matrix quantities described here are characterized by the *overbar* symbol. The details of the coordinates are described in Chapter 2. We have chosen the BSA system for the implementation section of this book as that is the system usually used in radar experiments; indeed the data produced by currently operating imaging radar polarimeters are produced in this context. Conversion from one coordinate system to another is an algebraic procedure, which has been described in the introductory material in Chapter 2.

The material presented in this chapter contrasts with the material in the other chapters of this book in that a different convention is used for the order of the elements of the scattering matrix $\overline{\boldsymbol{S}}$. Here $\overline{S}_{11}$, the upper left entry in the scattering matrix, represents the complex scattering coefficient for horizontal transmitting, horizontal receiving, rather than vertical polarization for both. This convention is used because it is consistent with the data provided by NASA to the user community, and is implicitly assumed in certain of the data processing and calibration algorithms presented below.

Finally, the definition of the Stokes parameters and therefore that of the *Stokes scattering operator* are different (see also Section 2-5.2). The term Stokes scattering operator in this book refers to the quantity denoted as Stokes matrix by van Zyl *et al.*, (1987). This form of the Stokes matrix is calculated by using the following definition of the Stokes vector, expressed in the BSA convention:

$$\mathbf{F} = \begin{pmatrix} \langle |E_h|^2 \rangle + \langle |E_v|^2 \rangle \\ \langle |E_h|^2 \rangle - \langle |E_v|^2 \rangle \\ \langle 2\mathrm{Re}(E_v E_h^*) \rangle \\ \langle 2\mathrm{Im}(E_v E_h^*) \rangle \end{pmatrix} \tag{6.1}$$

The resulting Stokes matrix elements are given in detail below.

[1]The official name of the Earth Observing System is Eos, the Greek goddess of dawn.

### 6-1.3 Polarimetric Measurements

The basic quantity measured by a polarimeter is a complex (amplitude and phase) scattering matrix for each resolution element of the radar image. This differs from conventional approaches in that a vector quantity must be determined, implying that a multichannel radar system be designed that is capable of maintaining phase coherence between antennas of different polarization states. Typical implementations use horizontally and vertically polarized antennas to achieve the necessary polarization diversity; however, any two different antennas could be used (Boerner *et al.*, 1981). We will describe here the horizontal-vertical ($hv$) implementation.

If a radar system is configured to measure all possible combinations available from the $h$- and $v$-antennas (that is, $h$-transmitting, $h$-receiving, and so forth), then the complete scattering matrix for a resolution element may be determined. Knowledge of the scattering matrix $\overline{S}$ permits calculation of the received power for any possible combination of transmitting and receiving antennas; this process is called polarization synthesis (van Zyl *et al.*, 1987; Zebker *et al.*, 1987). The technique has been described in Chapter 2 and is illustrated by several examples in Chapter 7. This important technique gives polarimetry its great advantage over conventional fixed-polarization radars — more information may be inferred about the surface if the full polarimetric properties are known.

### 6-1.4 The Radar Equation and Signal-to-Noise Ratio

Design of a useful polarimetric radar system requires that the desired signal, the radar echo from an area of interest, be distinguished from all other signals present in the radar receiver. These other signals, often called "noise," are present in all real systems and are minimized by careful choice of design parameters. The most important sources of noise in radar design are *thermal noise* generated within the radar hardware itself, thermal noise due to radiometric emission of the target area itself, *ambiguities* (spurious images generated by sampling of the radar return), and *quantization noise* generated when the analog radar signal is digitized for computer processing. Each of these sources may be characterized by a *signal-to-noise ratio* (SNR), a figure of merit representing the comparison of signal power to power in each of the noise terms. Although setting acceptable numerical values for this quantity is difficult without taking into account the exact application of the radar system, a rule of thumb is to set the total level from all noise sources to a few percent of the signal level. This requirement is typically stated as a total noise level of between −10 and −20 dB relative to the signal level.

The radar equation permits calculation of the absolute signal level received by a radar system. When combined with an estimate of the total noise power, the signal-to-noise ratio may be calculated. The radar equation has been previously introduced in Chapter 5, but we briefly redefine terms here so that the following calculations of signal-to-noise ratio for typical implementations are apparent.

If we begin with a transmitter capable of transmitting a signal of power $P_t$, and a transmitting antenna with gain $G_t$, the power density $P_s$ at the scatterer is

$$P_s = \frac{P_t G_t}{4\pi R^2}\,, \tag{6.2}$$

where $R$ is the distance from the transmitter to the scatterer. This follows from the total signal power spread over an area $4\pi R^2$, the area of a sphere of radius $R$, and with the signal distribution determined by the antenna pattern.

The scatterer itself is represented by a quantity denoted *radar cross section*, $\sigma$, which represents an effective scattering area subtended by the scatterer. The units of radar cross section are area, or $m^2$. The power reflected by the scatterer in the direction of the receiving antenna is then the incident power density $P_s$ times the radar cross section $\sigma$. For a monostatic radar, the signal is propagated back to the transmitting antenna, and in the process is subject to a second propagation loss of $4\pi R^2$. The total power received by the radar is then this received power density multiplied by the area of the receiving antenna $A_r$, leading to the following expression for the received power $P_r$:

$$P_r = \frac{P_t G_t}{4\pi R^2} \times \sigma \times \frac{A_r}{4\pi R^2}\,. \tag{6.3}$$

The radar cross section $\sigma$ is usually a function of polarization and viewing geometry, in addition to factors such as dielectric properties and geometric structure. Because a polarimeter is capable of synthesizing any polarization for each of the transmitting and receiving antennas, the radar cross section may be determined for any polarization state by using equation (6.3) above.

In the most general sense, $\sigma$ depends on the detailed viewing geometry, the incident and scattered wave directions as a function of both incidence angle and azimuthal direction. For monostatic polarimeters, such as all those described in this chapter, the incident and scattered directions are the same, and therefore the only angle that matters is the angle of incidence, which is defined as the angle between the wave propagation direction and a unit normal to the imaged surface.

The quantity $\sigma$ measured by the polarimeter is a function of the spatial resolution dimensions of the radar system. In other words, as the resolution of the radar system becomes coarser, the resolved patch of surface grows in size and so does the radar cross section. To eliminate this dependence on resolution, the cross section is usually presented in terms of a normalized radar cross section, called the backscattering coefficient $\sigma^0$. It is defined as the ratio of the measured radar cross section $\sigma$ to the actual area resolved on the surface. For a reasonably flat surface, then, the *backscattering coefficient* is given by

$$\sigma^0 = \frac{\sigma}{A} \tag{6.4}$$

where $A$ is the resolution area of the polarimeter. For a radar system with azimuth resolution $r_a$, slant range resolution $r_r$, and incidence angle $\theta$,

$$\sigma^0 = \frac{\sigma \sin\theta}{r_a r_r} \,. \tag{6.5}$$

Calculation of signal-to-noise ratio now follows from determination of the noise power terms. The most common method of expressing the thermal noise level in a radar system is in terms of an effective "noise temperature" for the receiver-antenna combination, measured in kelvin units (K). The noise power at the antenna for the system is

$$P_n = \mathrm{k} T_n B \tag{6.6}$$

where k is Boltzmann's constant = $1.38 \times 10^{-23}$ (J / K), $T_n$ is the effective noise temperature (K), and $B$ is the bandwidth of the receiver (Hz). If this is the only significant source of noise in the system, the signal-to-noise ratio is

$$\mathrm{SNR} = \frac{P_r}{P_n} \,. \tag{6.7}$$

All of the noise and signal terms may be modified by the data processor and exhibit various dependences on system parameters (this is common to any synthetic aperture radar and not unique to polarimetric systems). We again refer the interested reader to general radar texts for more complete treatments of signals and noise.

The major sources of noise other than thermal noise found in polarimetric synthetic aperture imaging radar systems are ambiguities and quantization noise. Ambiguous signals in range occur when the pulse repetition frequency (PRF) is very high, resulting in interpulse periods shorter than the echo duration, while ambiguous signals in azimuth occur when the doppler frequency spectrum of the echoes is undersampled. Thus, an engineering trade-off must be examined to optimize the system design. Typical systems are characterized by 25 dB suppression of both of these ambiguous echoes. The remaining significant noise term, the quantization noise, is mostly a function of the number of bits available in the digitizing subsystem. Early implementations used only a few bits of precision and this was an important noise source; now 8-bit quantization is commonplace. If the receiver gains are set correctly, the quantization noise terms for 8-bit quantization are suppressed 35 dB or more. As the quantization noise term is similar in nature and level to the corresponding term in conventional synthetic aperture design, we will not discuss it in detail here. However, the ambiguities in particular are often higher in level than in conventional implementations. We now discuss the causes for this phenomenon.

In the polarimeter implementations presented here, polarization diversity is obtained by transmitting two pulses for every corresponding pulse in a single channel system. For example, if a single-channel radar operates at a PRF of 1000 Hz in order to achieve a certain level of ambiguity suppression, a polarimeter with the same performance must transmit pairs of pulses, one $h$-polarized and one $v$-polarized, at the same rate. Because each of these two signals may appear ambiguously in the echo of the other, the effective PRF is doubled and the range ambiguity levels will increase while the azimuth ambiguity level remains constant. If, however, the rate of transmission of pulse pairs is reduced to 500 Hz to keep the range ambiguity level constant, the additional undersampling due to the lower PRF will result in a higher azimuth ambiguity level. (This trade-off will be made more apparent in the next section by examining Fig. 6.2, which illustrates a system timing diagram.) There is therefore no method to obtain identical performance with both systems unless one of the ambiguity levels is already zero. While this situation can occur in airborne systems if the data rate is adequate, it is very difficult to obtain for spaceborne geometries. The ambiguity level trade-off is thus a very important design consideration for spaceborne polarimetric systems.

## 6-1.5 Example Design Control Table

The radar equation may be used to monitor and assess the performance of a given radar implementation. This involves arranging the terms of the equation in a format that permits association of each quantity with its physical interpretation, and providing the means to calculate the signal levels at the input and output terminals of the various subsystems in the overall design. Thus, we can illustrate a design control table that may be used to estimate the performance of a polarimetric radar system. There are many ways to implement such a table, and the approach presented here is not the only applicable method. To compare the designs of two different systems is, however, instructive. In Table 6.1, we compare an aircraft and a spacecraft implementation. The system components used are typical, but do not necessarily reflect existing polarimeters. All values are given in dB, so that they may be added to form the total signal level. Each addition therefore corresponds to a multiplication in the preceding radar and noise equations.

**Table 6.1** Table 6.1 Aircraft and Spacecraft Design Control Table

| | Aircraft case | Spacecraft case |
|---|---|---|
| Transmit power | 37 dBW | 35 dBW |
| Antenna gain | 19 dB | 39 dB |
| $1/4\pi$ | -11 dB | -11 dB |
| $1/R^2$ | -84 dBm$^{-2}$ | -112 dBm$^{-2}$ |
| Ground area | 67 dBm$^2$ | 75 dBm$^2$ |
| $\sigma^0$ | -20 dB | -20 dB |
| $1/4\pi$ | -11 dB | -11 dB |
| $1/R^2$ | -84 dBm$^{-2}$ | -112 dBm$^{-2}$ |
| Antenna area | -4 dBm$^2$ | 15 dBm$^2$ |
| Received Power | -91 dBW | -102 dBW |
| Thermal noise | -126 dBW | -126 dBW |
| SNR | 35 dB | 24 dB |

The transmitted powers are comparable and in the 5 kW range for both the airborne and spaceborne designs — this follows from similar technology available in both environments. However, the antenna gain in the spaceborne case exceeds that of the airborne case by 20 dB, as much bigger antennas may be supported in space than on an aircraft. This increase is mitigated, however, by the increased distance to the surface, hundreds *versus* ten kilometers. Even though more area is illuminated on the ground, for the same backscatter coefficient, the total received power is typically at least 10 dB higher for the aircraft case. Hence, higher performance may be expected in the aircraft implementation, although at a loss of ground area imaged. In the above examples, the thermal signal-to-noise ratios are 35 dB and 24 dB for the aircraft and spacecraft cases, respectively.

We note that for both of the above systems the final performance will be somewhat less than predicted, as these examples do not consider the other noise sources described previously.

# 6-2 SYSTEM DESIGN

## 6-2.1 Scattering Matrix Measurement Approach

The first practical imaging radar polarimeter utilizing the full power of polarization synthesis was the NASA Jet Propulsion Laboratory CV-990 airborne synthetic aperture radar system ( Zebker *et al.*, 1987). This system was configured in the polarimetric mode from May through July 1985. The system was unfortunately destroyed along with the aircraft in a fire during an aborted takeoff attempt in July 1985. The CV-990 radar operated at a wavelength of 24.5 cm (L-band) and had a 4-look resolution of about 10 m by 10 m. The output images represent areas roughly 10 km by 10 km in size. Data were acquired over a variety of sites throughout the United States (see Thompson (1986) for a description of all available data.) This system has recently been replaced by a multifrequency system (aboard a NASA DC-8 aircraft) that operates simultaneously at wavelengths of 68 cm (P-band), 24 cm (L-band), and 6 cm (C-band). The operating characteristics in terms of resolution, number of looks, and output are very similar to the previous L-band radar and all three frequencies are fully polarimetric. This system has been flown over a variety of targets in the United States, including Alaska, and will be deployed to Europe and again in the United States in 1989.

The CV-990 radar has served as the prototype for all other currently operating imaging polarimeters and for the spaceborne systems proposed by NASA for operation in the 1990s. While other implementation approaches are feasible, to date all have utilized horizontal and vertical linearly polarized antennas, transmitting alternately and receiving simultaneously in order to measure all four elements of the scattering matrix. We therefore will describe this system in detail, and we will follow with a discussion of how the systems operated by other institutions differ in implementation.

A system block diagram of the CV-990 imaging radar polarimeter is shown in Fig. 6.1. The four scattering matrix elements were measured by first transmitting a pulse through the horizontally polarized feed of the antenna and subsequently receiving signals simultaneously on both feeds. The next transmission utilized the vertically polarized antenna feed for transmitting, again followed by simultaneous recording from both antenna ports. This approach is shown schematically in Fig. 6.2. With this implementation, each horizontally and vertically polarized transmitted pulse is offset in time from the other by half the interpulse period, typically 0.5 – 1 ms. In the data processor, a synthetic aperture is formed by coherently integrating many pulses; however, pulses corresponding to each different polarization combination are isolated and integrated independently from the other pulses. The total coherent integration required to achieve 3 m single-look resolution consists of approximately 1500 pulses over a 2 s period; therefore the interspersed sets of pulses very nearly correspond to the same synthesized spatial array.

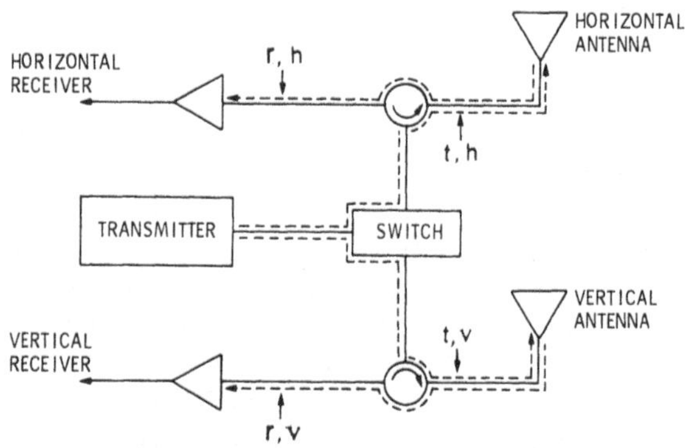


**Figure 6.1** Block diagram showing the major components of the JPL CV-990 SAR system. Also shown are the phase paths encountered by signals in the radar system. The path denoted $t$, $h$ is the phase path of the horizontally polarized transmitted signal, and so forth. These paths are measured from the received signals and compensated in the data processor.

Because the four synthesized spatial arrays are not measured simultaneously, the signals are partially decorrelated. Decorrelation results from the spatial coherence properties of the surface; unless each patch on the surface is viewed from nearly the same direction, the signals will be unrelated. This is the so-called "speckle" effect. In the aircraft implementations discussed here, this effect is of little consequence as the distance traveled by the aircraft during the time interval between pulses is small compared to the system resolution. Consequently, the reduction of cross-correlation intensity in this system for a shift of one interpulse time period out of a coherent integration of 1500 pulses is negligible compared to the total measured intensity, and we can consider to a good approximation that the

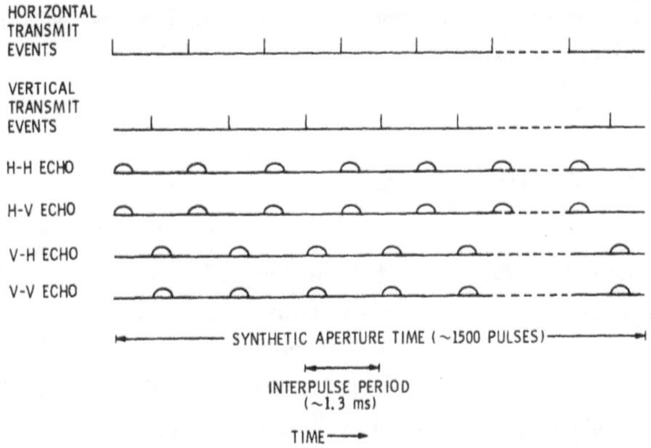


**Figure 6.2** System timing diagram showing that near simultaneity is achieved from overlapping sets of pulses.

processed signals arise from synthesized arrays that are identical in spatial position.

For the spaceborne geometries applicable to the SIR-C and Eos radars, the decorrelation distance is much closer in length to the distance covered by the antenna during the interpulse period. Therefore, significant decorrelation can occur as the spacecraft moves between the times of the horizontal and vertical transmissions. However, as each antenna has traveled along the same path in the sky, shifting the location of the sample points through interpolation allows the different polarization-channel images to be precisely overlayed. Therefore, along-track changes in aspect angle between the two polarization events may be corrected through resampling of one channel with respect to the other, albeit at a cost of additional processing time and complexity.

We note that the decorrelation can also be considered in terms of the doppler bandwidth of the received echo; in the spaceborne case, the pulse repitition frequency is commensurate with the processed bandwidth and no oversampling is possible. So that no significant decorrelation occurs, there must be very little change in doppler shift on the signal from one resolution cell between the times of horizontal and vertical transmissions. In other words, the doppler bandwidth must be rather oversampled. Thus, in the critically sampled situation applicable to spaceborne sensors, the processed synthetic aperture images must be resampled such that the various polarization combinations precisely overlay one another. The resampling is not generally required for the oversampled aircraft case. While optimum performance of even the airborne implementations requires resampling, reasonable performance may be achieved by oversampling the echoes, as is done on the NASA aircraft radars.

## 6-2.2 Calibration Tones

As will be discussed in greater detail below, calibration is much more critical to the design of radar polarimeters than to corresponding single-channel radar systems. This follows from the requirement to implement polarization synthesis (see Chapter 2) by forming linear combinations of the received and processed signals from the various radar channels. Errors in the amplitude and phase balance of the polarization channels will result in errors in synthesized polarization states, making comparison of observations with theory difficult, if not impossible within reasonable bounds. Phase balance may be readily applied by analyzing the received echoes themselves (see Section 6.4), but amplitude balance is not obviously exhibited in properties of the radar signals.

One approach to ensuring amplitude balance throughout most of the system is to inject calibration "tones" into the receiving and processing subsystems. This approach was used successfully in the NASA CV-990 and DC-8 radar polarimeters, and is planned for several future systems. The tones are introduced by adding a small, coherent signal to the received echo at the antenna port. Thus, every subsystem following the antenna in the receiving chain will process the calibration tone together with the signal, and any variations in gain may be normalized by the known level of the tone.

Several strategies have been used for determining the location within the received spectrum and the level of the tone. The CV-990 radar used a tone near the center of the passband of the receiver, resulting in stable estimation of receiver gains, closely related to the receiver transfer function at the frequencies of interest in the echoes. However, the signal level was of an amplitude that interfered with the echo, and therefore limited the system performance. The data processor was modified to reject the exact calibration tone frequency, but this results in a null in the system range spectral response. The end result is that the data are defocused somewhat in the range dimension, leading to poorer resolution capability. The DC-8 implementation uses a different approach; the calibration tone is injected just outside the transmitted signal bandwidth, but still within the receiver bandwidth. Because the signal is located closer to the edge of the receiver response, the estimated signal level is not quite as stable and not as well related to the receiver gain at the frequencies of interest. However, no degradation of the range impulse response occurs.

Either of these strategies may be used for polarimeter design, and each has its advantages and disadvantages. Again, a trade-off study is required, depending on the experiments to be implemented for any proposed system.

# 6-3 POLARIMETRIC RADAR HARDWARE

## 6-3.1 Antennas

The component that drives system design perhaps more than any other is the antenna, as more of the radar performance characteristics derive from the antenna than from any other single component of the flight electronics. There are many technologies available for the construction of radar antennas. All the approaches we list here are capable of supporting polarimetric radar work, although some have found much more widespread use than have others. These technologies include waveguide antennas, microstrip antennas, reflector antennas, and distributed antennas. Each type may be body-mounted or attached to gimbaled systems.

Slotted waveguide antennas are quite efficient at high frequencies as compared to the other antenna types, and can be the antenna of choice for many centimeter-wavelength radar systems. Antennas utilizing slotted waveguides as radiating elements were among the first aircraft antennas used for polarization-sensitive radar systems. NASA operated such a system in the 1970s and early 1980s, but true polarimetry, characterized by the ability to synthesize any polarization, was not possible with this system. While waveguides can be constructed to radiate in the vertical and horizontal polarization states, the NASA system had the antennas configured one above the other on the baggage door of the aircraft. In this configuration, the $h$- and $v$-antennas form an interferometer, as a result of which the phase difference between the received signals becomes a function of range, thereby degrading the measurement accuracy of the polarization state of the backscattered waves. We note that if the antennas had been offset in azimuth rather than range, resampling of the return data stream would have permitted polarization synthesis, as no phase gradient would be present in such a configuration.

Microstrip antennas are the most common types for polarimeters operating today. The NASA CV-990 and DC-8 radar systems use this approach as does the NADC/ERIM P-3 radar. In a microstrip antenna, an array of radiators is etched onto a substrate somewhat like a printed circuit board, and the patches are properly phased and weighted to form a beam of the desired pattern. A significant advantage of microstrip antennas for polarimetric work is that each radiating patch may be driven by two feeds, one of which causes the antenna to emit horizontally polarized radiation and the other causes the antenna to emit vertically polarized radiation. The phase centers for each of these modes are in nearly the same physical location; therefore, the undesired phase gradients in the range direction may be avoided. The great flexibility of the technology permitted by the variability of phasing and weighting of the patches also allows virtually any beam shape for the main pattern. Finally, microstrip antennas are relatively efficient at lower frequencies (less than $\approx$ 10 GHz) if the antennas are not too large.

Reflector antennas have been employed in certain single-channel radar systems; the NASA Magellan Venus radar as well as the Soviet Venera 15 and 16 systems utilize reflector technology. In each of these cases, the antenna feed components for the reflectors were such that linearly polarized radiation was transmitted. Polarimetric operation from this approach could be obtained if two polarization-diverse feed systems were employed. However, care would need to be taken to ensure that the phase centers of the antenna subsystems were at equal positions across the entire range swath; otherwise, the range phase gradient problem would occur.

One of the newest technologies developed for polarimetric radar antennas is known as distributed array technology. The name derives from the characteristic that the antenna radiating surface is actually an active device, where each of many radiating subelements of the antenna is driven by an independent *transmit-receive* (or TR) module. Each module consists of a small power amplifier for transmitting and a low-noise amplifier for receiving, and may drive a radiating patch such as that described above for the microstrip antenna. Thus, the advantages of collocation of the $h$- and $v$-phase centers is realized, along with the ability to combine the individual signals arbitrarily to form many different beam shapes. Another advantage is that, if the power amplifiers are located very close to the radiating elements, we have a very efficient feed network to the patches because the high-power signal paths are quite short. This increase in radiation efficiency is very important at the higher microwave frequencies.

The distributed amplifier approach is being used in the SIR-C radar design. This approach is also likely to be used by the EOS polarimeter, which, due to its higher orbit altitude, requires even greater efficiency than does SIR-C.

### 6-3.2 Flight Electronics

There are a number of competing hardware technologies in current polarimetric radar systems. In particular, the transmitter subsystems differ substantially, depending on the requirements of the individual implementation. Any implementation that allows coherence between two transmissions may be used; this important point is discussed next in this section. However, the majority of criteria determining the choice of transmitter technology are not specific to polarimeters, as opposed to single-channel radars; hence, we only list a few examples of possible approaches.

As noted above, successful polarimetric measurements are permitted only if phase coherence is maintained from pulse to pulse in the end-to-end radar system. This is a very different situation from the conventional synthetic aperture radar technique. In the conventional case, coherence must only be maintained over a time period extending from when the signal is transmitted to when it is received, because only the relationship of each transmission to its received phase is needed to form the synthetic aperture. However, polarimetry requires measurement of phase of one echo with respect to another; therefore, the phase coherence must be maintained over a period corresponding to several transmitted pulses. In theory, coherence must be maintained only for one extra echo, but for a practical system to have excellent coherence for a short time, phase coherence must be obtained for a much longer time.

Existence of some phase noise on the received echo may be tolerated if the purpose of the radar experiment is simply to map a given area, as little defocusing of the signal occurs until a large amount of phase noise is present. For polarimetric analysis, however, phase noise will be indistinguishable from variation of the target scene and hence may be misinterpreted as a geophysical effect in the terrain under investigation. For example, if polarization responses of target areas are analyzed, pixel-to-pixel variations in observed scattering properties will exhibit themselves as increases in the pedestal height of the responses (see Chapters 2 and 7 for a description and illustration of pedestal heights). These variations may be due to high-order surface scattering terms, volume scattering, or actual diversity in scattering mechanism from point to point. In any case, these geophysical phenomena may be important in understanding the radar echo characteristics. Proper design of a useful polarimetric radar system must minimize sources of phase noise so that any observed decorrelation or response pedestals may be ascribed to properties of the surface rather than those of the sensor.

The implication of the preceding statements is that the polarimetric radar system must be designed to much more strict phase-coherence requirements than conventional single-channel radar systems. The first requirement, then, is that a very stable local oscillator subsystem be used to generate all timing and reference signals throughout the system. The stability must be such that the noise spectrum of the oscillator is very low for time differences on the order of the round-trip time for each echo and for the set of echoes forming the full polarimetric measurement. Available technology can provide oscillators with a stability of 1 part in $10^{10}$ to $10^{12}$ over this range.

The next requirement is that the phase of the transmitted signal be controlled to the same level of stability. This is achieved quite readily by solid-state amplifiers and traveling wave tube amplifiers, and these are the transmitters of choice for polarimetric radar systems. Other common radar transmitters, such as magnetrons and klystrons, are characterized by much higher phase noise and may not be suitable for use in polarimeters. The receiver systems must also meet the same strict stability criteria, but this is not difficult with today's technology if a single low-noise oscillator is used to generate all the intermediate frequencies present in the receiving chain.

Finally, the data recording, storage, and processing system must maintain the same high degree of coherence from channel to channel in the polarimeter. Early implementations of synthetic aperture radar systems used optical data recording and processing techniques. This approach is quite satisfactory for single-channel systems as illumination by plane waves of laser light permit data associated with all the points on a section of film to be combined with known phases (that is, coherently), and synthetic apertures may be produced. However, for polarimetric operation, several different channels must be combined with arbitrary phase and amplitude weighting, which is very difficult to implement using standard optical techniques. Hence, digital recording and processing is the desired technology for polarimetric data systems. In addition, digital data acquisition and storage are much more suited to computer analysis of the observed radar scattering characteristics.

The digital systems first must accurately quantize the received signals in time and amplitude. Use of the stable oscillator to derive the analog-to-digital converter (ADC) clock provides accurate time records for the digitized signal. Modern technology easily permits eight or more bits of quantization; typical ADC design now can routinely quantize signals to 8 bits at 50 MHz or higher rates. These data are then either processed in real-time on-board the sensor or recorded for later playback and processing. In this portion of the digital electronics subsystem, the major considerations are the speed and number of digitized samples to be recorded.

### 6-3.3 Data Processing

Early synthetic aperture radar systems were designed with much higher data rates than could be reasonably accommodated by existing computers and therefore the image products were typically produced by optical processors. Sets of lenses and film provided the needed Fourier transform and convolution operations to convert the measured signals into high-resolution images. Optical techniques remained the only low-cost approach to data processing until the late 1970s when digital data acquisition systems and computers became fast and sufficiently cost-efficient.

The advent of digital systems also made possible the easy implementation of complex analysis procedures, including the ability to synthesize polarizations from the measured scattering or Stokes scattering operator elements. To measure the phase of signals in the optical processors is much more cumbersome, particularly when the fields to be compared are not measured exactly simultaneously, the situation encountered for all of the polarimetric radars described previously.

We can illustrate modern radar processing by describing the JPL DC-8 P-, L-, and C-band radar data processor. The signals from the four polarization channels ($hh$, $hv$, $vh$, $vv$) are each sampled at a 45 MHz rate and stored on high-density digital tapes. Each tape holds approximately 15 minutes of data from each of the three radars at all four channels. The total volume of data on a tape is about 72 Gb. Clearly, not all of this data may be processed at one time by the computer; therefore, on the ground, a small portion (4 Gb) is transferred to the disk storage system for further processing. This corresponds to a 10 by 10 km fully polarimetric frame at each frequency.

The data processor reads these data and "compresses" the echo in the range direction by a frequency-domain convolution with the range modulation function. The data are next presummed by a factor of four for further reducing data volume. The presumming operation must be implemented after the range compression — the processor must operate at nonzero squint angles because the antennas on the airplane are body-fixed and not gimbaled. After presumming, the data set for each polarization channel and each frequency is Fourier transformed in the azimuth direction. Here the signal may be interpolated along the curves of the range history for targets at each desired output range; this is the "range migration correction." Finally, the interpolated data set is multiplied by the proper azimuth reference function and inverse transformed to produce the complex high-resolution images for each of the canonical polarization channels. A data flow diagram is illustrated in Fig. 6.3.

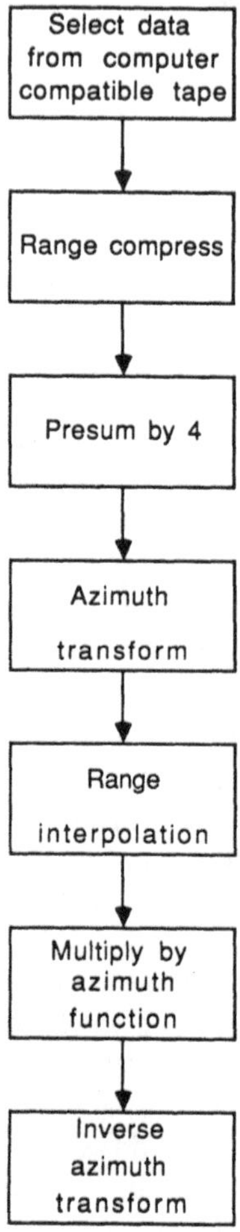


**Figure 6.3** Processor data flow illustration.

Once the data have been processed to form complex, high-resolution, multipolarization scattering matrices for each resolution element in a scene, the data may be converted to Stokes scattering operator form. The procedure used by the JPL processor is given below. As discussed in Chapter 2, the Stokes scattering operator elements $\overline{M}_{mn}$ may be defined in terms of the cross-fields of the original scattering matrix elements $\overline{S}_{ij}$. Note, however, that the coordinate system used in this chapter produces a slightly different set of equations than do the coordinates used for the modeling as described elsewhere in the book. We list here the equations used by JPL for the delivery of polarimetric data:

$$
\begin{aligned}
\overline{M}_{1,1} &= \frac{1}{4}[\,\overline{S}_{hh}\cdot\overline{S}^*_{hh} + \overline{S}_{vv}\cdot\overline{S}^*_{vv} + 2\overline{S}_{hv}\cdot\overline{S}^*_{hv}\,] \\
\overline{M}_{1,2} &= \frac{1}{4}[\,\overline{S}_{hh}\cdot\overline{S}^*_{hh} - \overline{S}_{vv}\cdot\overline{S}^*_{vv}\,] \\
\overline{M}_{1,3} &= \frac{1}{2}\,\mathrm{Re}[\,\overline{S}_{hh}\cdot\overline{S}^*_{hv}] + \frac{1}{2}\mathrm{Re}[\,\overline{S}_{hv}\cdot\overline{S}^*_{vv}] \\
\overline{M}_{1,4} &= -\frac{1}{2}\,\mathrm{Im}[\,\overline{S}_{hh}\cdot\overline{S}^*_{hv}] - \frac{1}{2}\mathrm{Im}[\,\overline{S}_{hv}\cdot\overline{S}^*_{vv}] \\
\overline{M}_{2,2} &= \frac{1}{4}[\,\overline{S}_{hh}\cdot\overline{S}^*_{hh} + \overline{S}_{vv}\cdot S^*_{vv} - 2S_{hv}\cdot S^*_{hv}\,] \\
\overline{M}_{2,3} &= \frac{1}{2}\,\mathrm{Re}[\,\overline{S}_{hh}\cdot\overline{S}^*_{hv}] - \frac{1}{2}\mathrm{Re}[\,\overline{S}_{hv}\cdot\overline{S}^*_{vv}] \\
\overline{M}_{2,4} &= -\frac{1}{2}\,\mathrm{Im}[\,\overline{S}_{hh}\cdot\overline{S}^*_{hv}] + \frac{1}{2}\mathrm{Im}[\,\overline{S}_{hv}\cdot\overline{S}^*_{vv}] \\
\overline{M}_{3,3} &= \frac{1}{2}\,\overline{S}_{hv}\cdot\overline{S}^*_{hv} + \frac{1}{2}\,\mathrm{Re}[\,\overline{S}_{hh}\cdot\overline{S}^*_{vv}] \\
\overline{M}_{3,4} &= -\frac{1}{2}\,\mathrm{Im}[\,\overline{S}_{hh}\cdot\overline{S}^*_{vv}] \\
\overline{M}_{4,4} &= \frac{1}{2}\,\overline{S}_{hv}\cdot\overline{S}^*_{hv} - \frac{1}{2}\,\mathrm{Re}[\,\overline{S}_{hh}\cdot\overline{S}^*_{vv}]\,. 
\end{aligned}
\tag{6.8}
$$

The remaining elements are obtained by using the assumption that the matrix is symmetric.

### 6-3.4 Data Compression

The data rates associated with high-resolution radar polarimetry are very high, and must be reduced in order to facilitate the widespread use of the data. We illustrate a data reduction technique by describing the method used by the Jet Propulsion Laboratory to produce user products for aircraft polarimeter systems. The same techniques will likely be utilized by the SIR-C and Eos instruments.

The volume of data produced by the NASA/JPL DC-8 data processor in the form of the high-resolution complex images is 372 Mbytes for a three-frequency fully polarimetric image. For practical uses, this must be reduced further; hence, a data compression scheme is employed that results in a 30 Mbyte data set. Consider a symmetrical Stokes scattering operator **M** of the form given in Chapter 2. As pointed out before, **M** may contain up to nine independent real numbers. Conventional computer storage would require 36 bytes to store these numbers on disk or tape, nine distinct floating point values of four bytes each. At present, JPL uses a quantization technique whereby those 36 bytes may be reduced to only ten. In addition, the data are further reduced to four-look format, yielding an additional factor of four improvement.

The compression procedure is implemented as follows. Each data set is scaled by a single value for all pixels such that the dynamic range of the total power elements in the Stokes scattering operators ( the $\overline{M}_{11}$ element ) falls within values between $2^{-128}$ and $2^{127}$. Because radar images typically exhibit a dynamic range less than 30 dB, this condition is easily met. The total power for each matrix is then coded into two bytes, one for the exponent in the above range and one for the mantissa. The remaining eight elements are then normalized to the $\overline{M}_{11}$ element in the Stokes scattering operator.

Four of these elements (those related to the cross-products of co- and cross-polarized channels) are observed to be much smaller than $\overline{M}_{11}$; many theories predict these elements to be small or zero for most natural surfaces. Therefore, before encoding, the square root of each of the four normalized elements is calculated. Each of the eight values is next truncated to eight bits (one byte) and the resulting eight bytes are saved. These eight bytes plus two for the total power are then stored, requiring only ten bytes per pixel.

The equations for the compression and decompression operations are given here for reference. For data encoding from the Stokes scattering operator to the compressed 10 byte format, the following apply:

$$\mathrm{byte}(1) = \mathrm{Int}\left(\frac{\log \overline{M}_{11}}{\log 2}\right) , \tag{6.9}$$

where Int is the integer part of the subsequent quantity, and

$$\mathrm{byte}(2) = \mathrm{Int}\left[254 \cdot \left(\frac{\overline{M}_{11}}{2^{\mathrm{byte}(1)}} - 1.5\right)\right] . \tag{6.10}$$

We then compute a normalization factor $x$, approximately equal to $\overline{M}_{11}$.

Choosing this factor for normalization results in slightly smaller errors than does using $\overline{M}_{11}$ directly:

$$x = \left(\frac{\mathrm{byte}(2)}{254} + 1.5\right) \cdot 2^{\mathrm{byte}(1)} . \tag{6.11}$$

The remaining eight bytes are coded as follows:

$$
\begin{aligned}
\text{byte}(3) &= 127 \cdot \overline{M}_{12}/x \\
\text{byte}(4) &= 127 \cdot \text{sign}(\overline{M}_{13}/x) \cdot \sqrt{|\overline{M}_{13}/x|} \\
\text{byte}(5) &= 127 \cdot \text{sign}(\overline{M}_{14}/x) \cdot \sqrt{|\overline{M}_{14}/x|} \\
\text{byte}(6) &= 127 \cdot \text{sign}(\overline{M}_{23}/x) \cdot \sqrt{|\overline{M}_{23}/x|} \\
\text{byte}(7) &= 127 \cdot \text{sign}(\overline{M}_{24}/x) \cdot \sqrt{|\overline{M}_{24}/x|} \\
\text{byte}(8) &= 127 \cdot \overline{M}_{33}/x \\
\text{byte}(9) &= 127 \cdot \overline{M}_{34}/x \\
\text{byte}(10) &= 127 \cdot \overline{M}_{44}/x,
\end{aligned} \tag{6.12}
$$

where sign $= \pm\ 1$, reflecting the sign of the quantity in brackets.

To reconstruct the Stokes scattering operator from the reduced data, the following inverse operations are required:

$$
\begin{aligned}
\overline{M}_{11} &= \left(\frac{\text{byte}(2)}{254} + 1.5\right) \cdot 2^{\text{byte}(1)} \\
\overline{M}_{12} &= \text{byte}(3) \cdot \frac{\overline{M}_{11}}{127} \\
\overline{M}_{13} &= \text{sign}(\text{byte}(4)) \cdot \left(\frac{\text{byte}(4)}{127}\right)^2 \cdot \overline{M}_{11} \\
\overline{M}_{14} &= \text{sign}(\text{byte}(5)) \cdot \left(\frac{\text{byte}(5)}{127}\right)^2 \cdot \overline{M}_{11} \\
\overline{M}_{23} &= \text{sign}(\text{byte}(6)) \cdot \left(\frac{\text{byte}(6)}{127}\right)^2 \cdot \overline{M}_{11} \\
\overline{M}_{24} &= \text{sign}(\text{byte}(7)) \cdot \left(\frac{\text{byte}(7)}{127}\right)^2 \cdot \overline{M}_{11} \\
\overline{M}_{33} &= \text{byte}(8) \cdot \frac{\overline{M}_{11}}{127} \\
\overline{M}_{34} &= \text{byte}(9) \cdot \frac{\overline{M}_{11}}{127}
\end{aligned}
$$

**Figure 6.4** Results of synthesizing a set of linear, co-polarized antennas and applying them to data collected over San Francisco, California. This figure consists of twenty images; the upper left image corresponds to horizontally polarized antennas used for transmission and reception of signals. Scanning from left to right and from top to bottom, the ellipse orientation angle of the linear polarization is advanced 2.5° per image and the square root of the resulting scattering cross section is displayed. The lower right image thus corresponds to a linear polarization with an ellipse orientation angle of 47.5°. For a complete discussion, see Zebker *et*

$$\overline{M}_{44} = \text{byte}(10) \cdot \frac{\overline{M}_{11}}{127}$$
$$\overline{M}_{22} = \overline{M}_{11} - \overline{M}_{33} - \overline{M}_{44}. \tag{6.13}$$

### 6-3.5 Examples of Polarimetric Observations

Detailed observations made by imaging radar polarimeter systems form the subject of Chapter 7. Here, however, we illustrate that the above CV-990 radar implementation may observe polarization-sensitive effects. Typical image products from the JPL CV-990 polarimeter are discussed in detail by Zebker *et al.* (1987). They present, for example, a figure consisting of 20 radar images, each at a slightly different synthesized linear polarization angle, varying from horizontal to linear at 47.5°. This set of images is also shown in Fig. 6.4. These images yield evidence for different types of scattering occurring in different parts of the surface. This portion of San Francisco includes the Pacific Ocean, an urban portion of the city, and the vegetated Golden Gate Park. Here we only note that each area has its own variation of brightness with angle, as discussed in more detail by Zebker *et al.* (1987).

## 6-4 POLARIMETRIC RADAR CALIBRATION

### 6-4.1 Background

Polarimetric radar systems differ from conventional implementations in the degree to which the measurements must be calibrated before they are useful for interpretation of scattering behavior. While for many applications single-channel radars need no calibration, polarimetric systems require that data from the various polarization-diverse channels be combined with the proper amplitude and phase weights to synthesize known polarization states. Good design practice can result in channel amplitudes that are balanced adequately, within one or two dB, but special techniques are required for phase balance and correction of nonideal antenna effects such as crosstalk between horizontal and vertical radiation patterns. In addition, the utility of the data is greatly improved if some form of absolute radiometric calibration is used. In this section, we present techniques to implement these calibration needs, and we detail algorithms used by existing polarimeter systems.

No single aspect of imaging radar system design is currently in flux to a greater degree than system calibration. One motivation for the increasing interest in this activity is due to the precise measurements of system parameters required for accurate polarization synthesis. Thus, a number of approaches to calibration are currently under development, including work at the Massachusetts Institute of Technology (Barnes, 1986), the University of Michigan (Sarabandi *et al.*, 1990; Whitt *et al.*, 1990), the Environmental Research Institute of Michigan (Sheen *et al.*, 1989), and the Jet Propulsion Laboratory (van Zyl, 1990; Zebker and Lou, 1990; Freeman *et al.*, 1990; Freeman and Curlander, 1990). Early work in the field of polarimetric calibration focused primarily on the development of calibration devices that may be deployed to aid in assessing system performance.

Although the development of specialized targets in principle permits measurement and characterization of a polarimetric radar system, practical implementation is hindered by a variety of constraints. These devices rely on extremely precise measurements of system performance. For example, the correction of antenna crosstalk impurities by using ground targets such as active transponders requires that the amplitude and phase of each interfering term be known to a precision 10 or 100 times better than that of the term, so that it may be nearly nulled. This sets unattainably high limits on the required cross section of calibration targets. In addition, such devices only measure the performance at one position in the radar swath, and many devices will be required to correct even one image.

Consequently, the emphasis now is on development of techniques that permit the data to provide many of the necessary parameters for characterizing the system performance and correcting the observations for deficiencies. We will concentrate on those algorithms in the section below. We note that absolute radiometric calibration still entails known targets in the scene to be imaged; however, the level of precision required for radiometric calibration is much less than for crosstalk removal. Relatively small and simple trihedral corner reflectors have proved effective for this task.

## 6-4.2 Phase Calibration

After standard image formation and associated data processing, phase calibration is the next required step in utilization of imaging radar polarimeter data. Errors in the phase relationship between polarization channels result in incorrect synthesis of polarization states, and may lead to false conclusions about the nature of scatterers on the surface. The Stokes scattering operators measured by an imaging radar polarimeter provide in themselves adequate information for accurate phase calibration of the observed polarimetric characteristics of a surface. This property allows the data to be reduced in volume in an operational synthetic aperture radar correlator with no prior knowledge of the conditions at the surface. No ground calibration equipment is necessary, as all important parameters are directly derived from the data.

Consider the system model shown in Fig. 6.1, where we illustrate the signal phase paths. The true scattering matrix for a resolution element $\overline{\mathcal{S}}$ is given by

$$\overline{\mathcal{S}} = \begin{pmatrix} \overline{S}_{hh} & \overline{S}_{hv} \\ \overline{S}_{vh} & \overline{S}_{vv} \end{pmatrix} , \tag{6.14}$$

where $\overline{S}_{ij}$ refers to the complex scattering amplitude for $j$-transmitting, $i$-receiving polarization, due to the as yet uncorrected path differences in the radar hardware. Hence, what we actually measure is a receiving matrix $\overline{\mathcal{R}}$:

$$\overline{\mathcal{R}} = \begin{pmatrix} \overline{S}_{hh} \exp j(\phi_{t,h} + \phi_{r,h}) & \overline{S}_{hv} \exp j(\phi_{t,v} + \phi_{r,h}) \\ \overline{S}_{vh} \exp j(\phi_{t,h} + \phi_{r,v}) & \overline{S}_{vv} \exp j(\phi_{t,v} + \phi_{r,v}) \end{pmatrix} , \tag{6.15}$$

where the phase factors correspond to the paths shown in Fig. 6.1. For notational simplicity, we can factor out the phase of the $vv$ term, resulting in

$$\overline{\mathcal{R}} = \exp j(\phi_{t,v} + \phi_{r,v}) \begin{pmatrix} \overline{S}_{hh} \exp j(\phi_t + \phi_r) & \overline{S}_{hv} \exp j\phi_r \\ \overline{S}_{vh} \exp j\phi_t & \overline{S}_{vv} \end{pmatrix} , \tag{6.16}$$

where $\phi_t = \phi_{t,h} - \phi_{t,v}$ and $\phi_r = \phi_{r,h} - \phi_{r,v}$.

Now, in the backscatter case, reciprocity dictates that $\overline{S}_{hv} = \overline{S}_{vh}$. Thus, we can solve for the difference $\phi_t - \phi_r$ by examining the complex product $\overline{R}_{hv}\overline{R}^*_{vh} = \overline{S}_{hv}\overline{S}^*_{vh} \exp j(\phi_r - \phi_t)$ averaged over the image. For each pixel, the argument of this product will vary slightly as the system noise in each of the channels is uncorrelated, but we may average over many pixels to obtain the mean value, which depends mainly on the phase path difference in which we are interested.

Finally, if we know the dominant scattering mechanism of at least one point in the image, we can determine the rest of the factors necessary for phase calibration. Typically, a looser constraint is required: knowledge of the predicted phase difference of the $hh$ and $vv$ signals suffices for determination of $\phi_t + \phi_r$. These two sum-and-difference equations may now be solved for the individual phases $\phi_t$ and $\phi_r$. These factors relate the measured matrix $\overline{\mathcal{R}}$ to $\overline{\mathcal{S}}$, which we may determine from

$$\overline{\mathcal{S}} = \begin{pmatrix} \overline{R}_{hh} \exp -j(\phi_t + \phi_r) & \overline{R}_{hv} \exp -j\phi_r \\ \overline{R}_{vh} \exp -j\phi_t & \overline{R}_{vv} \end{pmatrix} . \tag{6.17}$$

Note that this determination omits the leading phase factor of equation (6.16), as only the relative phases are preserved.

The Stokes scattering operators, rather than the individual scattering matrices, may also be used in the phase calibration procedure. This step is important as it allows averaging ("looks") of the radar data prior to distribution, which makes feasible many more applications of polarimetric radar data. We can ignore the leading factor $\exp j(\phi_{t,v} + \phi_{r,v})$ in equation (6.16) as calculation of the Stokes matrix removes this term. The Stokes scattering operator elements are defined only in terms of the relative phases of the polarimetric channels (see (6.7)). The first step in data volume reduction is that the scattering matrices be made symmetrical. We repeat the estimation of $\overline{R}_{hv}\overline{R}^*_{vh}$ as above, and store the value of the average phase difference over the full image in a header that accompanies the data set. We then apply equation (6.15) and obtain a scattering matrix for each pixel in which the off-diagonal elements possess the same phase factor. Because the off-diagonal elements are now in phase, they may be averaged without introducing errors. We thus modify $\overline{\mathcal{R}}$ to get a new matrix $\overline{\mathcal{Z}}$:

$$\overline{\mathcal{Z}} = \begin{pmatrix} \overline{S}_{hh} \exp j(\phi_t + \phi_r) & \overline{S}_{hv} \exp j\phi_r \\ \overline{S}_{vh} \exp j\phi_t \exp -j(\phi_t - \phi_r) & \overline{S}_{vv} \end{pmatrix} \tag{6.18}$$

or

$$\overline{\mathcal{Z}} = \begin{pmatrix} \overline{S}_{hh} \exp j(\phi_t + \phi_r) & \overline{S}_{hv} \exp j\phi_r \\ \overline{S}_{vh} \exp j\phi_r & \overline{S}_{vv} \end{pmatrix} . \tag{6.19}$$

We next obtain the symmetric matrix $\overline{\mathcal{Y}}$ from $\overline{\mathcal{Z}}$:

$$\overline{\mathcal{Y}} = \frac{1}{2}(\overline{\mathcal{Z}} + \tilde{\overline{\mathcal{Z}}}) , \tag{6.20}$$

where $\tilde{\overline{\mathcal{Z}}}$ is the transpose of $\overline{\mathcal{Z}}$.

Knowledge of $\overline{\mathcal{Y}}$ allows us to calculate the Stokes scattering operator elements by using (6.8). The values of $\overline{\mathcal{Y}}$ and $\overline{\mathcal{S}}$ do not yet agree because the data have not yet been phase calibrated, which must be accomplished for correctly decoding the polarizations. These elements of $\overline{\mathcal{Y}}$ are the data values produced by the standard data processor, before phase calibration:

$$\begin{aligned}
\overline{Y}_{hh}\overline{Y}^*_{hh} &= \overline{S}_{hh}\overline{S}^*_{hh} \\
\overline{Y}_{vv}\overline{Y}^*_{vv} &= \overline{S}_{vv}\overline{S}^*_{vv} \\
\overline{Y}_{hv}\overline{Y}^*_{hv} &= \overline{S}_{hv}\overline{S}^*_{hv} \\
\overline{Y}_{hh}\overline{Y}^*_{vv} &= \overline{S}_{hh}\overline{S}^*_{vv} \exp j(\phi_t + \phi_r) \\
\overline{Y}_{hh}\overline{Y}^*_{hv} &= \overline{S}_{hh}\overline{S}^*_{hv} \exp j\phi_t \\
\overline{Y}_{vh}\overline{Y}^*_{vv} &= \overline{S}_{vh}\overline{S}^*_{vv} \exp j\phi_r \ .
\end{aligned} \tag{6.21}$$

We again must select an area in the image for which the scattering phase difference between $hh$ and $vv$ is known; it is usually chosen from an area where the phase difference theoretically is near zero. Note that from (6.21) the measurement of $\overline{Y}_{hh}\overline{Y}^*_{vv}$ yields an estimate of the sum $\phi_t + \phi_r$. Recall that we previously saved the value $\phi_t - \phi_r$ in the header of the data file; these two equations in two unknowns are solved for the individual values.

Thus, what remains is to apply the newly derived phases $\phi_t$ and $\phi_r$ to the stored Stokes scattering operators. This results in phase-calibrated matrix cross-products related to those of $\overline{\mathcal{Y}}$ by

$$\begin{aligned}
\overline{O}_{hh}\overline{O}^*_{hh} &= \overline{Y}_{hh}\overline{Y}^*_{hh} \\
\overline{O}_{vv}\overline{O}^*_{vv} &= \overline{Y}_{vv}\overline{Y}^*_{vv} \\
\overline{O}_{hv}\overline{O}^*_{hv} &= \overline{Y}_{hv}\overline{Y}^*_{hv} \\
\overline{O}_{hh}\overline{O}^*_{vv} &= \overline{Y}_{hh}\overline{Y}^*_{vv} \exp -j(\phi_t + \phi_r) \\
\overline{O}_{hh}\overline{O}^*_{hv} &= \overline{Y}_{hh}\overline{Y}^*_{hv} \exp -j\phi_t \\
\overline{O}_{vh}\overline{O}^*_{vv} &= \overline{Y}_{vh}\overline{Y}^*_{vv} \exp -j\phi_r \ .
\end{aligned} \tag{6.22}$$

Note that these cross-products correspond to the correctly phase-calibrated Stokes matrix (compare with (6.21) above), as the cross-products associated with $\overline{\mathcal{O}}$ accurately depict the original scattering matrix elements with no auxiliary phase factors.

We note that there is a formal ambiguity in determining $\phi_t$ and $\phi_r$ from the two equations as each of the sum and difference may only be measured modulo $2\pi$. Thus, $\phi_t$ and $\phi_r$ may each possess errors of $\pi$; that is, each may be the negative of the true value. The effect of this in practical situations is minimal, as it merely causes the component of linearly polarized radiation at 45° to be interpreted as polarized at 135°, and the reverse. For azimuthally isotropic scatterers, then, no change would be observed; this is the case for most natural surfaces. However, for applications in which the distinction between 45° and 135° is important, this possible ambiguity must be considered.

### 6-4.3 Crosstalk Calibration and Removal

We have already noted the importance of estimating amplitude and phase balance of the polarization channels in a high-resolution polarimeter to permit accurate polarization synthesis. A remaining error source is due to imperfections in the radar antenna, resulting in impure polarization states for the individual polarization combinations measured directly by the hardware.

The technique presented here is currently applied to certain scenes of compressed data produced by the JPL aircraft SAR processor. In the process of compressing the data, the measured scattering matrices are symmetrized, and so we may consider only symmetrical matrices when modeling the resulting compressed measurements. Thus, if $\overline{\mathcal{Y}}$ is the symmetrized measured scattering matrix, then

$$\overline{\mathcal{Y}} = \tilde{\overline{\mathcal{Y}}}. \tag{6.23}$$

Now consider the effects of the transmitting and receiving systems on the symmetrized scattering matrix. Let $\overline{\mathcal{S}}_R$ represent the receiving system, and $\tilde{\overline{\mathcal{S}}}_T$ represent the transmitting system. We want to use a system model of the form:

$$\overline{\mathcal{Y}} = A\, e^{i\phi}\, \overline{\mathcal{S}}_R \overline{\mathcal{S}}\, \overline{\mathcal{S}}_T\ , \tag{6.24}$$

where $\overline{\mathcal{S}}$ is the desired scattering matrix, $A$ is an overall absolute amplitude factor and $\phi$ is an overall absolute phase. As discussed previously, the absolute phase $\phi$ is lost when received power is calculated during polarization synthesis. Therefore, calibrating $\phi$ is not necessary. In the backscatter case, reciprocity dictates that

$$\overline{\mathcal{S}} = \tilde{\overline{\mathcal{S}}}\ . \tag{6.25}$$

From (6.23), (6.24), and (6.25), we then find that we must model

$$\overline{\mathcal{S}}_R = \tilde{\overline{\mathcal{S}}}_T\ . \tag{6.26}$$

Physically, this means that the system must be modeled as reciprocal. While this may not be strictly true, the data being made symmetrical

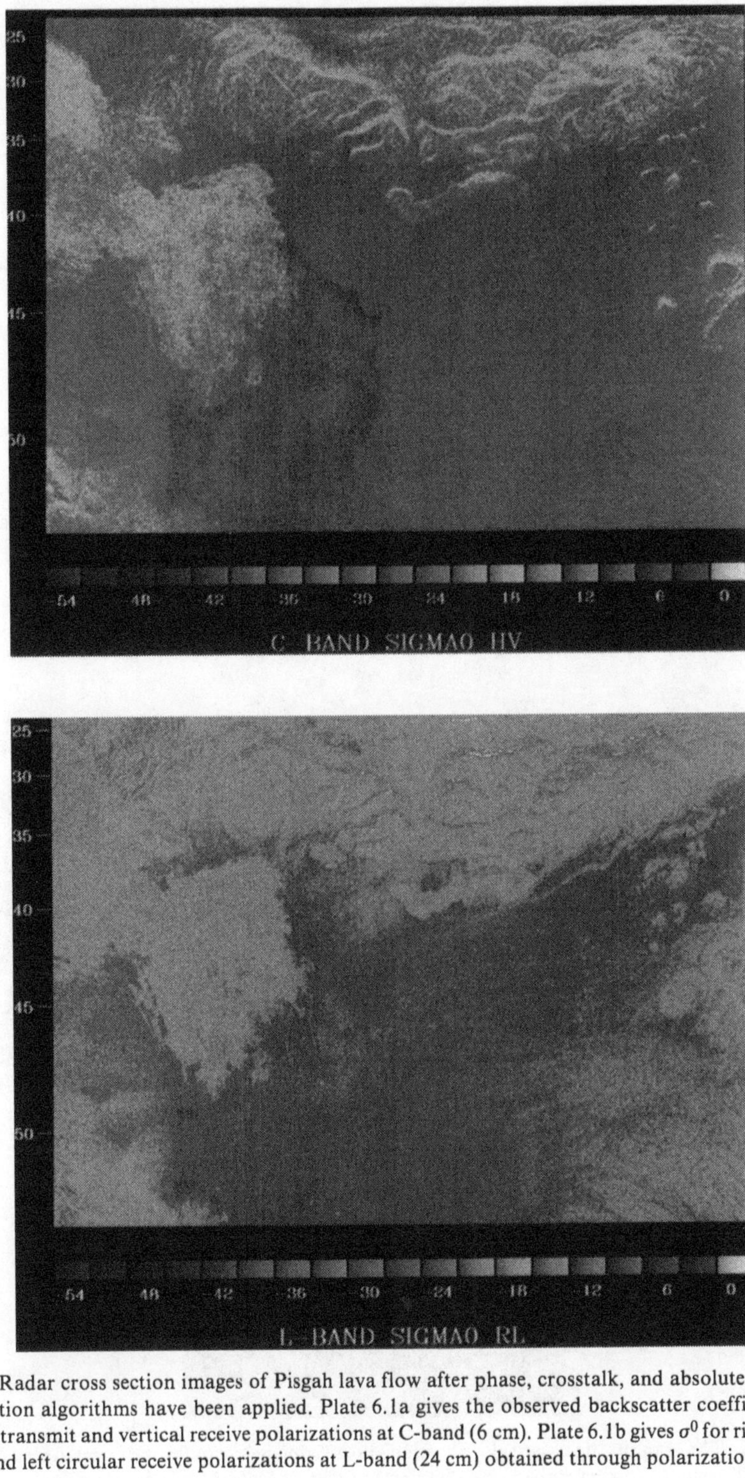


**Plate 6.1.** Radar cross section images of Pisgah lava flow after phase, crosstalk, and absolute calibration and correction algorithms have been applied. Plate 6.1a gives the observed backscatter coefficient $\sigma^0$ for horizontal transmit and vertical receive polarizations at C-band (6 cm). Plate 6.1b gives $\sigma^0$ for right circular transmit and left circular receive polarizations at L-band (24 cm) obtained through polarization synthesis.

**Plate 7.1.** Results of the unsupervised classification scheme of scattering behavior in the San Francisco Bay area, California. Pixels colored blue are areas with polarization characteristics similar to those predicted by odd numbers of reflections, and those colored red are areas which behave similar to even numbers of reflections. Green pixels are areas which exhibit a large amount of diffuse scattering, and yellow pixels are those that do not belong to one of the three classes mentioned. Radar illumination is from the top and aircraft motion is from left to right. The image represents an area of about 11 km on a side.

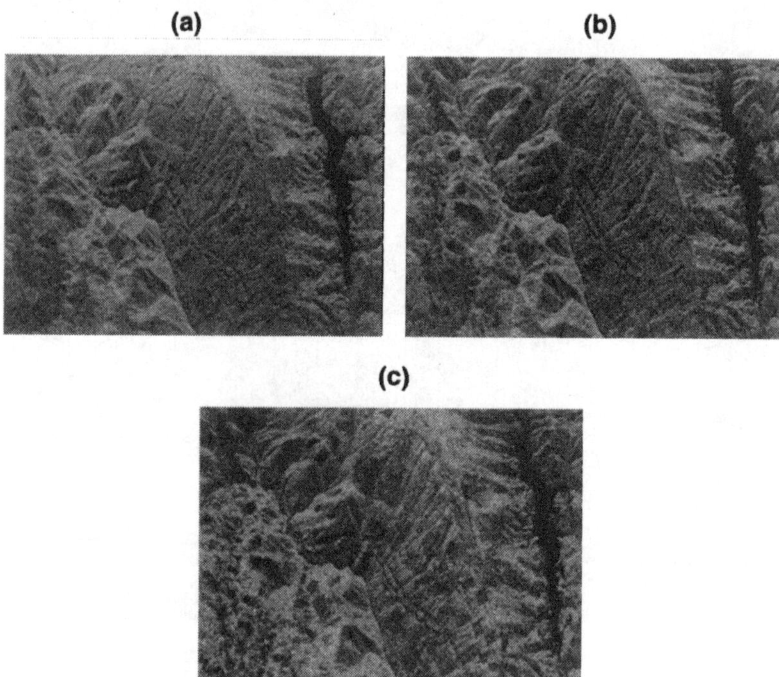


**Plate 7.2.** Results of the unsupervised classification of scattering behavior for the forested area near Mount Shasta, California. Colors have the same meaning as those in Plate 7.1. Radar illumination is from the top and aircraft motion is from the right to the left. Images represent an area of about 5 km from top to bottom and about 12 km from left to right.

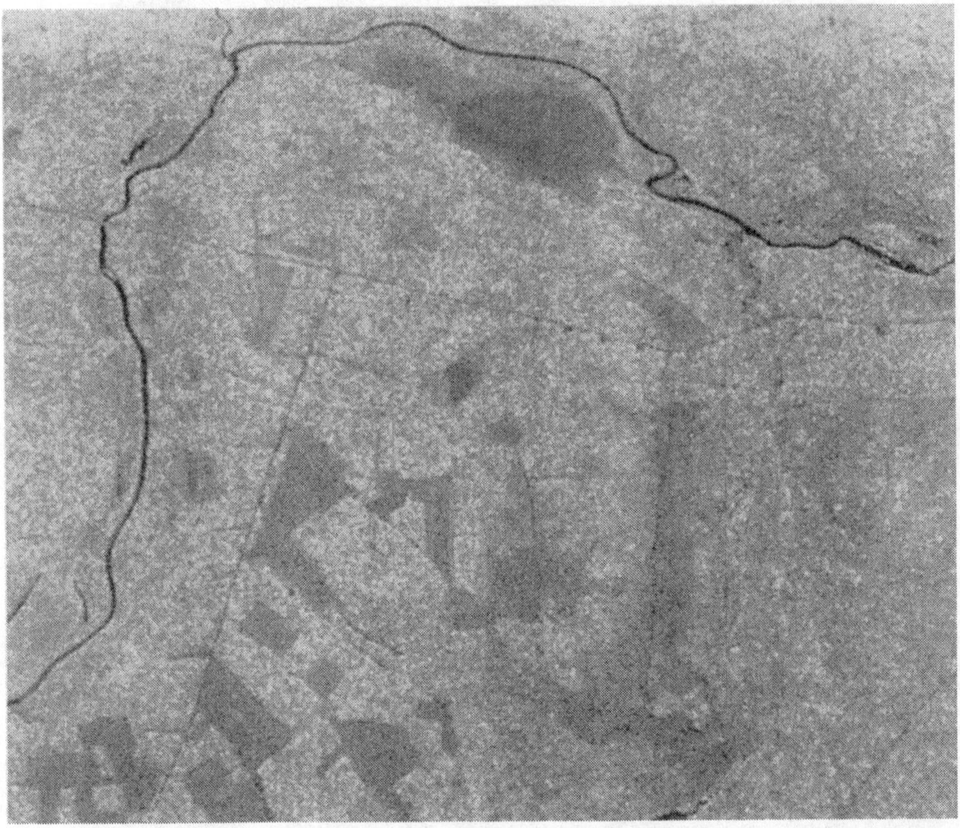

**Plate 7.3.** Results of the unsupervised classification of scattering behavior in a forested area near Moosehead Lake, Maine, L-band data. Colors have the same meaning as in Plate 7.1. Radar illumination is from the top and aircraft motion is from left to right. The image represents an area of about 11 km on a side.

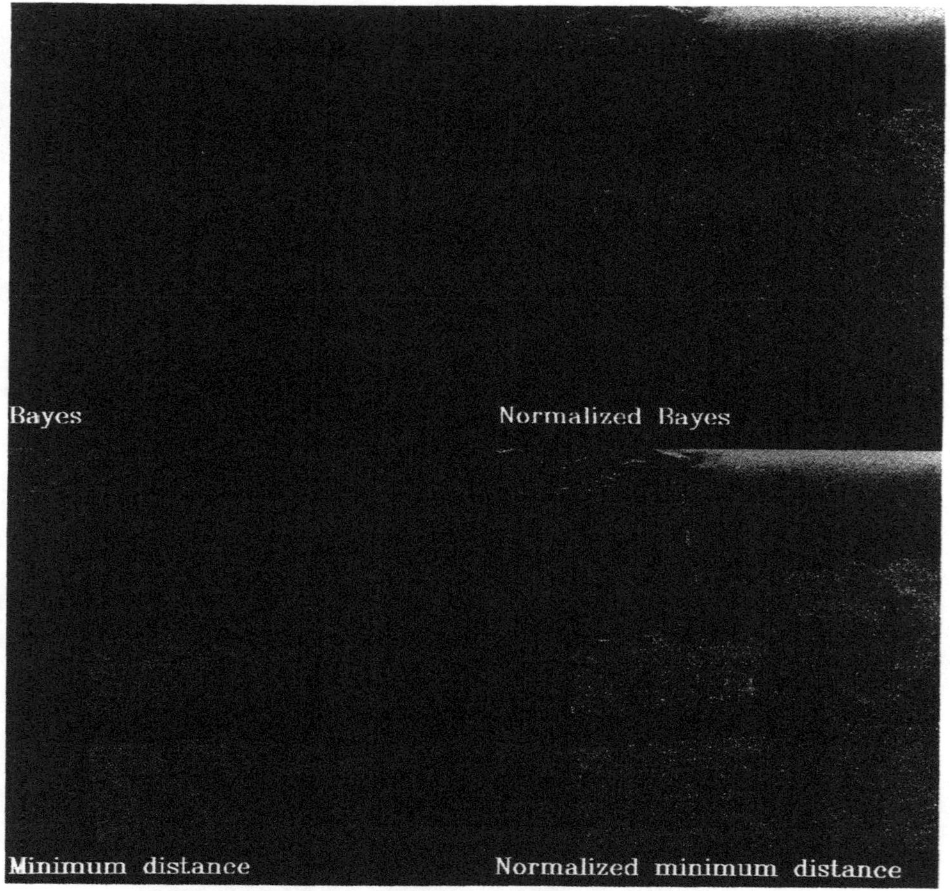


**Plate 7.4.** Supervised classifications, (a) unnormalized maximum likelihood, (b) normalized maximum likelihood, (c) unnormalized minimum distance, and (d) normalized minimum distance, of the San Francisco Bay area, based on the training areas shown in Figure 7.20. Pixels colored red are classified as urban areas, those colored green as vegetated areas, and those colored blue as ocean. Some misclassification, due to high specular return, is visible in the unnormalized classification

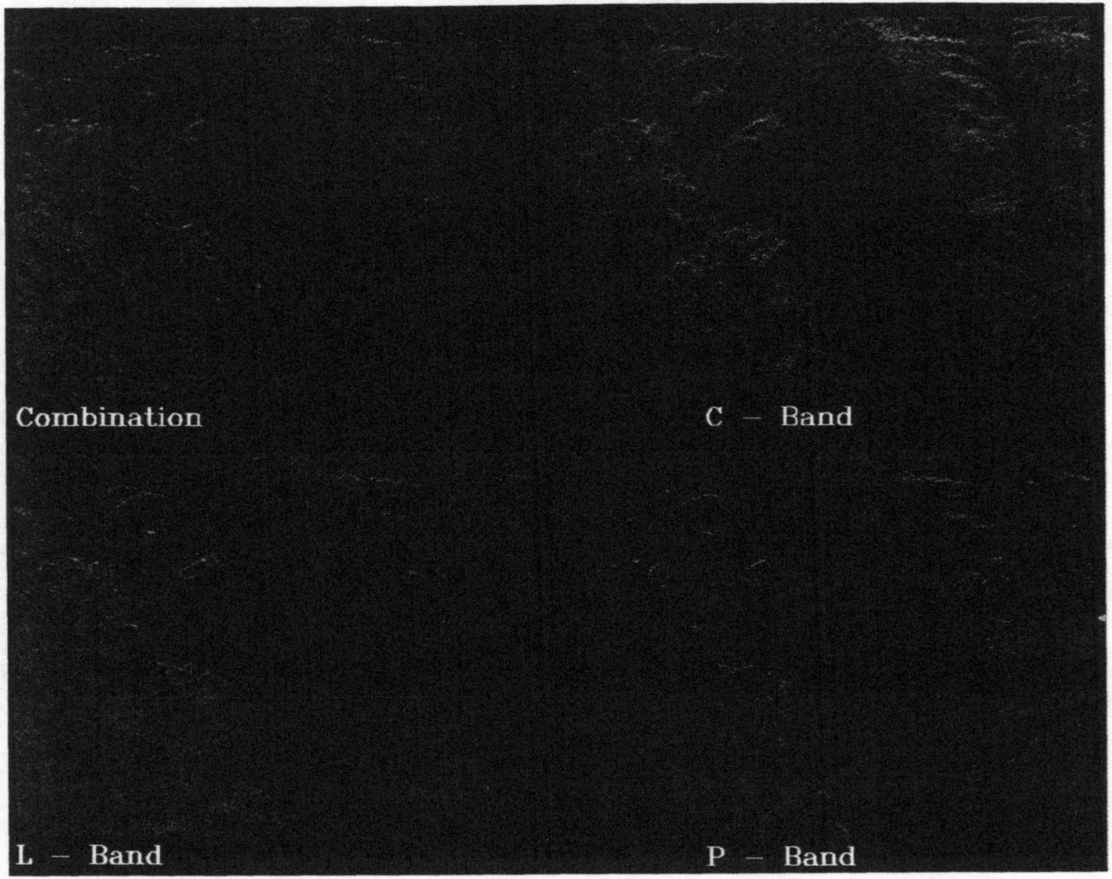


**Plate 7.5.** Normalized maximum likelihood classifications of a forested area near Mount Shasta, California, based on the training areas shown in Figure 7.22 and C-band, L-band, P-band and all three frequency data sets. Pixels colored red are classified as burned areas, those colored blue as open water. Consistent with the results of Plate 7.2, the burned and unburned areas are almost indistinguishable in the C-band results. The full three-frequency result is significantly less noisy than any of the single-frequency results.

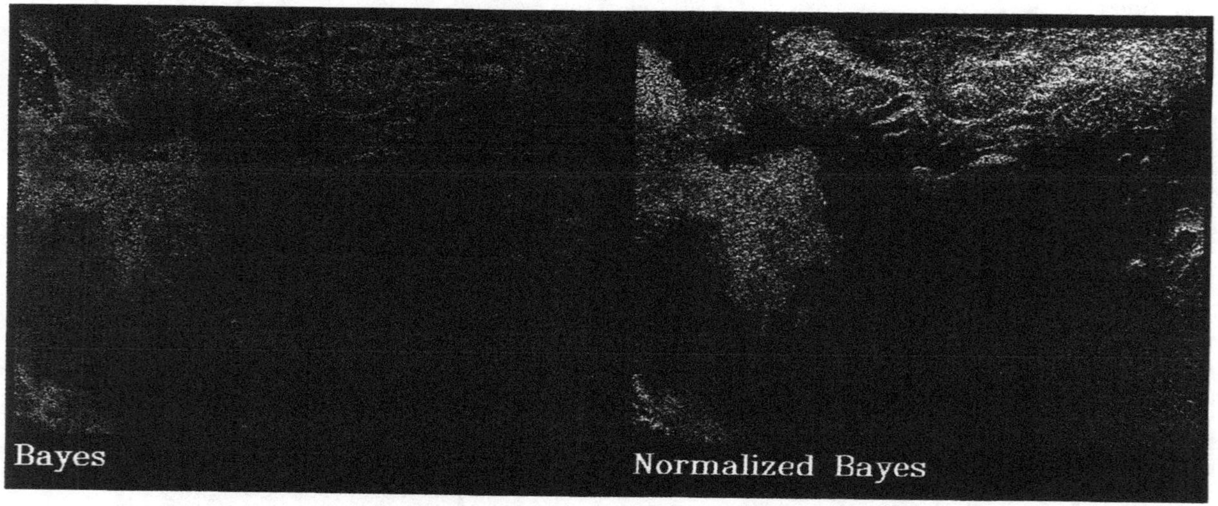


**Plate 7.6.** Normalized and unnormalized Bayes classification of the C-band data of the Pisgah lava flow area, California, based on the training areas shown in Figure 7.23. Pixels colored blue are classified as Phase I (pahoehoe) lava, those colored red as Phase II (a'a) lava, those colored green as Phase III (pahoehoe) lava, those colored yellow as alluvial surfaces, and those colored purple as dry lake bed. The results of the normalized classifier are significantly more noisy than that of the unnormalized classifier, suggesting that, for this scene, the overall brightness carries much of the information.

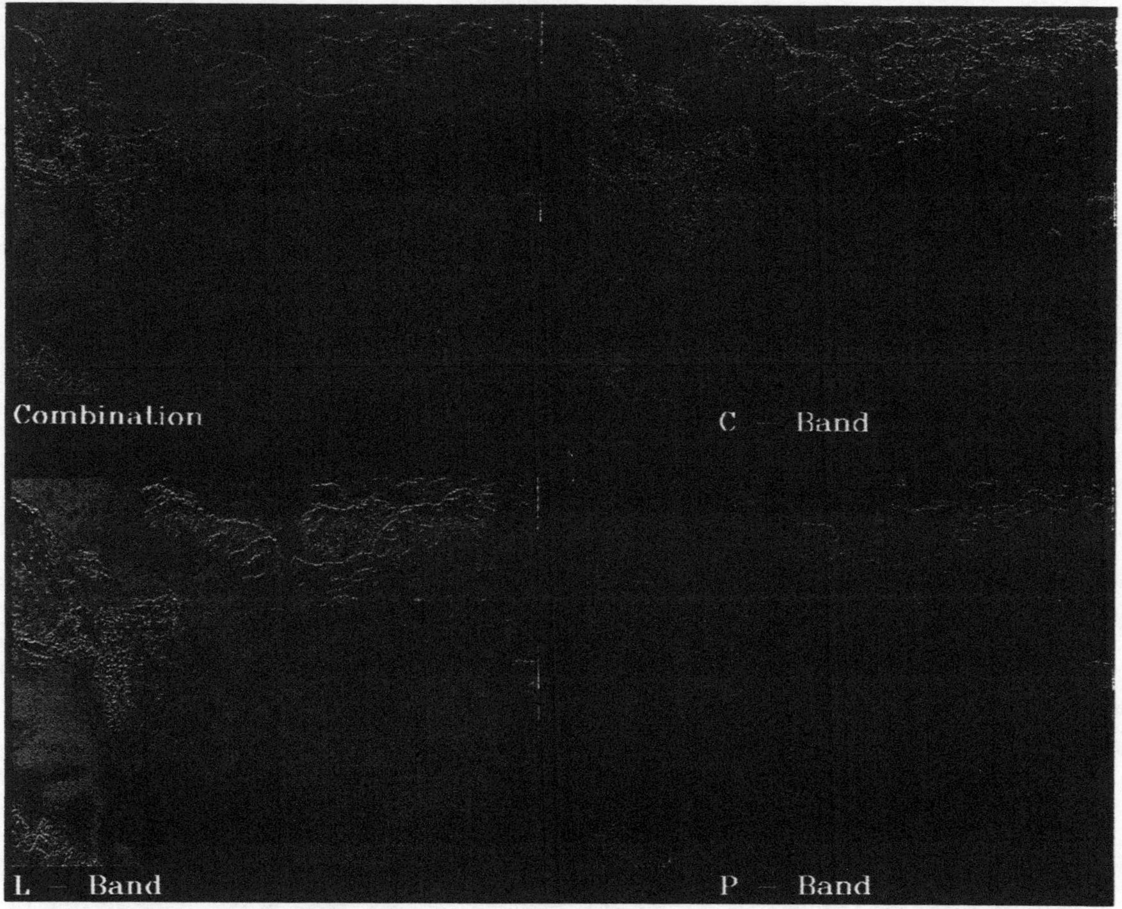


**Plate 7.7.** Unnormalized maximum likelihood classifications of the Pisgah lava flow area, California, based on the training areas shown in Figure 7.23 and C-band, L-band, P-band, and all three frequency data sets. Colors have the same meaning as in Plate 7.6. The C-band results are the noisiest. The box-shaped areas in the L-band and the vertical bars in the P-band results are due to interference signals from unknown sources. The full three-frequency result is significantly less noisy than any of the single-frequency results.

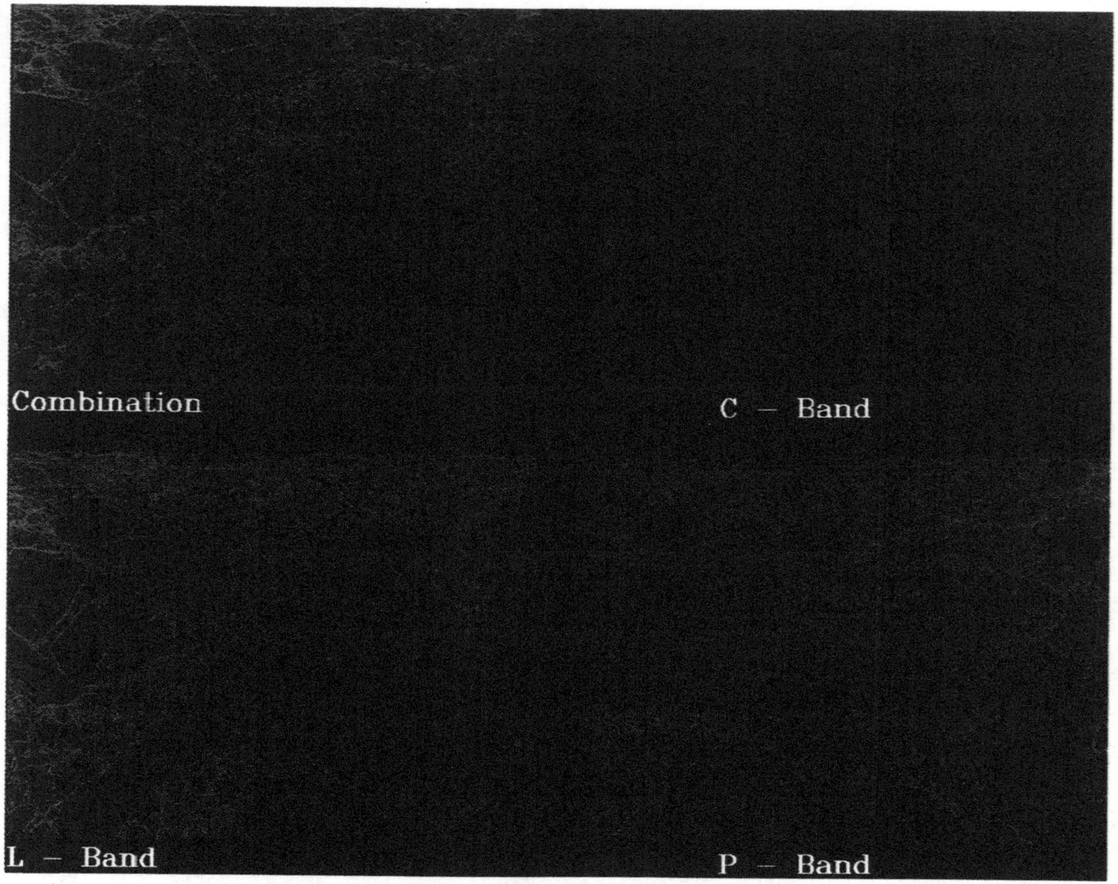


**Plate 7.8.** Unnormalized maximum likelihood classifications of part of the Beaufort Sea, Alaska, based on the training areas shown in Figure 7.24 and C-band, L-band, P-band, and all three frequency data sets. Pixels colored red are classified as multiyear ice, those colored green as first-year ice, and those colored blue as pressure ridges. Unlike all other cases considered so far, the C-band

during compression forces us to adopt this system model. A consequence is that the system parameters estimated may not have any simple relationship to real physical system parameters after the measured scattering matrix is made symmetrical. As will be shown below, however, the data can still be calibrated by using the estimated system parameters. This, indeed, is what we seek; knowledge of detailed attenuation paths in the radar system is not required.

The measured symmetrized scattering matrix $\overline{\mathcal{Y}}$ may now be written as

$$\begin{pmatrix} \overline{Y}_{hh} & \overline{Y}_{hv} \\ \overline{Y}_{hv} & \overline{Y}_{vv} \end{pmatrix} = A\, e^{i\phi} \begin{pmatrix} 1 & \delta_2 \\ \delta_1 & f \end{pmatrix} \begin{pmatrix} \overline{S}_{hh} & \overline{S}_{hv} \\ \overline{S}_{hv} & \overline{S}_{vv} \end{pmatrix} \begin{pmatrix} 1 & \delta_1 \\ \delta_2 & f \end{pmatrix} . \tag{6.27}$$

The elements of these matrices are complex numbers, and $\delta_1$ represents the crosstalk when vertically polarized electric fields are transmitted or received, $\delta_2$ represents the crosstalk when horizontally polarized electric fields are transmitted or received, and $f$ represents the one-way co-polarized channel imbalance in amplitude and phase.

To show how the crosstalk calibration can be separated from the channel imbalance and radiometric calibration, we notice that we can write

$$\begin{pmatrix} 1 & \delta_1 \\ \delta_2 & f \end{pmatrix} = \begin{pmatrix} 1 & 0 \\ 0 & f \end{pmatrix} \begin{pmatrix} 1 & \delta_1 \\ \delta_2/f & 1 \end{pmatrix} \tag{6.28}$$

or

$$\overline{\mathcal{S}}_T = \overline{\mathcal{S}}_{Tc} \overline{\mathcal{S}}_{Tx} , \tag{6.29}$$

and

$$\begin{pmatrix} 1 & \delta_2 \\ \delta_1 & f \end{pmatrix} = \begin{pmatrix} 1 & \delta_2/f \\ \delta_1 & 1 \end{pmatrix} \begin{pmatrix} 1 & 0 \\ 0 & f \end{pmatrix} \tag{6.30}$$

or

$$\overline{\mathcal{S}}_R = \overline{\mathcal{S}}_{Rx} \overline{\mathcal{S}}_{Rc} . \tag{6.31}$$

Notice that $\overline{\mathcal{S}}_{Rc}$ and $\overline{\mathcal{S}}_{Tc}$ are independent of the system crosstalk. Thus, by using (6.29) and (6.31) in (6.24), we can write

$$\begin{aligned} \overline{\mathcal{Y}} &= \overline{\mathcal{S}}_{Rx} \overline{\mathcal{S}}_{Rc} \overline{\mathcal{S}}\, \overline{\mathcal{S}}_{Tc} \overline{\mathcal{S}}_{Tx} \\ &= \overline{\mathcal{S}}_{Rx} \overline{\mathcal{J}}\, \overline{\mathcal{S}}_{Tx} , \end{aligned} \tag{6.32}$$

where we have introduced the intermediate scattering matrix:

$$\overline{\mathcal{J}} = \begin{pmatrix} \overline{J}_{hh} & \overline{J}_{hv} \\ \overline{J}_{hv} & \overline{J}_{vv} \end{pmatrix} = A\, e^{i\phi} \begin{pmatrix} \overline{S}_{hh} & f\overline{S}_{hv} \\ f\overline{S}_{hv} & f^2\overline{S}_{vv} \end{pmatrix} = A\, e^{i\phi} \overline{\mathcal{S}}_{Rc} \overline{\mathcal{S}}\, \overline{\mathcal{S}}_{Tc} \tag{6.33}$$

on which only channel imbalance and radiometric calibration steps remain. Note that $\overline{\mathcal{S}}_{Rx}$ and $\overline{\mathcal{S}}_{Tx}$ both contain the element $\delta_2/f$. As will be shown below, only this ratio must be estimated and we need not know $f$ and $\delta_2$ separately to correct for the crosstalk. Thus, the crosstalk effects can be corrected separately from the channel imbalance and the radiometric calibration. This is important, because, as will be shown below, the crosstalk calibration can be performed without relying on external calibration devices. Thus, even if no calibration targets were deployed before imaging, the system crosstalk effects may still be removed from the data.

We shall show below that the way in which we have separated the matrices $\overline{\boldsymbol{S}}_{Rx}$ and $\overline{\boldsymbol{S}}_{Rc}$, and $\overline{\boldsymbol{S}}_{Tx}$ and $\overline{\boldsymbol{S}}_{Tc}$, means that we must correct the crosstalk before correcting the channel imbalance. This permits correction of crosstalk effects in images without any external calibration devices, and where balancing the channels would not be possible. A simple exercise is to separate these matrices in such a way that channel balance is implemented first. While this will lead to slightly different equations for estimating the crosstalk parameters to use in the correction, the two methods are similar enough that the second will not be described here.

To linearize the problem, we make the following two assumptions:

1. The magnitudes of $\delta_1$ and $\delta_2$ are small compared to 1. This simply means that one assumes a reasonably well isolated system.

2. The co- and cross-polarized components of $\overline{\boldsymbol{S}}$ are uncorrelated for natural distributed targets.

This is true to second order for rough surfaces as predicted by Rice's small perturbation model, and is also predicted to be true to second order by the random medium model (see, for example, Tsang *et al.*, 1985, chapter 5). Stated mathematically, this means that

$$\langle \overline{S}_{hh}^{*} \overline{S}_{hv} \rangle = 0, \tag{6.34a}$$

and

$$\langle \overline{S}_{vv}^{*} \overline{S}_{hv} \rangle = 0. \tag{6.34b}$$

Notice that (6.34a) and (6.34b) are also true for the elements of $\overline{\mathcal{J}}$. Here, $\langle\ \rangle$ means the ensemble average, which in terms of image parameters means an average over an extended area.

By computing the matrix multiplications in (6.32) and using the assumptions stated above, we have

$$\begin{aligned}
&\langle \overline{Y}_{hh} \overline{Y}_{hh}^{*} \rangle \approx \langle \overline{J}_{hh} \overline{J}_{hh}^{*} \rangle \\
&\langle \overline{Y}_{vv} \overline{Y}_{vv}^{*} \rangle \approx \langle \overline{J}_{vv} \overline{J}_{vv}^{*} \rangle \\
&\langle \overline{Y}_{vv} \overline{Y}_{hh}^{*} \rangle \approx \langle \overline{J}_{vv} \overline{J}_{hh}^{*} \rangle \\
&\langle \overline{Y}_{hv} \overline{Y}_{hh}^{*} \rangle \approx \delta_1 \langle \overline{J}_{hh} \overline{J}_{hh}^{*} \rangle + \delta_2/f \langle \overline{J}_{vv} \overline{J}_{hh}^{*} \rangle + 2(\delta_2/f)^{*} \langle \overline{J}_{hv} \overline{J}_{hv}^{*} \rangle \\
&\langle \overline{Y}_{hv} \overline{Y}_{vv}^{*} \rangle \approx \delta_1 \langle \overline{J}_{hh} \overline{J}_{vv}^{*} \rangle + \delta_2/f \langle \overline{J}_{vv} \overline{J}_{vv}^{*} \rangle + 2\delta_1^{*} \langle \overline{J}_{hv} \overline{J}_{hv}^{*} \rangle\ .
\end{aligned} \tag{6.35}$$

Hence,

$$\delta_1 = \frac{\Delta P + 2P^* \langle \overline{J}_{hv}\overline{J}^*_{hv}\rangle \langle \overline{Y}_{vv}\overline{Y}^*_{hh}\rangle}{\Delta^2 - 4\langle \overline{J}_{hv}\overline{J}^*_{hv}\rangle^2 \langle \overline{Y}_{vv}\overline{Y}^*_{hh}\rangle \langle \overline{Y}_{hh}\overline{Y}^*_{vv}\rangle} \tag{6.36}$$

and

$$\delta_2/f = \frac{\Delta Q + 2Q^* \langle \overline{J}_{hv}\overline{J}^*_{hv}\rangle \langle \overline{Y}_{hh}\overline{Y}^*_{vv}\rangle}{\Delta^2 - 4\langle \overline{J}_{hv}\overline{J}^*_{hv}\rangle^2 \langle \overline{Y}_{vv}\overline{Y}^*_{hh}\rangle \langle \overline{Y}_{hh}\overline{Y}^*_{vv}\rangle} , \tag{6.37}$$

where

$$\begin{aligned}
\Delta &= \langle \overline{Y}_{hh}\overline{Y}^*_{hh}\rangle \langle \overline{Y}_{vv}\overline{Y}^*_{vv}\rangle - \langle \overline{Y}_{vv}\overline{Y}^*_{hh}\rangle \langle \overline{Y}_{hh}\overline{Y}^*_{vv}\rangle - \langle \overline{J}_{hv}\overline{J}^*_{hv}\rangle \langle \overline{J}_{hv}\overline{J}^*_{hv}\rangle \\
P &= \langle \overline{Y}_{hv}\overline{Y}^*_{hh}\rangle \langle \overline{Y}_{vv}\overline{Y}^*_{vv}\rangle - \langle \overline{Y}_{vv}\overline{Y}^*_{hh}\rangle \langle \overline{Y}_{hv}\overline{Y}^*_{vv}\rangle - \langle \overline{J}_{hv}\overline{J}^*_{hv}\rangle \langle \overline{Y}_{vv}\overline{Y}^*_{hv}\rangle \\
Q &= \langle \overline{Y}_{hh}\overline{Y}^*_{hh}\rangle \langle \overline{Y}_{hv}\overline{Y}^*_{vv}\rangle - \langle \overline{Y}_{hv}\overline{Y}^*_{hh}\rangle \langle \overline{Y}_{hh}\overline{Y}^*_{vv}\rangle - \langle \overline{J}_{hv}\overline{J}^*_{hv}\rangle \langle \overline{Y}_{hh}\overline{Y}^*_{hv}\rangle \, .
\end{aligned} \tag{6.38}$$

Notice that, to estimate $\delta_1$ and $\delta_2/f$, we need to know $\langle \overline{J}_{hv}\overline{J}^*_{hv}\rangle$. For ocean and bare land surfaces, this term has been observed to be small compared to the co-polarized cross-products. However, for heavily vegetated areas, this term may become comparable to the co-polarized cross-products, and may even become larger than the magnitude of $\langle \overline{Y}_{vv}\overline{Y}^*_{hh}\rangle$.

Now consider the problem of finding the cross-products involving the elements of $\overline{\mathcal{J}}$ from knowledge of the elements of $\overline{\mathcal{Y}}$ and the estimated values of $\delta_1$, $\delta_2/f$. From (6.32),

$$\overline{\mathcal{J}} = \overline{\mathcal{S}}^{-1}_{Rx}\overline{\mathcal{Y}}\,\overline{\mathcal{S}}^{-1}_{Tx} \, . \tag{6.39}$$

Therefore, assuming that we know the values of $\delta_1$ and $\delta_2/f$ and using the assumptions stated earlier, we find that

$$\begin{aligned}
\langle \overline{J}_{hh}\overline{J}^*_{hh}\rangle &= \langle \overline{Y}_{hh}\overline{Y}^*_{hh}\rangle \\
\langle \overline{J}_{vv}\overline{J}^*_{vv}\rangle &= \langle \overline{Y}_{vv}\overline{Y}^*_{vv}\rangle \\
\langle \overline{J}^*_{hh}\overline{J}_{vv}\rangle &= \langle \overline{Y}^*_{hh}\overline{Y}_{vv}\rangle \\
\langle \overline{J}^*_{hh}\overline{J}_{hv}\rangle &= \langle \overline{Y}_{hv}\overline{Y}^*_{hh}\rangle - \delta_1\langle \overline{Y}_{hh}\overline{Y}^*_{hh}\rangle - (\delta_2/f)\langle \overline{Y}_{vv}\overline{Y}^*_{hh}\rangle \\
\langle \overline{J}^*_{hv}\overline{J}_{vv}\rangle &= \langle \overline{Y}_{vv}\overline{Y}^*_{hv}\rangle - \delta_1^*\langle \overline{Y}_{vv}\overline{Y}^*_{hh}\rangle - (\delta_2/f)^*\langle \overline{Y}_{vv}\overline{Y}^*_{vv}\rangle \\
\langle \overline{J}_{hv}\overline{J}^*_{hv}\rangle &= \langle \overline{Y}_{hv}\overline{Y}^*_{hv}\rangle + |\delta_1|^2\langle \overline{Y}_{hh}\overline{Y}^*_{hh}\rangle + |\delta_2/f|^2\langle \overline{Y}_{vv}\overline{Y}^*_{vv}\rangle \\
&\quad - \delta_1\langle \overline{Y}_{hh}\overline{Y}^*_{hv}\rangle - \delta_1^*\langle \overline{Y}^*_{hh}\overline{Y}_{hv}\rangle - (\delta_2/f)\langle \overline{Y}_{vv}\overline{Y}^*_{hv}\rangle \\
&\quad - (\delta_2/f)^*\langle \overline{Y}^*_{vv}\overline{Y}_{hv}\rangle + \delta_1(\delta_2/f)^*\langle \overline{Y}_{hh}\overline{Y}^*_{vv}\rangle \\
&\quad + \delta_1^*(\delta_2/f)\langle \overline{Y}^*_{hh}\overline{Y}_{vv}\rangle \, .
\end{aligned} \tag{6.40}$$

From (6.35) to (6.40) we can now illustrate the technique for estimating $\delta_1$ and $\delta_2/f$. First, we use (6.35) with $\delta_1 = 0$ and $\delta_2/f = 0$ to estimate an initial guess of the crosstalk parameters by (6.38), (6.39), and (6.40). This initial guess is then used to improve our estimate of $\langle \overline{J}_{hv} \overline{J}^*_{hv} \rangle$. This improved estimate of $\langle \overline{J}_{hv} \overline{J}^*_{hv} \rangle$ is then used to improve our estimate of the crosstalk parameters. This procedure is repeated until the values of $\langle \overline{J}_{hv} \overline{J}^*_{hv} \rangle$, $\delta_1$, and $\delta_2/f$ have reached stable solutions.

An example of the crosstalk removal procedure is shown in Fig. 6.5, where we plot corner reflector responses before and after the correction procedure. Figure 6.6 depicts an L-band image acquired over the Pisgah lava flow area in the Mojave Desert, California, on June 3, 1988. Prior to imaging, five trihedral corner reflectors, each measuring 1.8 m on a side, were placed in the Pisgah area. These corner reflectors were spaced such that they would cover the whole swath in range. As the experiment involved imaging the scene with parallel, partially overlapping swaths, all corner reflectors were oriented to place the boresight at 45° incidence angle. Careful measurements of the orientations of the corner reflectors were made before and after the flight. Images of the area were acquired simultaneously at three radar wavelengths, P-band (67 cm), L-band (24 cm), and C-band (5.6 cm).

Figure 6.7 shows the theoretical co-polarized response (van Zyl, 1985) of an ideal trihedral corner reflector, and Fig. 6.5a shows the measured co-polarized response of the five corner reflectors at L-band after only phase calibration has been performed. The responses show noticeable distortion, with the distortion becoming worse in the near range (smaller incidence angles) in the L-band image.

By using the procedure described above, crosstalk parameters were estimated from the image data. Because $\delta_1$ and $\delta_2/f$ should only be a function of range, we could obtain a better estimate of the crosstalk at a particular range were we to use the average of the estimates over an entire range line. A plot of the estimated crosstalk parameters is shown in Fig. 6.8a. We notice that the estimated crosstalk parameters $\delta_1$ and $\delta_2/f$ at L-band are significantly different and they are a function of the incidence angle. The parameter $\delta_1$ for L-band is only about $-12$ dB in the near range. This crosstalk leads to the distortion of the L-band polarization response observed in Fig. 6.5.

Figure 6.5b shows the measured co-polarized responses of the five corner reflectors at L-band after crosstalk calibration. We notice a dramatic improvement when these responses are compared to the theoretical response of a trihedral corner reflector (Fig. 6.7). This confirms that most of the distortion evident in the responses in Fig. 6.5 has been introduced by the crosstalk.

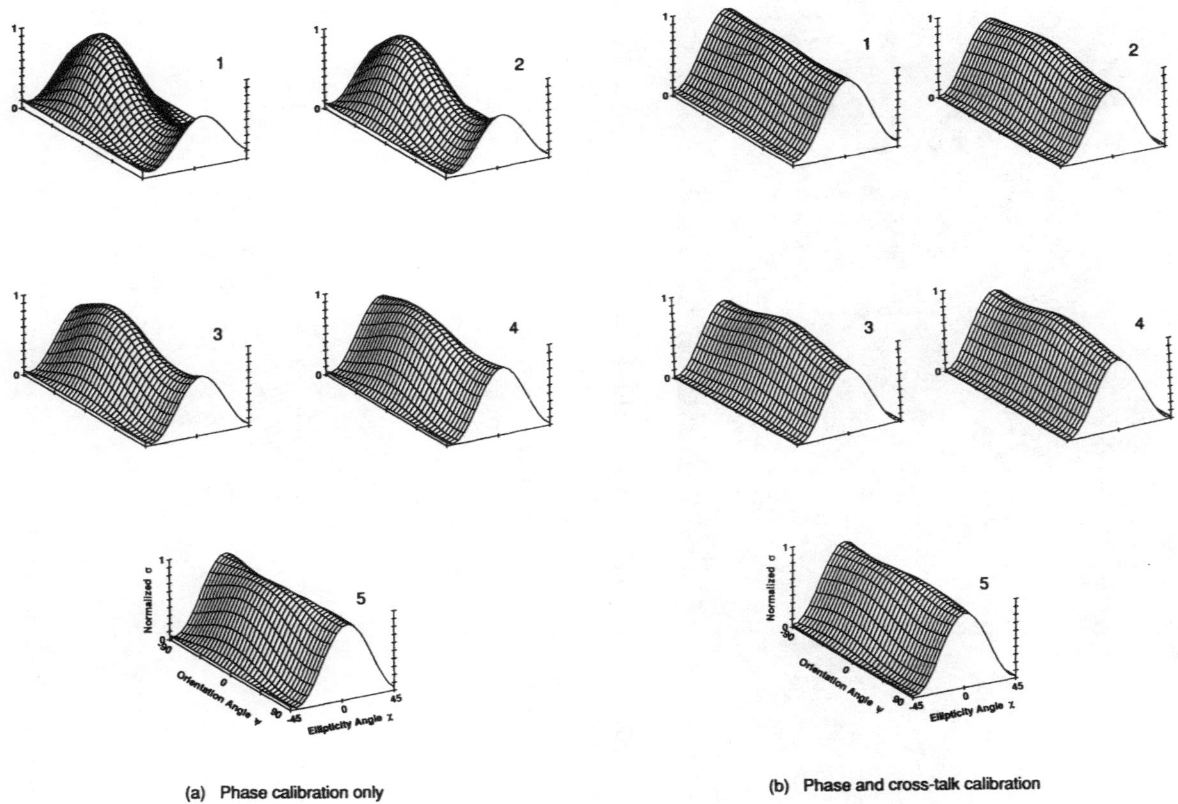


**Figure 6.5** (a) Measured L-band polarization responses of the corner reflectors after phase calibration only. Note that the amount of distortion decreases with increasing incidence angle. (b) The same data after crosstalk correction. Observed responses are now symmetrical, but still exhibit channel imbalance.

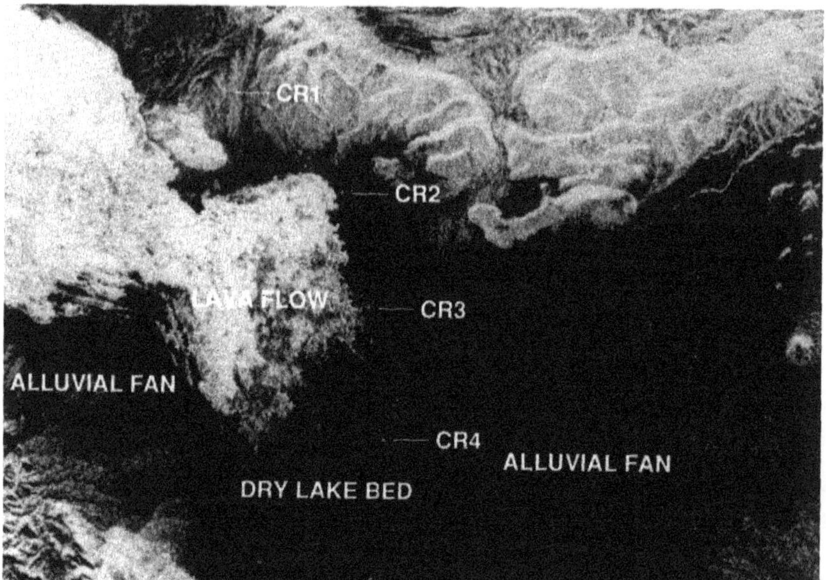


**Figure 6.6** Total power image of the Pisgah lava flow area, California, obtained with the NASA/JPL DC-8 airborne SAR at L-band. The image covers an area of about 12 km by 5 km and each pixel represents approximately 12 m by 7 m on the ground. The radar illumination is from the top and the aircraft movement is from right to left.

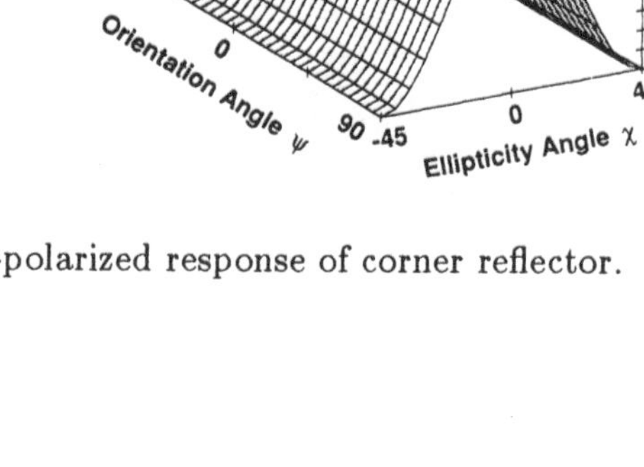


**Figure 6.7** Co-polarized response of corner reflector.

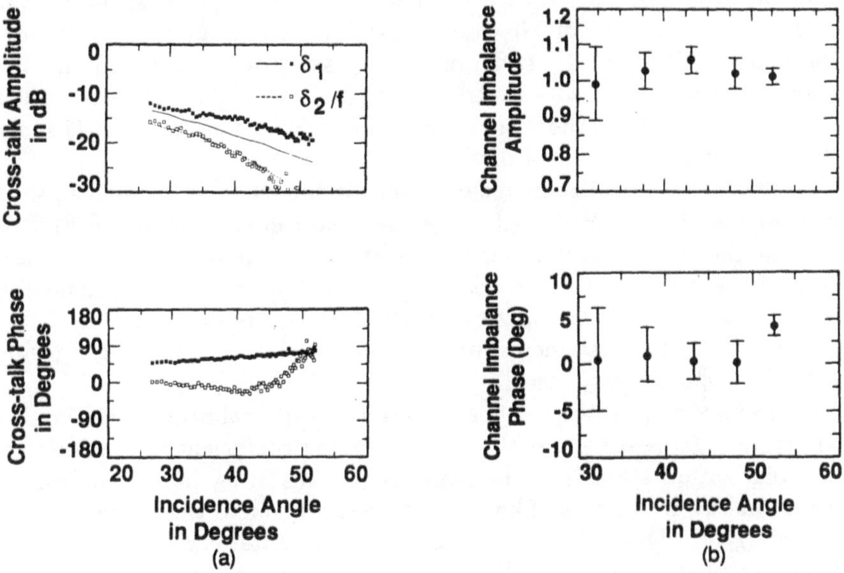


**Figure 6.8** (a) Estimated crosstalk parameters as a function of range from the crosstalk correction algorithm. (b) Estimated amplitude and phase of the co-channel imbalance factor $f$ using the corner reflector signals at L-band. The estimated channel imbalance phase is close to zero, confirming the accuracy of the previous phase calibration.

### 6-4.4 Absolute Amplitude Calibration and Channel Gain Balance

At this point the data have been corrected for all of the parameters that may be inferred without the use of test targets or detailed system performance knowledge. What remains is to utilize the backscatter measurements from each of the corner reflectors to correct the residual amplitude offsets in the various polarization channels, and to correct for the absolute gain of the radar system.

We illustrate these procedures with data from the same Pisgah images as above. To estimate the value of $f$ for the Pisgah images, we used the measured cross-products for the five trihedral corner reflectors. Figure 6.8b shows the estimated amplitude and phase of $f$ as a function of incidence angle for the five corner reflectors. The measured amplitudes at L-band vary between 1.01 and 1.06, a total spread of 0.05. The estimated phase of $f$ is close to zero for all corner reflectors, confirming the accuracy of the phase calibration applied from analysis of the background signals as described above. We note that if we set $f$ equal to 0.31 dB, the resulting ratios measured from the corner reflectors nearly satisfy the SIR-C goals of channel imbalance of $\pm$ 0.4 dB.

To correct the entire image data set for the channel imbalance, we approximated the curves described by the measurements shown in Fig. 6.8b by straight-line segments. Figure 6.9 shows the polarization response of each of the five corner reflectors after crosstalk correction and channel imbalance calibration. Comparison with the ideal polarization response in Fig. 6.7 indicates that the calibrated responses are in excellent agreement with the expected response.

The final step is that of absolute radiometric calibration, determining the system gains such that the backscattering coefficient for each point at all polarization states may be measured. This is again accomplished by using the several targets of known cross section across the radar swath and recording the observed absolute cross sections. After corrections have been made for phase and crosstalk, and approximate corrections have been applied to compensate partially for range propagation loss, antenna patterns, different receiver channel gains, and so forth, the resulting measurements will vary only slowly as a function of range. Therefore, a relatively simple matter is to fit a smooth correction curve to the measured values of the known target data and apply it to the full data set. This procedure results in the final image.

We illustrate the final product with several images of the Pisgah lava field area in the Mojave Desert of California. Visible in the scene are the five calibration trihedral corner reflectors. We plot absolute cross section images of the area for horizontal transmitting, horizontal receiving, a canonical polarization, in Plate 6.1a. Also shown is a synthesized image of left circularly polarized transmitting, right circularly polarized receiving energy where the synthesis technique of Chapter 2 has been used (Plate 6.1b). Each image displays absolute radar cross section in dB.

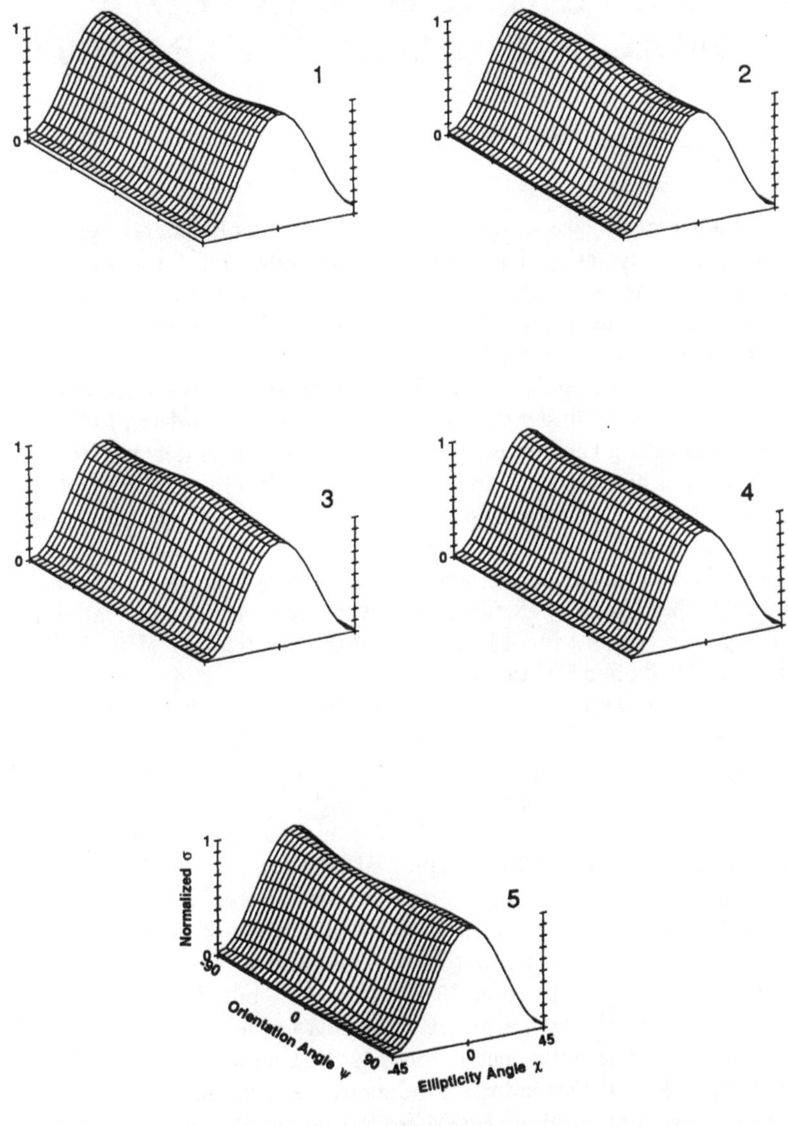


**Figure 6.9** Measured L-band polarization responses of the corner reflectors after channel imbalance, crosstalk and phase calibrations were performed. Note the excellent agreement with the ideal response in Fig. 6.7.

# 6-5 EXISTING AND PLANNED IMAGING RADAR POLARIMETERS

## 6-5.1 Airborne Systems

In this final section, we describe imaging radar polarimeter systems that are currently in use as well as some of those under development. All follow the same basic design approach as described above for the NASA CV-990 radar system and utilize similar architecture. The CV-990 radar remains the prototype for polarimeter design.

Several additional polarimetric imaging systems have since been developed or are currently under development. The Jet Propulsion Laboratory is currently operating the successor to the CV-990 polarimeter, a new system simultaneously operating at three wavelengths (6 cm, 24 cm, and 68 cm). This instrument is installed on a DC-8 aircraft based at Ames Research Center in Mountain View, California. All three polarimeters on the DC-8 have similar resolution and image parameters to the CV-990 radar. This system, which became operational in the spring of 1988, collected data over many types of terrain, including lava fields, forests, agricultural areas, ice regions, and the open ocean.

The Environmental Research Institute of Michigan, supported by NADC, has also tested a new airborne imaging polarimeter (Kozma *et al.*, 1986). This instrument operates at three wavelengths, 3 cm, 6 cm, and 24 cm (X-, C-, and L-band), at a higher resolution than the JPL instrument, 1.6 or 3.2 m in range resolution and 2.2 m single-look in azimuth (the JPL single-look resolution is 3.2 m). The platform for the NADC/ERIM radar is a P-3 aircraft. This system also measures the complex scattering matrix $\boldsymbol{S}$ for each single-look resolution element in a scene of swathwidth 6 or 12 km. We note that both this system and the above-mentioned JPL system can image swaths greater than 20 km in width, but data rate limitations preclude polarimetric data acquisition in this mode.

Another airborne polarimetric radar is the Canadian Centre for Remote Sensing (CCRS) CV-580 radar (Goodenough and Livingstone, 1985). This is an improved version of an ERIM SAR that CCRS operated from 1981–1984. The current system consists of a 6-cm wavelength radar and is being augmented with a 3-cm wavelength radar. Both of these radars now employ digital data recording, which is required for subsequent polarization synthesis. The CCRS implementation also incorporates on-board range processing, with optional on-board azimuth data processing. The ability to process signals in near real-time greatly aids in the utility of the instrument, and is a subject of great interest in the area of new system design.

### 6-5.2 Spaceborne Systems

The next step in technology development for polarimetric radar systems is the implementation of a spaceborne sensor, providing wide swaths and global imaging capabilities. The first of this new class of polarimeters will be the NASA SIR-C, scheduled for launch in 1992. This sensor will have dual-wavelength (24 cm and 6 cm) polarimetric capability. An X-band system, built by a German and Italian team, will be flown simultaneously, but will not have polarimetric capability. Data will be acquired over sites distributed around the world on three space shuttle flights. The signals will be recorded on-board the shuttle on digital tape and will be processed on the ground to produce the polarimetric data sets.

The SIR-C radar forms the basis of a prototype for the next in NASA's continuing remote sensing program, a radar designed to fly on the Eos satellite, scheduled for launch in the late 1990s. This will be a free-flying satellite dedicated to continual long-term monitoring of the earth's surface. Although exact wavelengths and modes remain to be defined, multiwavelength polarimetry is one of the areas of highest interest for the investigators. The final radar design will likely include L-, C-, and X-band wavelengths (24, 6, and 3 cm, respectively), and produce data at 10–50 m resolution, with nearly global coverage capability. Such massive coverage requirements will entail novel approaches to hardware design and, especially, data processing and distribution networks. In addition, the Eos sensor will incorporate a very wide swath mode, denoted SCANSAR, as the beam is scanned across- track over a wide range of incidence angles in several increments. This will result in the capability for nearly global coverage of the earth with polarimetric multifrequency imaging radar, although, due to data system limitations, to utilize all frequencies and all polarizations simultaneously may not be possible. NASA foresees widespread use of all instruments on Eos platforms over a 10 to 15 year period, implying many new challenges to the design and high throughput requirements of such a system.

# References


Barnes, R.N., "Antenna Polarization Calibration Using In-Scene Reflectors," Project Report TT-65, MIT Lincoln Laboratory, September 1986.

Barton, D.K., *Modern Radar System Analysis,* Artech House, Norwood, MA, 1988.

Boerner, W.M., M.B. El-Arini, C.Y. Chan, and P.M. Mastoris, "Polarization Dependence in Electromagnetic Inverse Problems," *IEEE Trans. on Antennas and Propagation,* Vol. AP-29, pp. 262–271, 1981.

Born, M., and E. Wolf, *Principles of Optics,* Sixth Edition, Pergamon Press, New York, 1980, pp. 24–32.

Brookner, E., *Aspects of Modern Radar,* Artech House, Norwood, MA, 1988.

Freeman, A., Y. Shen, and C.L. Werner, "Polarimetric SAR Calibration Experiment Using Active Radar Calibrators," *IEEE Trans. Geosci. Rem. Sens.,* Vol. GE-28, pp. 224-240, 1990.

Freeman, A., and J. Curlander, "Radiometric Correction and Calibration of SAR Images," *Photo. Eng. Rem. Sens.,* 1990.

Goodenough, D.C., and C. Livingstone, "Airborne Multifrequency, Multipolarization SAR, the CCRS Experience," *International Geoscience and Remote Sensing Symposium Digest,* 1985.

Hagfors, T., "A Study of the Depolarization of Lunar Echoes," *Radio Science,* Vol. 2, pp. 445–465, 1967.

Jackson, J.D., *Classical Electrodynamics,* John Wiley and Sons, New York, 1976.

Kozma, A., A.D. Nichols, R.F. Rawson, S.J. Shackman, C.W. Haney, and J.J. Shanne, "Multifrequency-Polarimetric SAR for Remote Sensing," *Proceedings of the IGARSS'86 Symposium,* ESA SP-254, pp. 715–720, 1986.

Sarabandi, K., F.T. Ulaby, and M.A. Tassoudji, "Calibration of Polarimetric Radar Systems with Good Polarization Isolation," *IEEE Trans. Geosci. Rem. Sens.,* Vol. GE- 28, pp. 70–75, January 1990.

Sheen, D.R., A. Freeman, and E. Kasischke, "Phase Calibration of Polarimetric Radar Images," *IEEE Trans. Geosci. Rem. Sens.*, Vol. GE-27, pp. 719–731, 1989.

Thompson, T.W., "A User's Guide for the NASA/JPL Synthetic Aperture Radar and the NASA/JPL L- and C- band Scatterometers," JPL publication no. 83-38, 1986.

Tsang, L., J.A. Kong, and R.T. Shin, *Theory of Microwave Remote Sensing,* John Wiley and Sons, New York, 1985.

Van de Hulst, H.C., *Light Scattering by Small Particles,* Dover, New York, 1981, pp. 28–42.

van Zyl, J.J., On the Importance of Polarization in Radar Scattering Problems, Ph.D. Thesis, Antenna Laboratory Report No. 120, 152 pp., California Institute of Technology, Pasadena, CA, 1985.

van Zyl, J.J., H.A. Zebker, and C. Elachi, "Imaging Radar Polarization Signatures: Theory and Observation," *Radio Science,* Vol. 22, pp. 529–543, 1987.

van Zyl, J.J., "A Technique to Calibrate Polarimetric Radar Images Using Only Image Parameters and Trihedral Corner Reflectors," *IEEE Trans. Geosci. Rem. Sens.,* Vol. GE-28, 1990.

Whitt, M.W., F.T. Ulaby, P.F. Polatin, and V. Liepa, "A General Polarimetric Radar Calibration Technique," *IEEE Trans. Ant. and Prop.,* Vol. AP-38, 1990.

Zebker, H.A., J.J. van Zyl, and D.N. Held, "Imaging Radar Polarimetry from Wave Synthesis," *Journal of Geophysical Research*, Vol. 92, No. B1, pp. 683–701, 1987.

Zebker, H.A., and Y. Lou, "Phase Calibration of Imaging Radar Polarimeter Stokes Matrices," *IEEE Trans. Geosci. Rem. Sens.,* Vol. GE-28, pp. 246-252, 1990.

# Chapter 7

# Polarimetric SAR Applications

**Jakob J. van Zyl, Howard A. Zebker and Charles Elachi**

A number of polarimetric SAR analysis techniques and applications have been reported in the recent literature (Evans *et al.*, 1988; Swartz *et al.*, 1988; Lim *et al.*, 1989; van Zyl, 1989a). These include the measurement and interpretation of the polarization response, techniques for image contrast enhancement based on polarimetric filtering, and classification according to physical scattering mechanisms. In this chapter, those techniques will be described and subsequently illustrated for several diverse applications by analyzing the observed polarimetric scattering properties of different types of earth terrain.

## 7-1 POLARIZATION RESPONSE

Measurement of the full polarization properties of a scatterer allows very complete testing of scattering models and ultimately derivation of surface geophysical parameters. To be useful for geophysical applications, a model must correctly predict the scattering coefficients not only as a function of incidence angle or electrical properties of the scatterer, but also as a function of the polarization of the transmitting and receiving antenna. In order to describe completely the polarization behavior of a given scattering situation, we need to calculate the complete average Stokes scattering operator, Mueller matrix, or covariance matrix. This information may be summarized in graphical form through the polarization response introduced in Chapter 2. A comparison of the measured and predicted polarization responses provides a simple and efficient way to test the polarization behavior of the model. In this section, we shall compare the observed polarization responses and those predicted by models for terrain types varying from (i) very smooth, bare playa (dry lake bed) surfaces, (ii) different types of lavas, and (iii) heavily vegetated surfaces. These comparisons will be only qualitative; calibrated polarimetric SAR data necessary for detailed quantitative comparison currently exist for only a few areas on the earth's surface (van Zyl, 1989b).

The power of the polarization response to analyze scattering behavior was clearly illustrated when Zebker and Norikane (1987) inferred the orientation of a trihedral corner reflector from its observed response. When analyzing the polarization responses of five trihedral corner reflectors deployed on Goldstone dry lake bed, four of the corner reflectors exhibited responses that were found to be consistent with a three-bounce geometry (see Fig. 2.6 for the expected response of a trihedral corner reflector). One of the corner reflectors exhibited a response more consistent with a two-bounce geometry like that of a dihedral corner reflector (see Fig. 2.5), but the response appeared to be rotated, i.e., the dihedral reflector trough appeared not to be parallel to the $h$-$v$ coordinates. The measured responses of the dry lake bed, a "regular" trihedral corner reflector, and the rotated trihedral corner reflector are shown in Fig. 7.1. By applying a rotation to the coordinate axes similar to what was done for the short, thin cylinder in Chapter 2, Zebker and Norikane (1987) found that the dihedral trough was rotated by 52°. They further showed that the actual two-bounce scattering observed was due to a fortuitous alignment of the corner reflector, and was the result of a rotation about the axis of symmetry of the trihedral corner reflector such that one side of the reflector was aligned with the wave direction of incidence, and thus that side was invisible to the radar (see Fig. 7.2).

Rice's small perturbation model was used by van Zyl *et al.* (1987) to explain the observed polarization response of the ocean surface in San Francisco Bay, California. The observed and model responses are shown in Fig. 7.3. The data were acquired in the summer of 1985 at L-band with the NASA/JPL CV-990 polarimetric radar. There is excellent agreement between the observed and predicted responses at 50° incidence angle. However, the observed and predicted responses for 20° incidence angle do not agree as well. This is likely explained by the fact that the Bragg scattering mechanism, which, according to the first-order small perturbation model is the mechanism responsible for scattering, dominates mainly at larger incidence angles; at 20°, other mechanisms may be more important. Another difference visible in Fig. 7.3 is that the observed polarization responses appear to sit on a "pedestal." This pedestal is absent in the responses predicted by the first-order small perturbation model.

Rice's original model was extended to the full second order by van Zyl *et al.* (1987) to predict the observed pedestal. When the second-order fields are added, the model predicts a pedestal, and also that its height increases with increasing roughness.

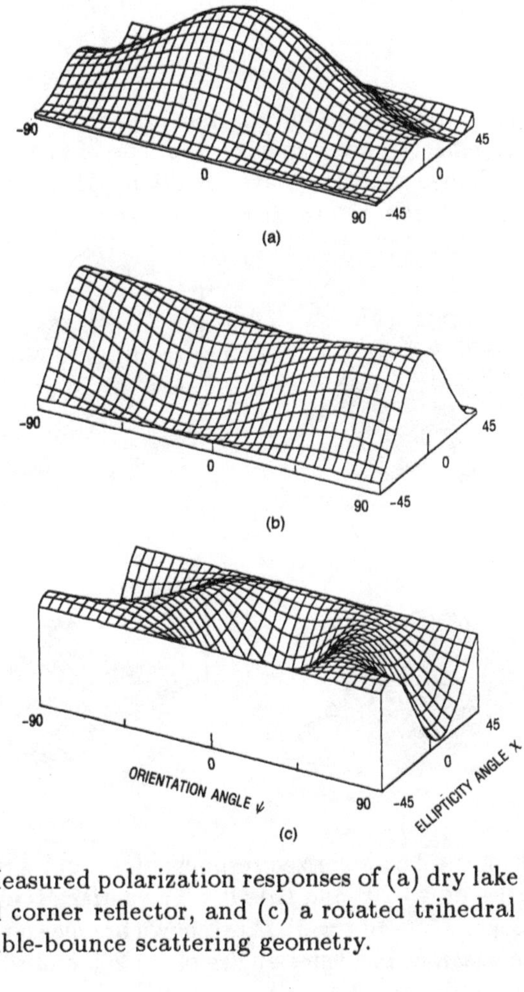


**Figure 7.1** Measured polarization responses of (a) dry lake bed, (b) a "regular" trihedral corner reflector, and (c) a rotated trihedral corner reflector exhibiting double-bounce scattering geometry.

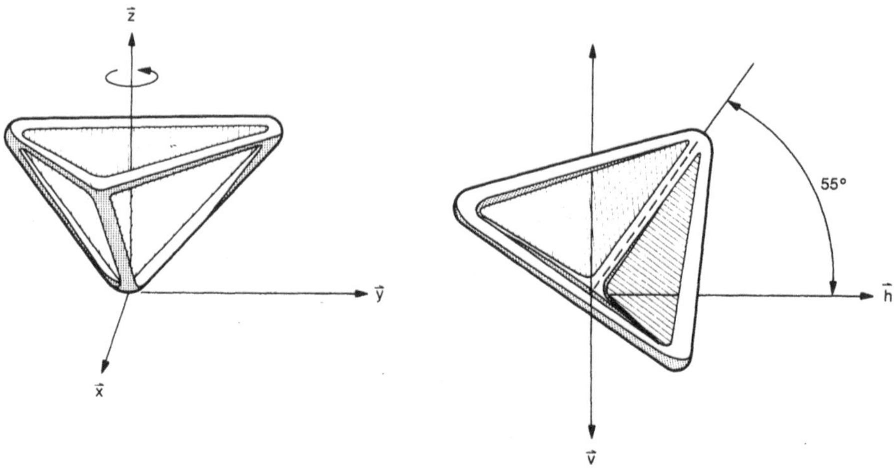


**Figure 7.2** A trihedral corner reflector may appear to the radar system as a dihedral corner reflector when spun on the axis of symmetry (left) such that one side is aligned with the wave direction of incidence, and is thus invisible when viewed from the radar look direction (right).

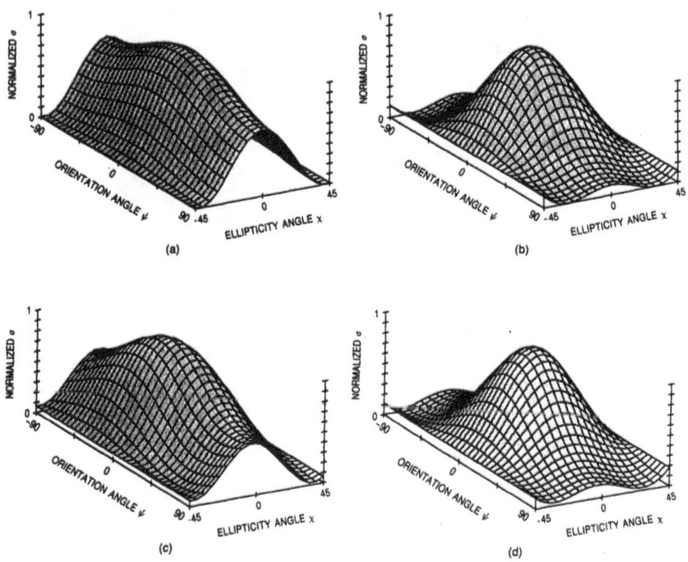


**Figure 7.3** Theoretical co-polarized responses for rough surface scattering at incidence angles of (a) 20° and (b) 50°. The surface is characterized by $kh = 0.1$, rms slope = 5° and $\epsilon = 81$. Also shown are measured co-polarized responses of the ocean at incidence angles of (c) 20° and (d) 50°.

A second experiment was undertaken for better quantifying the ability of the second-order small perturbation model to explain the observed polarization responses of smoother areas. Data were acquired over the Pisgah lava flow in the Mojave desert in the summer of 1985 (van Zyl *et al.*, 1988). To avoid problems associated with possible subsurface scattering, the observed polarization response of a smooth part of a pahoehoe type of lava flow was selected. This observed polarization response is shown in Fig. 7.4(a). Since the data were not absolutely calibrated, the elements of the Stokes scattering operator were normalized by the $M_{11}$ element. The model input parameters, rms height, correlation length, and complex dielectric constant were then varied until the best fit between the observed and predicted polarization responses was found. The resulting predicted response is given in Fig. 7.4(b), and shows excellent agreement with the observed response. The inferred surface parameters are dielectric constant $6.75 + i0$, rms height 4.1 cm, and correlation length 3.0 cm (van Zyl *et al.*, 1988). *In situ* measurements of surface rms height using a helicopter mounted stereo camera pair yielded values ranging between 2.28 to 5.06 cm. L-band dielectric constant values ranging between 4 and 8 have been reported by Ulaby *et al.* (1988) for volcanic rocks. Although the inferred height and correlation length exceed the region of validity for the first-order small perturbation model (Ulaby, *et al.* 1982), the inferred values are in agreement with those measured for the pahoehoe surface. No validity conditions have yet been established for the second-order small perturbation method. In fitting the observed response, the second-order terms were required to be smaller than the first-order ones.

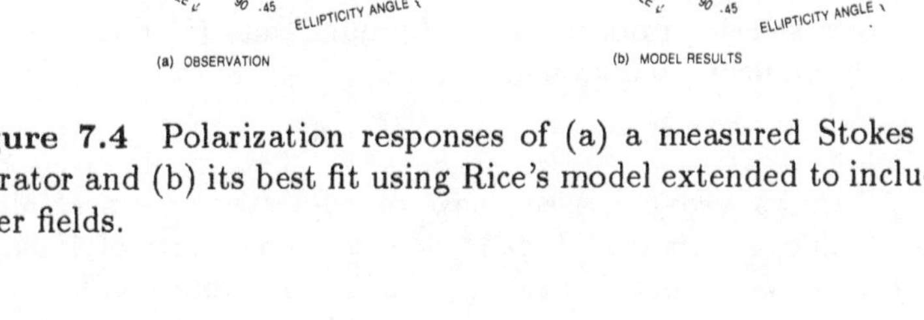


**Figure 7.4** Polarization responses of (a) a measured Stokes scattering operator and (b) its best fit using Rice's model extended to include second-order fields.

Response analysis has also been applied to forest ecology systems. To explain the observed L-band polarization responses for a forested area, Durden *et al.* (1989) constructed a three-layer model using the discrete scatterer approach. The forest floor is modeled as a rough dielectric surface above which is a layer of trunks and a layer of branches. Both trunk and branch layers are modeled by randomly oriented dielectric cylinders. The model does not include leaves; Durden *et al.* compared theoretical calculations of responses from trees with and without leaves and determined that the scattering from the leaves could be neglected at L-band. Five distinct scattering mechanisms are included in the model: backscatter from branches, backscatter from trunks, double-bounce scattering from branches and ground, double-bounce scattering from trunks and ground, and backscatter from ground. This model is shown schematically in Fig. 7.5. As demonstrated below, the leaf contributions are not required to match theory and observation.

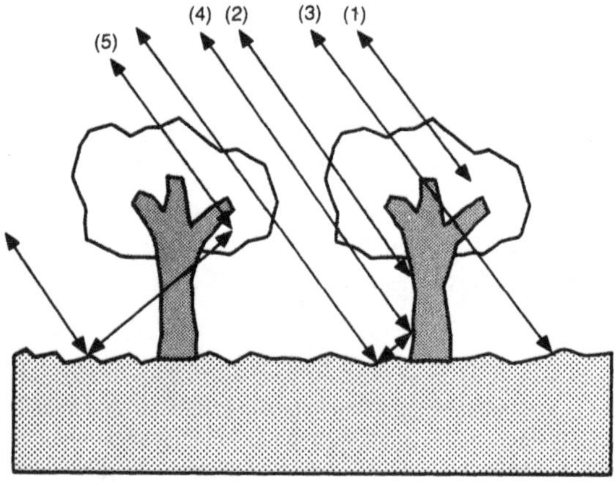


**Figure 7.5** Illustration of the scattering mechanisms included in the forestry model proposed by Durden *et al.* (1989): (1) backscatter from branches, (2) backscatter from trunks, (3) backscatter from ground, (4) double-bounce scattering from trunks and ground, and (5) double-bounce scattering from branches and ground.

Figure 7.6 shows a *vv*-polarized image of a forested area near Moosehead Lake, Maine, acquired at L-band during the summer of 1985 with the NASA/JPL CV-990 imaging radar polarimeter. Noted on this image are three sample areas corresponding to a heavily forested area, a swampy area and a clear-cut. In this image, the forested areas are primarily softwood trees such as white spruce, although some areas are mixed hardwood and softwood (Durden *et al.*, 1989). The clear-cut areas ranged from three to seven years in age and had shrubs and some softwood regrowth. The swampy areas consist of considerable standing water and relatively thin canopies of black spruce trees.

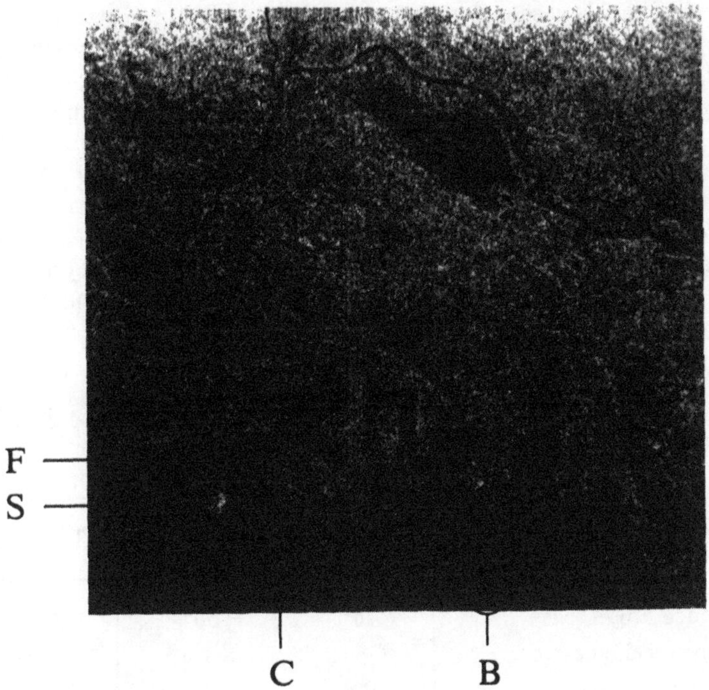


**Figure 7.6** Vertical polarization transmit and receive ($vv$) image of a forested area near Moosehead Lake, Maine, showing three of the four sites studied by Durden *et al.* (1989): (F) forest, (S) swamp, and (C) clear-cut.

By adjusting the model input parameters, Durden *et al.* (1989) duplicated the observed polarization responses. To explain the scattering from the clear-cut area, only a ground layer covered by one layer of randomly oriented dielectric cylinders is used. The model parameters that resulted in the best fit between model results and observations are shown in Table 7.1. Figure 7.7 shows the observed and model predicted responses for the forested area. Excellent agreement is evident. By comparing the relative amount of backscattered energy for each of the five mechanisms used in the model, the backscatter was found to be dominated by the scattering from the randomly oriented branches. This explains the high pedestal in the responses shown in Fig. 7.7. The observed and model-predicted responses for the swampy area are shown in Fig. 7.8. Although the agreement is not as good as in the case of the forested area, the responses exhibit similar characteristics. Both responses are somewhat similar to that of a dihedral corner reflector. Examination of the energy in the five mechanisms indicated that the scattering in this case is dominated by the double-bounce ground-trunk interaction, as is to be expected for an area with a thin canopy covering a layer of standing water. Although the double-bounce scattering dominates, the scattering from the branches was found to contribute significantly to the total observed backscatter from the swampy area. The observed and predicted responses of the clear-cut area are shown in Fig. 7.9. Although the shapes of these responses are similar to those of the forested area, a larger portion of the return signal was found to be polarized, and direct scattering from the ground was more important in this case. For a more complete description, the reader is referred to Durden *et al.* (1989).

**Table 7.1** Model Parameters Used to Predict the Response Shown in Figures 7.7 to 7.9

| Parameter | Forest | Swamp | Clear-cut |
|---|---|---|---|
| Incidence angle | 45° | 50° | 50° |
| Vegetation dielectric | $8+i2$ | $8+i2$ | $8+i2$ |
| Branch lengths (m) | 0.8 | 0.8 | 0.8 |
| Branch radii (cm) | 1.0 - 3.0 | 1.0 - 3.0 | 0.5 - 1.5 |
| Branch density ($m^{-3}$) | 5.0 | 1.0 | 10.0 |
| Branch layer thickness (m) | 7.5 | 7.5 | 1.0 |
| Trunk lengths (m) | 5.0 | 5.0 | 0. |
| Trunk radii (cm) | 12.5 - 14.5 | 12.5 - 14.5 | 0. |
| Trunk density ($m^{-3}$) | 0.1 | 0.1 | 0. |
| Trunk layer thickness (m) | 5.0 | 5.0 | 0. |
| Ground dielectric | $6+i1$ | $35+i4$ | $6+i1$ |
| rms Ground height (m) | 0.04 | 0.02 | 0.04 |
| rms Ground slope | 0.1 | 0.1 | 0.1 |

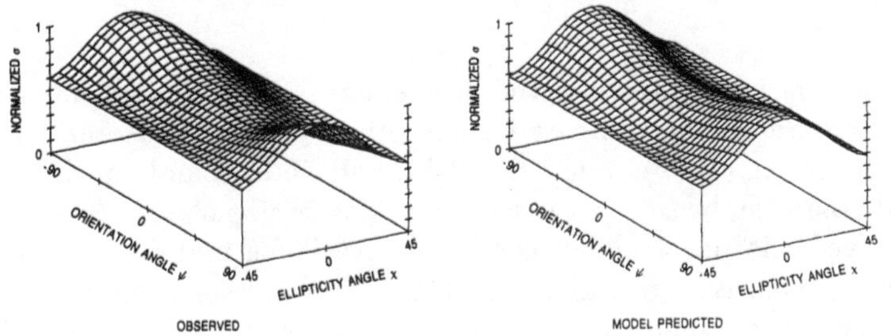


**Figure 7.7** Observed and model predicted polarization responses of the forested area outlined in Figure 7.6.

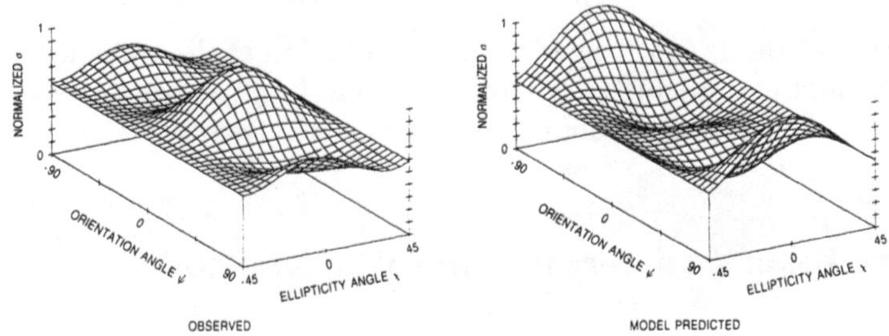


**Figure 7.8** Observed and model predicted polarization responses of the swampy area outlined in Figure 7.6.

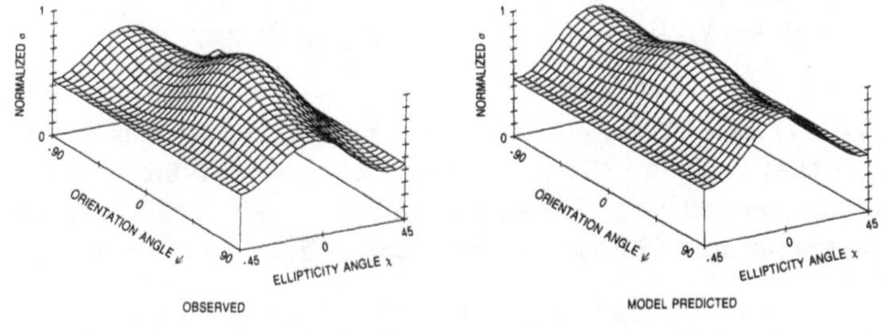


**Figure 7.9** Observed and model predicted polarization responses of the clear-cut area outlined in Figure 7.6.

# 7-2 CONTRAST ENHANCEMENT

Several authors have addressed the problem of finding the optimum polarizations to enhance the contrast between two types of scatterers. Ioannidis and Hammers (1979), van Zyl (1985), and Dubois and van Zyl (1989) proposed solutions based on the Stokes scattering operator. Swartz *et al.* (1988) solved the problem by using the covariance matrix approach. In all cases, the goal is to maximize the ratio of the power received from one scatterer (which we shall call the "target") relative to that received from a second scatterer (which we shall call the "clutter"). Mathematically, this means that we must find the maxima of

$$\mathrm{CR}(\psi_t, \chi_t, \psi_r, \chi_r) = \frac{P_{\mathrm{target}}(\psi_t, \chi_t, \psi_r, \chi_r)}{P_{\mathrm{clutter}}(\psi_t, \chi_t, \psi_r, \chi_r)}, \tag{7.1}$$

where $(\psi_t, \chi_t)$ and $(\psi_r, \chi_r)$ refer to the polarization of the transmitting and receiving antennas, respectively. Here, we shall briefly review two different methods, and then present some applications and results.

## 7-2.1 Stokes Scattering Operator Method

Using the Stokes scattering operator representations of the target and the clutter, we can rewrite (7.1) as (Ioannidis and Hammers, 1979; van Zyl, 1985; Dubois and van Zyl, 1989):

$$\mathrm{CR}(\psi_t, \chi_t, \psi_r, \chi_r) = \frac{\overline{\mathbf{A}}(\psi_r, \chi_r) \cdot \overline{\mathcal{M}}_{\mathrm{target}} \overline{\mathbf{A}}(\psi_t, \chi_t)}{\overline{\mathbf{A}}(\psi_r, \chi_r) \cdot \overline{\mathcal{M}}_{\mathrm{clutter}} \overline{\mathbf{A}}(\psi_t, \chi_t)}, \tag{7.2}$$

with $\overline{\mathbf{A}}(\psi_r, \chi_r)$ and $\overline{\mathbf{A}}(\psi_t, \chi_t)$ being the normalized Stokes vectors (see Chapter 2) that describe the polarization of the receiving and transmitting antennas, respectively. Because both antennas radiate fully polarized waves, the maximum of (7.2) must be found subject to the constraint that

$$\overline{\mathbf{A}} \cdot \begin{pmatrix} 1 & 0 & 0 & 0 \\ 0 & -1 & 0 & 0 \\ 0 & 0 & -1 & 0 \\ 0 & 0 & 0 & -1 \end{pmatrix} \overline{\mathbf{A}} = 0, \tag{7.3}$$

or, writing $\overline{\mathbf{A}}$ as

$$\overline{\mathbf{A}} = \begin{pmatrix} 1 \\ \mathbf{a} \end{pmatrix}, \tag{7.4}$$

where **a** is a real three-dimensional vector, we can rewrite (7.3) as

$$\mathbf{a} \cdot \mathbf{a} = 1. \tag{7.5}$$

This problem can now be solved by using the Lagrange multiplier method. Because efforts to solve the general problem led to complicated nonlinear equations, Dubois and van Zyl (1989) proposed a simpler, hybrid method. In this method, the optimum *receiving* polarization for an arbitrary transmitting polarization is found analytically. The optimum *transmitting* polarization is then found numerically by using this analytical expression for the optimum contrast ratio for a fixed transmitting polarization. As can be shown, for a given arbitrary transmitting polarization, the optimum contrast ratios are

$$\begin{aligned} \mathrm{CR}_{\mathrm{max}}(\psi_t, \chi_t) &= \frac{S_{01}S_{02} - \mathbf{s}_1 \cdot \mathbf{s}_2}{S_{02}^2 - \mathbf{s}_2 \cdot \mathbf{s}_2} \\ &\quad + \sqrt{\left(\frac{S_{01}S_{02} - \mathbf{s}_1 \cdot \mathbf{s}_2}{S_{02}^2 - \mathbf{s}_2 \cdot \mathbf{s}_2}\right)^2 - \left(\frac{S_{01}^2 - \mathbf{s}_1 \cdot \mathbf{s}_1}{S_{02}^2 - \mathbf{s}_2 \cdot \mathbf{s}_2}\right)} \end{aligned} \tag{7.6}$$

and

$$\begin{aligned} \mathrm{CR}_{\mathrm{min}}(\psi_t, \chi_t) &= \frac{S_{01}S_{02} - \mathbf{s}_1 \cdot \mathbf{s}_2}{S_{02}^2 - \mathbf{s}_2 \cdot \mathbf{s}_2} \\ &\quad - \sqrt{\left(\frac{S_{01}S_{02} - \mathbf{s}_1 \cdot \mathbf{s}_2}{S_{02}^2 - \mathbf{s}_2 \cdot \mathbf{s}_2}\right)^2 - \left(\frac{S_{01}^2 - \mathbf{s}_1 \cdot \mathbf{s}_1}{S_{02}^2 - \mathbf{s}_2 \cdot \mathbf{s}_2}\right)}, \end{aligned} \tag{7.7}$$

where

$$\begin{pmatrix} S_{01} \\ \mathbf{s}_1 \end{pmatrix} = \overline{\boldsymbol{\mathcal{M}}}_{\mathrm{target}} \overline{\mathbf{A}}(\psi_t, \chi_t) \tag{7.8}$$

and

$$\begin{pmatrix} S_{02} \\ \mathbf{s}_2 \end{pmatrix} = \overline{\boldsymbol{\mathcal{M}}}_{\mathrm{clutter}} \overline{\mathbf{A}}(\psi_t, \chi_t) \; . \tag{7.9}$$

Notice from (7.6) to (7.9) that the optimum contrast ratios are now only functions of the transmitting antenna polarization. To find the maximum contrast ratio, for example, the maximum of (7.6) is found numerically over all $(\psi_t, \chi_t)$. Once the maximum contrast is found, the optimum receiving antenna polarization can be found from

$$\mathbf{a}_r = \frac{\mathbf{s}(\mathrm{CR}_{\mathrm{opt}})}{\sqrt{\mathbf{s}(\mathrm{CR}_{\mathrm{opt}}) \cdot \mathbf{s}(\mathrm{CR}_{\mathrm{opt}})}}, \tag{7.10}$$

where

$$\mathbf{s}(\mathrm{CR}_{\mathrm{opt}}) = \mathbf{s}_1 - \mathrm{CR}_{\mathrm{opt}} \mathbf{s}_2 \tag{7.11}$$

and $\mathrm{CR_{opt}}$ is the maximum of $\mathrm{CR_{max}}$ in (7.6), optimized over all transmitting antenna polarizations, and $\mathbf{s}_1$ and $\mathbf{s}_2$ are given by (7.8) and (7.9), respectively.

## 7-2.2 Covariance Matrix Method

Using the covariance matrix (see Chapter 2), Swartz *et al.* (1988) have shown that the contrast ratio in the backscatter case can also be written as

$$\mathrm{CR} = \frac{\overline{\mathbf{B}}_b^* \cdot \overline{\mathcal{C}}'_{\text{target}} \overline{\mathbf{B}}_b}{\overline{\mathbf{B}}_b^* \cdot \overline{\mathcal{C}}'_{\text{clutter}} \overline{\mathbf{B}}_b}, \tag{7.12}$$

where $\overline{\mathbf{B}}_b$ is the antenna vector (see Chapter 2) which in the backscatter case can be reduced to (Swartz *et al.*, 1988):

$$\overline{\mathbf{B}}_b^* = \begin{pmatrix} p_v^t p_v^r \\ p_h^t p_v^r + p_v^t p_h^r \\ p_h^t p_h^r \end{pmatrix}; \tag{7.13}$$

and the $p_x^y, x = h, v; y = r, t$ are defined in Chapter 2. The quantity $\overline{\mathcal{C}}'$ is the covariance matrix for the backscatter direction, which can be reduced to a $3 \times 3$ Hermitian matrix (Swartz *et al.*, 1988):

$$\overline{\mathcal{C}}' = \begin{pmatrix} \overline{S_{vv}S_{vv}^*} & \overline{S_{vv}S_{hv}^*} & \overline{S_{vv}S_{hh}^*} \\ \overline{S_{hv}S_{vv}^*} & \overline{S_{hv}S_{hv}^*} & \overline{S_{hv}S_{hh}^*} \\ \overline{S_{hh}S_{vv}^*} & \overline{S_{hh}S_{hv}^*} & \overline{S_{hh}S_{hh}^*} \end{pmatrix}. \tag{7.14}$$

We note here that $\overline{\mathbf{B}}_b$ and $\overline{\mathcal{C}}'$ are reduced versions of the original antenna vector and covariance matrix as defined in Chapter 2. This reduction in dimensionality follows from the reciprocity relation (2.37) (i.e., $\overline{S}_{hv} = \overline{S}_{vh}$), which, if used in the definition of the covariance matrix (see (2.73)), results in two identical rows and two identical columns. Here we use the subscript $b$ on the antenna vector to denote the reduced version of the antenna vector and the prime superscript to denote the reduced covariance matrix.

Using the Lagrange multiplier approach, the optimum antenna vector is found to be the solution to

$$\overline{\mathcal{C}}'_{\text{target}} \overline{\mathbf{B}}_b = \lambda \overline{\mathcal{C}}'_{\text{clutter}} \overline{\mathbf{B}}_b. \tag{7.15}$$

Thus, to maximize the contrast ratio, we simply need to find the maximum eigenvalue $\lambda$ in (7.15). Also note that (7.15) is valid for both point and distributed targets.

We note that a comparison of the two methods suggests that the covariance matrix approach would require fewer computations than the Stokes scattering operator approach.

### 7-2.3 Examples

*Discrimination between man-made and natural targets.* The concept of maximizing target-to-clutter ratios by using polarizations can be interpreted as a polarization filter. Swartz *et al.* (1988) and Dubois and van Zyl (1989) applied the concept of polarization filtering to the problem of finding the filter that would maximize the contrast between man-made and natural targets.

In the case of Swartz *et al.* (1988), the man-made target was an area in the city from an image of San Francisco, and the natural target was an area in Golden Gate Park from the same image. They found that a transmitting antenna with $(\psi_t, \chi_t) = (-41.3°, -6.44°)$ and a receiving antenna with $(\psi_r, \chi_r) = (60.3°, 3.51°)$ would yield a maximum contrast ratio of 9.38 dB. Notice that the transmitting and receiving antennas are very nearly cross-polarized. Among the canonical polarizations, a horizontally polarized transmitting, horizontally polarized receiving combination gave the best contrast ratio: 7.3 dB.

To investigate the effect of the incidence angle and the type of natural target used as clutter on the parameters of the polarization filter, Dubois and van Zyl (1989) performed a different experiment. First, two areas to be discriminated are chosen, the first an urban area, and the second a vegetated area. The two areas are chosen to be similar in both size and incidence angle. The corresponding polarization filter is computed by using these two experimentally determined Stokes scattering operators. Figure 7.10 shows the maximum contrast ratio as a function of the transmitting antenna polarization parameters, calculated by using (7.6), for a typical pair of areas. The filter parameters are those that yield the maximum in Fig. 7.10. As pointed out by Dubois and van Zyl (1989), the robustness of the polarization filter is related to the variation of the maximum contrast ratio in the neighborhood of the optimum filter parameters. For example, if the maximum contrast ratio varies rapidly as a function of polarization, we could expect significant degradation in the filter performance for small errors in antenna polarizations. The contrast ratio response shown in Fig. 7.10 indicates that the optimum filter for detecting man-made targets in the presence of vegetation is very robust, because relatively large errors in transmitted polarization would not result in a significant degradation of filter performance. This procedure of estimating the polarization filter parameters is then repeated for twenty area pairs at different incidence angles. The resulting filter parameters are shown in the scatter plot in Fig. 7.11. Note that the resulting orientation and ellipticity angles cluster well around $(\psi_t, \chi_t) = (45°, 0°)$ and $(\psi_r, \chi_r) = (-45°, 0°)$, i.e., a 45° linear cross-polarized antenna pair. Note that this result agrees well with that obtained by Swartz *et al.* (1988). To quantify the performance of the polarization filter, Dubois and van Zyl (1989) introduced an *enhancement factor*, a measure of how radar polarization alone provides discrimination between two classes of scatterers, defined as

$$\text{Enhancement Factor} = \frac{\overline{\mathbf{A}}(\psi_r, \chi_r) \cdot \overline{\mathcal{M}}_{\text{target}} \overline{\mathbf{A}}(\psi_t, \chi_t)}{\overline{\mathbf{A}}(\psi_r, \chi_r) \cdot \overline{\mathcal{M}}_{\text{clutter}} \overline{\mathbf{A}}(\psi_t, \chi_t)} \frac{(M_{11})_{\text{clutter}}}{(M_{11})_{\text{target}}}. \tag{7.16}$$

The enhancement factor was found to be about 3 dB and independent of the incidence angle and size of the training areas used.

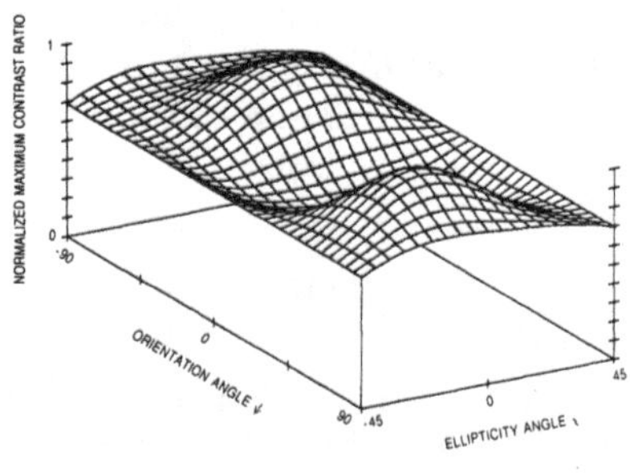


**Figure 7.10** Maximum contrast between an urban area and a forested area as a function of the transmitting antenna polarization parameters.

Dubois and van Zyl (1989) also applied the technique using the ocean as clutter and found that the resulting filter had essentially the same parameters as when vegetated areas were used as clutter. Figure 7.12(a) shows an L-band total power image, i.e., the (1,1) element of the Stokes scattering operator, image of San Francisco acquired in the summer of 1985 with the NASA/JPL CV-990 polarimetric SAR. Figure 7.12(b) shows the filter output using a 45° cross-polarized antenna pair. As expected, good contrast is observed between urban areas and vegetated areas, as well as urban areas and the ocean. Also notice the increased visibility of the lighthouse, visible off the northwest corner of the city.

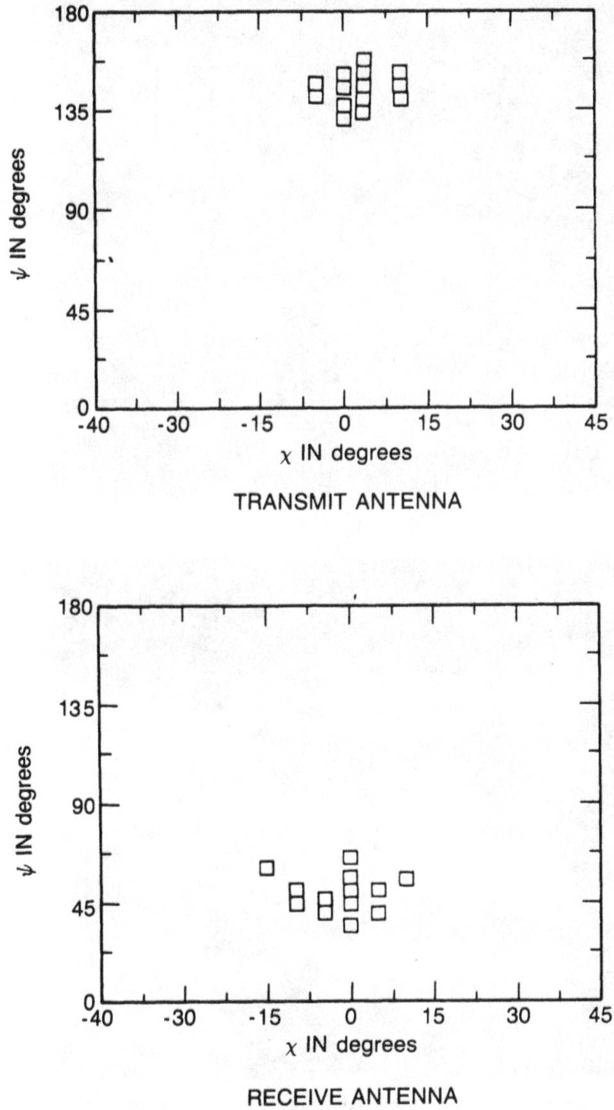


**Figure 7.11** Scatter plots showing the parameters of the polarization filters that would maximize the contrast between urban areas and forested areas. The results for 20 urban-forest pairs, each pair chosen at a different radar wave incidence angle, are shown.

(a)

(b)

**Figure 7.12** (a) L-band total power ((1,1) element of the Stokes scattering operator) image of San Francisco, and (b) image generated with polarization filter to maximize the contrast between the urban areas and the natural areas such as the parks and the ocean. Notice the increased visibility of the lighthouse in the ocean off the northwest corner of the city.

*Mapping of lava flow boundaries.* The previous example is a case where returns from two areas with different scattering mechanisms are compared. In the case of man-made targets, scattering is dominated by a double-bounce mechanism, while in the case of natural targets, scattering is either dominated by a single reflection mechanism for the ocean, or by diffuse scattering in the case of vegetated areas (van Zyl, 1989). We now consider the usefulness of this technique when the contrast between areas with similar scattering mechanisms must be maximized. As an illustration, we consider the case of mapping different lava flow units; from each unit there is rough surface scattering, where the main difference is in surface height variation, i.e., surface roughness.

Basaltic lava flows exhibit significant differences in surface roughness both between individual flows (MacDonald, 1953) and within a single flow as a function of the distance from the vent (Peterson and Tilling, 1980). Field work has shown that variations in surface textural characteristics of basaltic flows can be interpreted in terms of lava rheology and the emplacement characteristics of the flow (Basaltic Volcanism Study Project, 1981). The geometry of basalt flows has been shown to be related to the total erupted volume (Malin, 1980), the magma effusion rate (Walker, 1973), or the thermal history of the flow (Bologa and Pieri, 1986). In addition, the yield strength and viscosity of lava is directly related to the existence and geometry of channelized flows (Moore, 1987). The existence and variability of specific types of lava surface texture, and the extent and geometry of flows can thus be used to infer the eruptive style and the plausible range of geochemical properties of the magma. Regional-scale mapping of lava flow surface morphology is thus crucial to understanding the style and evolution of volcanism in a given region. Here, we present an analysis of the applicability of the polarization filter concept to the problem of mapping lava flow boundaries.

The test data consist of an L-band image, acquired in the summer of 1985 with the NASA/JPL polarimetric SAR, of a group of lava flows near the edge of the Craters of the Moon National Monument, Idaho. Figure 7.13 shows the *hh*, *vv* and *hv* images of this area, and Fig. 7.14 shows a map of the geologic units within the image, based on the geologic map by Kuntz *et al.*, (1988). The descriptions of the lava flows that follow are also taken from this reference. Here, we shall concentrate on mapping the boundary between the Blue Dragon pahoehoe (fa2) and the Carey Kipuka a'a flow (fd2), and the boundary between the Carey Kipuka a'a flow and the Little Park a'a hawaiite flow (fd3). The Blue Dragon flow is a tube- and surface-fed pahoehoe flow, dated at $2076 \pm 45$ years. Pressure ridges, pressure plateaus, and collapse depressions are common morphological features, indicating a high-volume, short-duration eruption. The Carey Kipuka a'a

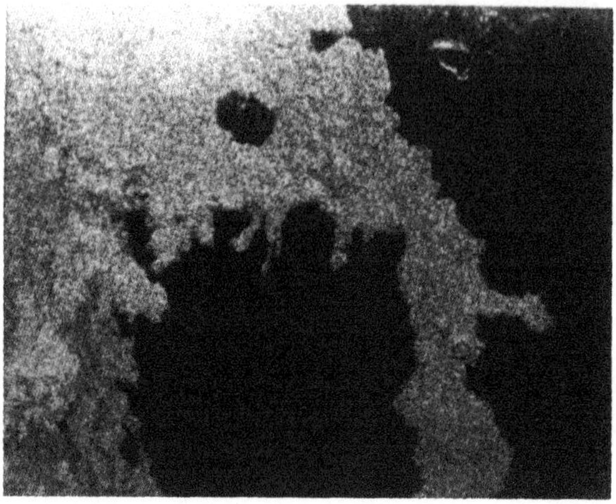

(a)

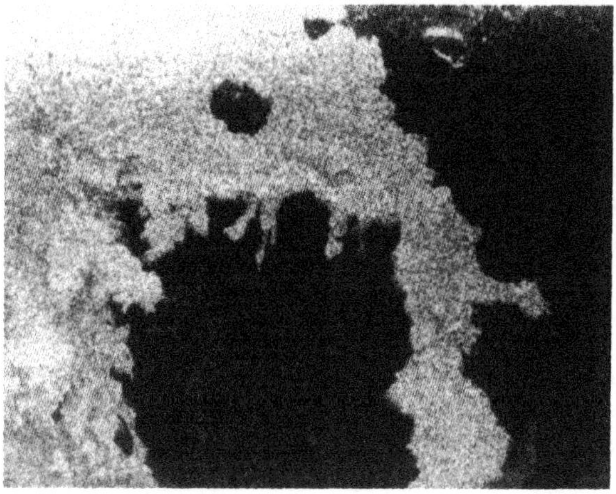

(b)

**Figure 7.13** (a) $hh$, (b) $vv$, and (c) $hv$ images of lava flows near the edge of the Craters of the Moon National Monument, Idaho. Notice that some of the lava flow boundaries visible in the $hv$ image cannot be seen in the $hh$ or $vv$ images.

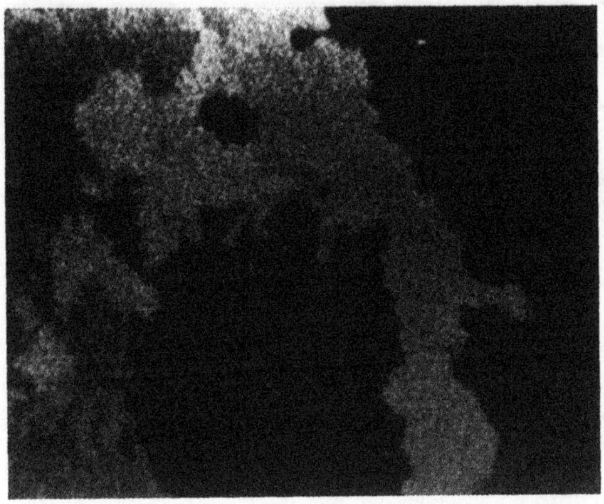

(c)

**Figure 7.13** (continued)

flow is a surface-fed, medium dark-gray a'a hawaiite flow with lobes and steep, bulbous flow fronts as high as 10 m. The lobes contain prominent flow ridges that are perpendicular to and convex toward the the flow direction, and cracks that are roughly parallel to the flow direction. The age of this unit is $6600 \pm 60$ years. The Little Park a'a flow is very similar to the Carey Kipuka a'a flow, except that the flow contains rafted blocks of bedded cinders as much as 100 m in the longest dimension and as high as 30 m. The age of the Little Park a'a flow is estimated at $6500 \pm 60$ years.

A comparison of the images in Fig. 7.13 and the geologic map in Fig. 7.14 shows that the flow boundaries are not easily recognized in the *hh* and the *vv* images, but the *hv* image shows a clear distinction between the Blue Dragon and the Carey Kipuka a'a flow. The boundary between the Little Park a'a flow and the Carey Kipuka a'a flow is not clearly visible in any of the radar images. Figure 7.15(a) shows the maximum contrast ratio responses for the Carey Kipuka a'a flow and the Blue Dragon pahoehoe flow, while the maximum contrast response for the Carey Kipuka and the Little Park a'a flows is shown in Fig. 7.15(b). These responses show that the maximum contrast ratio in both cases is almost independent of the transmitted polarization, unlike the case of the man-made target *versus*

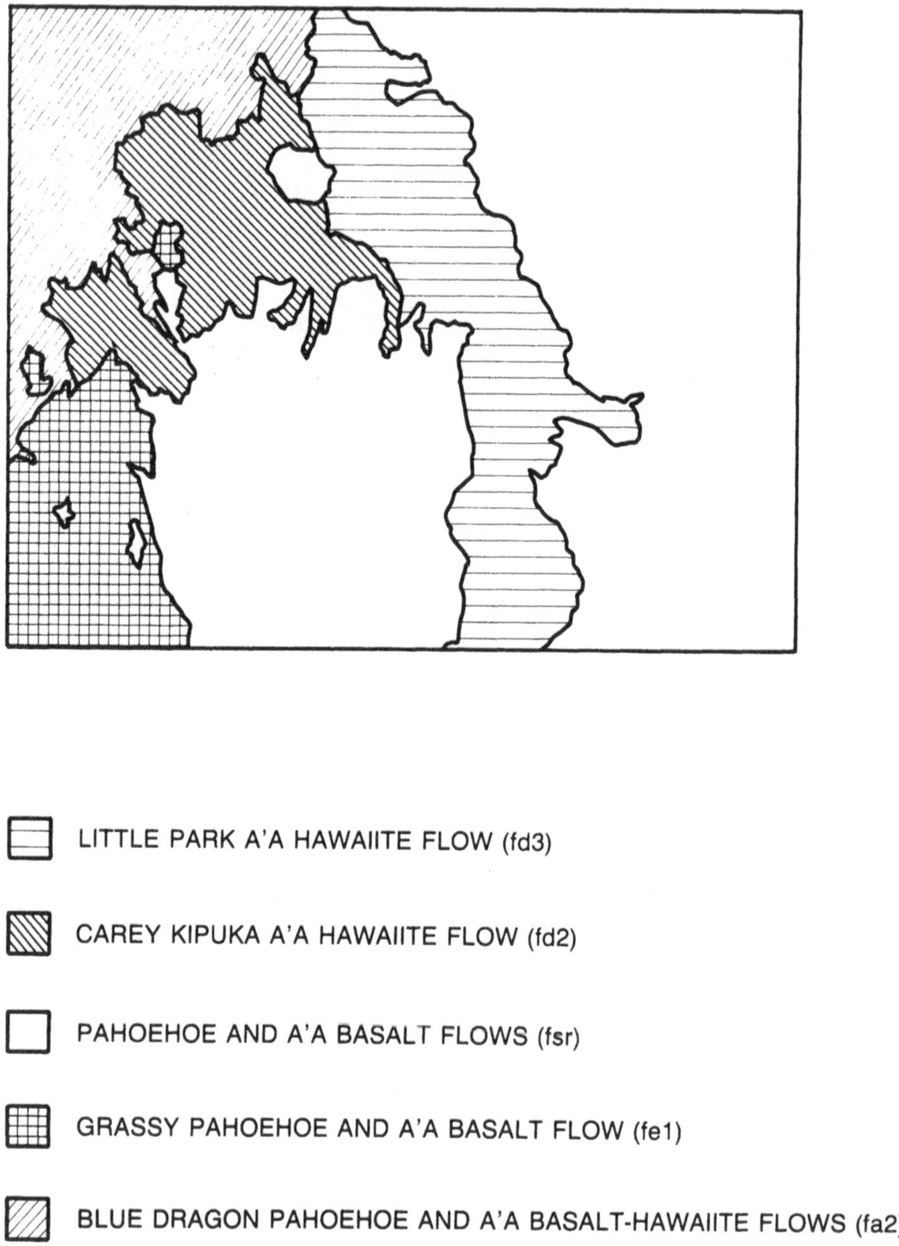


**Figure 7.14** Geologic map (after Kuntz *et al.* (1988)) of the area covered by the images shown in Figure 7.13.

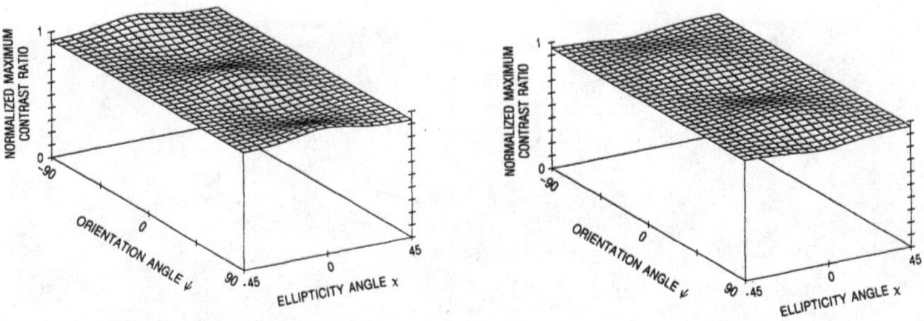


(a) CAREY KIPUKA A'A / BLUE DRAGON PAHOEHOE (b) CAREY KIPUKA A'A / LITTLE PARK A'A

**Figure 7.15** Maximum contrast responses for (a) the Carey Kipuka a'a flow and the Blue Dragon pahoehoe flow, and (b) the Carey Kipuka and Little Park a'a flows.

vegetation case shown in Fig. 7.10. In the case of the Carey Kipuka a'a flow *versus* the Blue Dragon pahoehoe flow, the polarization filter parameters are $(\psi_t, \chi_t) = (-90°, 0°)$ and $(\psi_r, \chi_r) = (3°, 4°)$, yielding an enhancement factor of 4.7 dB, while the parameters for the Carey Kipuka and Little Park a'a flows are $(\psi_t, \chi_t) = (-40°, 0°)$ and $(\psi_r, \chi_r) = (47°, 1°)$, with an enhancement factor of 1 dB. The images synthesized by using these two filters are shown in Figs. 7.16(a) and 7.16(b). Although subtle differences between the two figures can be found, the flow boundaries are not significantly easier to distinguish in one image than in the other, as expected from the results in Fig. 7.15. Also, notice that the filter parameters for the Carey Kipuka a'a and the Blue Dragon pahoehoe flows are very nearly those of horizontally polarized transmitting antenna and vertically polarized receiving antenna, i.e., the filter image is almost identical to the *hv* image. Also, as expected from the enhancement factor of only 1 dB, the boundary of the Carey Kipuka and Little Park a'a flows is not clearly visible, even if an optimum filter image is used. There is enough change in intensity across the boundary in the original images, however, to allow us to trace parts of the boundary between the Carey Kipuka and the Little Park a'a flows.

Using the filtered images, a photo-interpretation map of flow boundaries was constructed for the area, and is shown in Fig. 7.17. The flow assignments are based on the geologic map published by Kuntz *et al.* (1988). We note some important differences between the geologic map and the interpretation map. Whereas the difference in the Carey Kipuka and Little Park a'a flow boundaries can be attributed to uncertainties in tracing the flow boundary from the radar images, the big feature visible in the Grassy pahoehoe flow (fe1), outlined by dotted lines in the radar-derived map, is

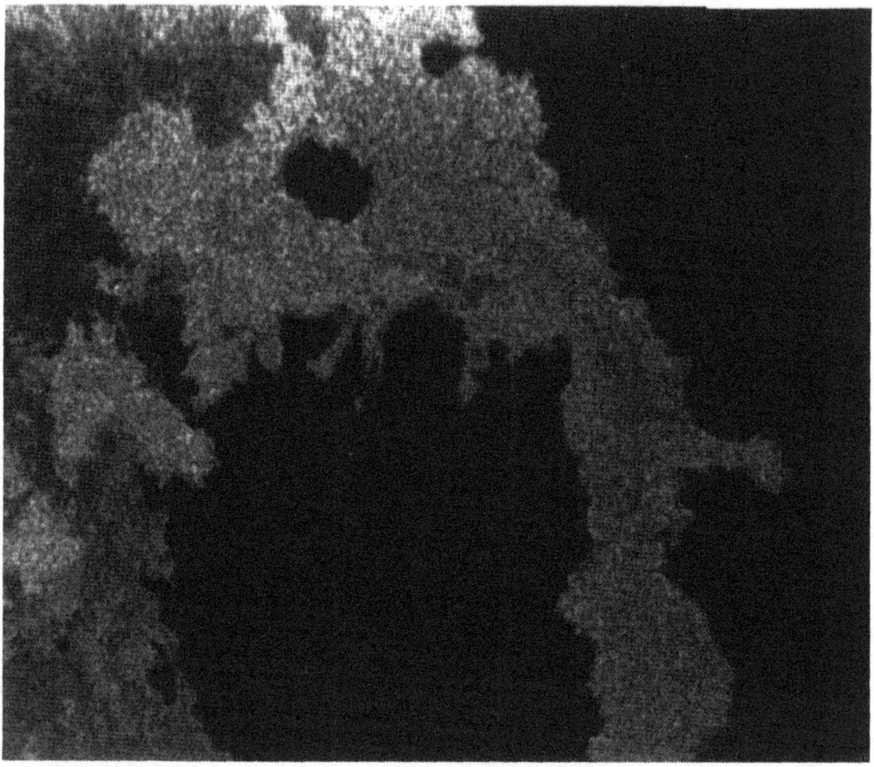

(a)

**Figure 7.16** Polarization filter images to maximize the contrast between the Carey Kipuka a'a flow and the Blue Dragon pahoehoe flow (a) $(\psi_t, \chi_t) = (-90°, 0°)$; $(\psi_r, \chi_r) = (3°, 4°)$ and between the Carey Kipuka a'a flow and the Little Park a'a flow (b) $(\psi_t, \chi_t) = (-40°, 0°)$; $(\psi_r, \chi_r) = (47°, 1°)$. Note that the boundaries are equally visible in both images, as expected from the flat contrast responses shown in Figure 7.15.

not shown in the map by Kuntz *et al.* (1988). They point out, however, that this flow contains large amounts of a'a in collapsed areas where lava flowed over steep slopes. Although this feature appears similar to the Carey Kipuka a'a flow in both radar brightness and texture, it may simply be one of the areas containing a'a in the Grassy flow.

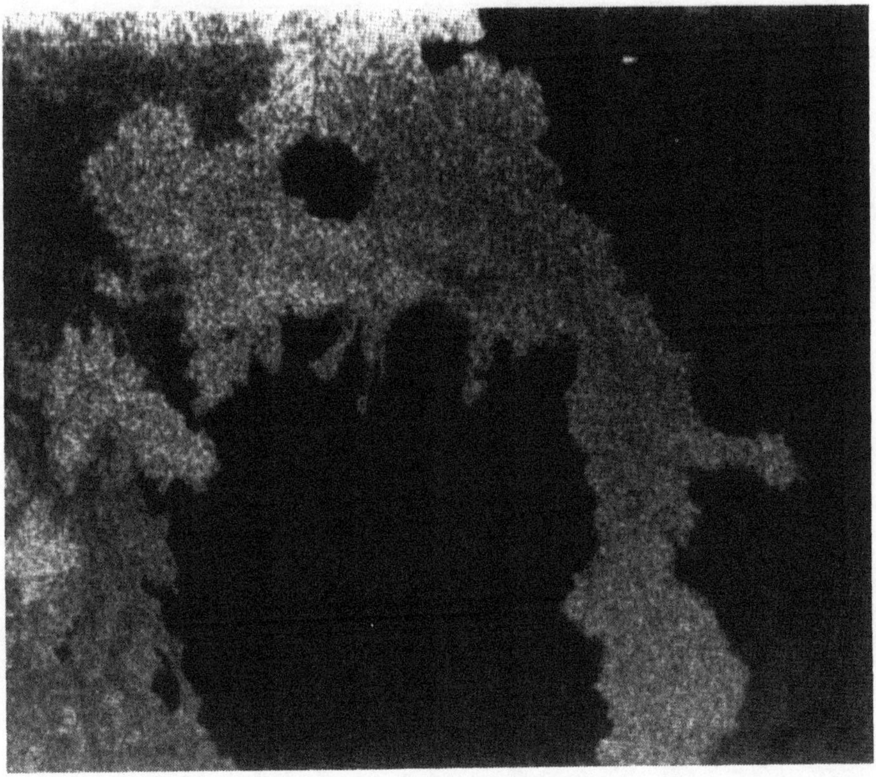

(b)

**Figure 7.16** (continued)

The two examples treated here show that as long as the areas to be contrasted have different scattering mechanisms, the polarizion filter can be expected to provide an excellent way to discriminate different areas in radar images. When scattering from different surfaces is due to the same mechanism, the usefulness of the polarization filter decreases. If the surfaces are sufficiently different physically, however, such as pahoehoe and a'a lava surfaces, the polarization filter provides a simple and effective way to map surface boundaries.

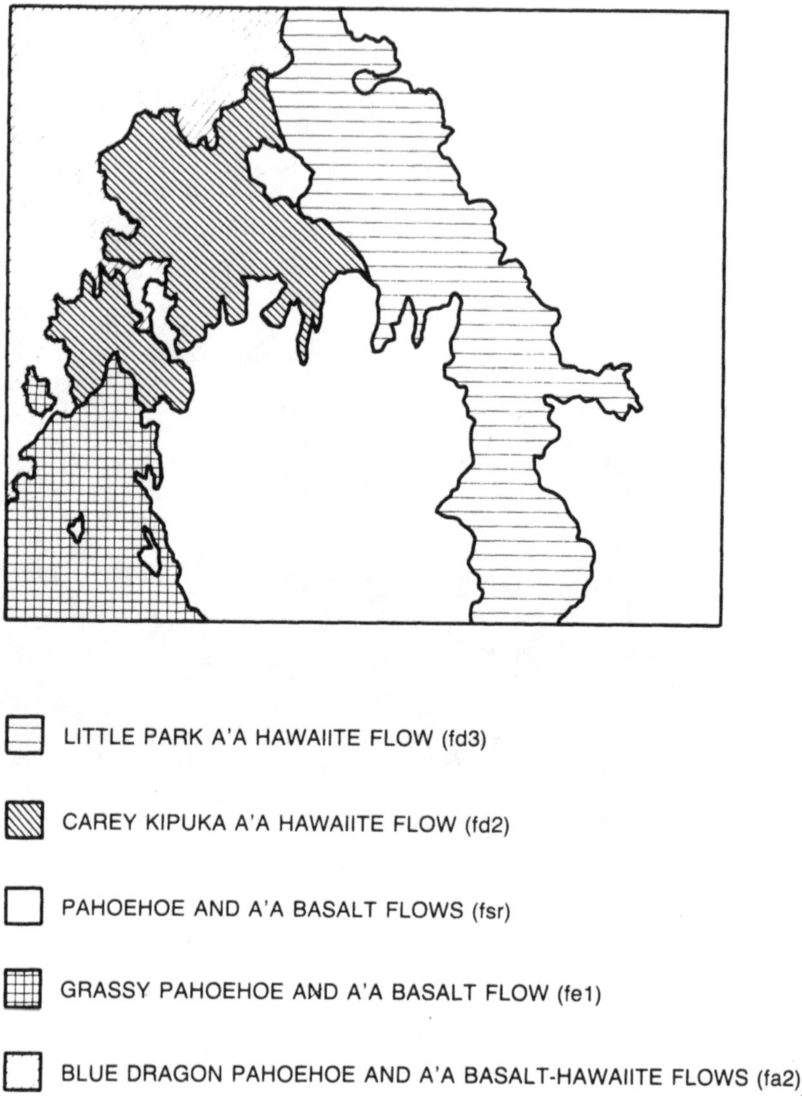


**Figure 7.17** Photo-interpretation map of flow boundaries constructed from the images in Fig. 7.16. Note the small differences in the boundary between this map and the geologic map shown in Fig. 7.14. The area outlined with the dotted line inside the Grassy pahoehoe flow is an area containing rougher lava; a consequence of the lava flowing over a steep slope.

# 7-3 UNSUPERVISED CLASSIFICATION OF SCATTERING BEHAVIOR

An algorithm to classify pixels according to the dominant scattering mechanisms observed in the polarimetric radar return was developed by van Zyl (1989a). The philosophy behind the classification algorithm is to group pixels in an image into large, general classes of scattering behavior, rather than a restricted classification based on a relatively small number of specific models.

The only information gathered in a real *remote* sensing situation (i.e., the case where no ground truth is available and chances for gathering ground truth are small) is the Stokes scattering operator, the Mueller matrix, or the covariance matrix describing the scattering properties of the area. An unfortunate reality is that more than one scattering medium may actually cause the same type of matrix to be measured. In fact, the same theoretical model, given different sets of input parameters, can predict identical matrices. To circumvent these difficulties, this algorithm classifies pixels into general classes. This should be viewed as a first step in the detailed analysis of the scattering properties of the imaged area, to be followed by an analysis using detail-model inversion and all the information available in the matrix elements.

The algorithm classifies scattering behavior by comparing the Mueller matrix for every pixel in an image with the polarization properties of three simple classes. These are: (1) odd number of reflections; (2) even number of reflections; and (3) diffuse scattering. The comparison consists of a two-part experiment relating the polarization of the scattered wave to that of the transmitted wave:

(i) First, linear polarizations are transmitted and the orientation angle of the scattered wave polarization ellipse (see Fig. 1.6), as viewed from the receiving antenna, is observed as the orientation angle of the transmitted wave is varied.

(ii) Next, any one of the circular polarizations is transmitted and the handedness of the scattered wave is compared to that of the transmitted wave.

First, consider the class of scattering events where the scattered wave suffered an odd number of reflections. As shown in Chapter 2, if scattering by a large flat plate at normal incidence is considered, for such a single scattering event, the resulting scattering matrix elements $S_{hh}$ and $S_{vv}$ are in phase. Strictly speaking, the phase difference between $S_{hh}$ and $S_{vv}$

depends, among other factors, on the frequency of the illumination and the (complex) dielectric constant of the scattering medium. As is easily shown, every scattering event adds approximately 180° phase shift between the vertically and horizontally polarized electric fields of the scattered wave. Thus, for an odd number of scattering events, the resulting phase angle of $\langle S^*_{hh} S_{vv} \rangle$ will in general be nearly zero. For this class of scatterers, van Zyl (1989a) showed that the following polarization properties will be observed:

(i) The orientation angle of the scattered wave polarization increases as the orientation angle of the transmitted wave polarization increases.

(ii) The handedness of the scattered wave is the opposite of that of the transmitted wave. Note that handedness is defined when the wave in question is observed *receding from* the observer (see Chapter 1).

Pixels in an image that exhibit these polarization characteristics are classified as *odd number of reflections.*

Next, consider the class of scattering events where the scattered wave suffers an even number of reflections before reaching the radar receiver. Because every scattering event adds 180° phase shift between the vertically and horizontally polarized electric fields of the scattered wave, it follows from the arguments of the previous paragraphs that the resulting phase angle of $\langle S^*_{hh} S_{vv} \rangle$ will in general be nearly 180° for this class. Therefore, for this class, the resulting average Mueller matrix has the following properties:

(i) The orientation angle of the scattered wave polarization increases as the orientation angle of the transmitted wave polarization decreases.

(ii) The handedness of the scattered wave is the same as that of the transmitted wave.

Pixels in an image that exhibit these polarization characteristics are classified as *even number of reflections.*

Finally, consider the case of diffuse scattering. As defined by van Zyl (1989a), diffuse scattering means that the *hh* and *vv* returns are almost completely uncorrelated. In practice this situation is expected to occur when a significant amount of multiple scattering is present, such as in the case of scattering from forested areas (Boerner *et al.*, 1987), or scattering by extremely rough surfaces or surfaces covered by blocky boulders. This means that the rapidly varying *hh* to *vv* phase difference will cause some of the individual Mueller matrices to exhibit the properties of an odd number and some to exhibit the properties of an even number of reflections. A Mueller matrix of this form will exhibit the following polarization properties:

(i) The orientation angle of the scattered wave polarization increases as the orientation angle of the transmitted wave polarization increases.

(ii) The handedness of the scattered wave is the same as that of the transmitted wave.

### 7-3.1 Examples

Three examples will be considered. The first example shows application of this method to classify land use in an urban area. In the second example, we consider the application of this technique to forest ecology, and use multifrequency polarimetric radar data to discriminate between burned and unburned areas in a forest. The third example shows how this technique could be used to map clear-cuts and wetlands in vegetated areas.

*Land use.* Plate 7.1 shows the results of classification of an image of the San Francisco Bay area in California, acquired with the NASA/JPL CV-990 polarimetric SAR during the summer of 1985. All pixels that exhibited polarization characteristics similar to that of the odd number of reflections class are colored blue, those that were similar to the even number of reflections class are colored red, and those that exhibit characteristics similar to areas with a significant amount of diffuse scattering are colored green. All pixels for which "unable to classify" was obtained are colored yellow. Generally, pixels in the ocean were classified as being similar to the odd number of reflections class. This is not surprising, because, as shown earlier (see also van Zyl, *et al.* (1987)), the polarization responses observed for the ocean are adequately explained by the second-order small perturbation model. Plate 7.1 also shows that most pixels in the urban area, as well as the lighthouse in the ocean (visible off the northwest corner of San Francisco) were classified as being similar to the even number of reflections class. In the case of the urban area, this is probably due to a double bounce off the surface of the streets onto the sides of the buildings before the signals return to the radar, whereas for objects on the ocean surface, a probable mechanism is a reflection off the water surface onto the side of the object and back to the radar. Most of the pixels in Golden Gate Park and other tree-covered parts of the city, including the wider tree-lined streets, were classified as being similar to areas with a large amount of diffuse scattering, possibly as a result of increased multiple scatter in the vegetation.

*Forest ecology* (Evans and van Zyl, 1989). A key element in understanding global climate change is the ability to quantify sources and sinks of greenhouse gasses such as $CO_2$, $CH_4$ and $N_2O$. For example, the magnitude of the carbon source and sink strength in forests depends largely on the successional stage. In addition, forests are rapidly being cleared for agriculture and pasture resulting in a relatively instantaneous (depending on whether the forest is burned or left to decompose) influx of $CO_2$ in the atmosphere. Information on (1) the biomass of the existing forests, (2) the areal extent of the cleared forests, and (3) the successional stage of the primary and secondary growth forests (and thus the carbon source or sink strength) are essential to understanding the role of forests in the global carbon cycle. Radar data, because they provide information about the volumetric properties of the canopy, may play a major role in determining the total forest biomass.

Figure 7.18 shows an L-band total power image of an area in Mt. Shasta in northern California. This area contains conifers and some hardwoods. The area at the lower left of this image is a burned area with somewhat smaller trees (35 m *versus* 45 m for the unburned trees). Plate 7.2 shows classification results based on the C-, L-, and P-band data of the same area. We can see in Plate 7.2(a) that much of the area is classified as dominated by odd numbers of reflections at C-band, indicating that the canopy is dense enough to result in single reflections from the branches, and the clear-cut areas allow single reflections from the ground. At L-band, the clear-cut areas are still dominated by single reflections, but there is much more pervasive diffuse scattering by the canopy, in both the burned and the unburned areas, and some double reflections. At P-band, the clear-cut areas are more visible, and the burned area in the lower left of the image exhibits significantly more double-bounce scattering. This increase could either be caused by the difference in the size of the trees as noted above, or some change in the canopy as a result of the fire that results in increase penetration into the canopy of the burned area. According to Durden *et al.* (1989), there is also a decrease in the P-band unpolarized power of the burned area relative to the unburned area. They conclude, based on model results, that this may be caused by a lower dielectric constant in the burned area, which explains the increased canopy penetration seen here.

*Mapping of clear-cut areas and wetlands.* Plate 7.3 shows the classification map for the Moosehead Lake, Maine image studied by Durden *et al.* (1989). We note that the forested areas are classified as being dominated by diffuse scattering, and the clear-cut areas are classified as odd numbers of reflections. The blue oval area in the center near the top of the image is an open bog, for which the scattering mechanism is classified as odd numbers of reflections, the same as the clear-cut areas. There are two other interesting areas in this image, both of which are classified as even numbers of reflections. The first is in the lower left corner, near the bend in the river. This area is near a beaver dam (van Zyl, 1989a), and the scattering is

**Figure 7.18** L-band total power image of an area near Mount Shasta in northern California.

dominated by double reflections. The second area is at almost the same incidence angle, but almost in the middle (left to right) in the image. This is a swampy area (Durden *et al.*, 1989) with fewer and shorter trees than the surrounding areas. Note that the results of this classification are consistent with the analysis performed by Durden *et al.* (1989) (see also Section 7.1) using a three-layer scattering model. They found that the scattering from the clear-cut areas is dominated by direct backscattering from the ground, i.e., single reflections are important in this case. For the swampy areas, the double reflections from the ground to the trunks and *vice versa* were found to be the dominant scattering mechanism, consistent with the even numbers of reflections classification. For the heavily forested areas, scattering from the randomly oriented branches was found to dominate, which supports the diffuse scattering classification.

The swampy and clear-cut areas are plainly visible in Plate 7.3. When Plate 7.3 is compared with the *vv* image of the same area shown in Fig. 7.7, it becomes evident that some of the clear-cuts are not discernable in the *vv* image. Also, based on the brightness in the *vv* image alone, the swampy areas are not distinguishable from specular reflections. Using classification maps such as those in Plate 7.3 acquired over a period of time, the rate and extent of clear-cutting, as well as the existence and change in areal extent of the swampy areas can easily be mapped.

# 7-4 SUPERVISED CLASSIFICATION OF EARTH TERRAIN

Several investigators have reported the use of radar data to map earth terrain types and land cover. These investigations can be grouped into four major approaches: (1) mapping based on the textural information in radar images, e.g., Stromberg and Farr (1986); (2) mapping based on multiple incidence angle images, e.g., Cimino *et al.* (1986); (3) mapping based on multipolarization parameters, such as amplitude (i.e., *hh*, *vv*, and *hv*) images, e.g., Evans *et al.* (1986) or Wu and Sader (1987), or the phase difference between the *hh* and *vv* elements of the scattering matrix, e.g., Boerner *et al.* (1987) or Ulaby *et al.* (1987); and (4) more recently, the use of fully polarimetric classifiers, e.g., Lim *et al.* (1989). Here, we shall discuss the use of fully polarimetric radar data to classify different types of earth terrain.

The optimum polarimetric classifier distance measure was derived by Kong *et al.* (1988) for unnormalized polarimetric data under the assumption that polarimetric data are characterized by a multivariate Gaussian distribution. In practice, in some cases, due to mainly incidence angle effects, the unnormalized polarimetric classifier was found to confuse classes (Lim *et al.* 1989). To circumvent this problem, the optimum normalized polarimetric classifier distance measure was derived by Yueh *et al.* (1988). Here, we shall compare these two methods with each other, and also with a simpler minimum-distance type of classifier.

## 7-4.1 Optimum (Bayes) Classification Distance

In this section we shall only consider the backscatter case. Using the Bayes maximum likelihood test, Kong *et al.* (1988) showed that the decision rule for classification can be written as

$$\overline{\mathbf{X}}_b \in w_i \text{ if } d_i(\overline{\mathbf{X}}_b) < d_j(\overline{\mathbf{X}}_b) \text{ for all } j \neq i, \tag{7.17}$$

where

$$\overline{\mathbf{X}}_b = \begin{pmatrix} \overline{S}_{vv} \\ \overline{S}_{hv} \\ \overline{S}_{hh} \end{pmatrix} \tag{7.18}$$

is the feature vector of the pixel to be classified, and $d_i$ is the distance measure for the $i$th class. Assuming that polarimetric data are characterized by a multivariate Gaussian distribution, Kong *et al.* (1988) derived the optimum distance measure for the $i$th class as

$$d_i(\overline{\mathbf{X}}_b) = \overline{\mathbf{X}}_b^* \cdot \left(\overline{\mathcal{C}}_i'\right)^{-1} \overline{\mathbf{X}}_b + \ln|\overline{\mathcal{C}}_i'| - \ln[P_a(i)], \tag{7.19}$$

where $\overline{\mathcal{C}}_i'$ is the covariance matrix of the $i$th class as defined in (7.14) and $P_a(i)$ is the *a priori* probability that a particular pixel belongs to class $i$.

In some cases, the overall brightness of one class in a certain part of the image may be closer to the brightness of the training data of a different class. This could happen, for example, when two classes of rough surfaces at angles significant away from normal incidence are used for training data, and one class is much rougher (and hence much brighter in the radar image) than the other. For angles near normal incidence, specular reflections could cause the smoother surface class to exhibit brightnesses close to that of the rougher surface class at larger incidence angles. The result is mis-classification due to the brightness variations. To overcome this kind of problem, Yueh *et al.* (1988) derived the optimum distance measure for normalized polarimetric radar data as

$$d_i(\overline{\mathbf{X}}_b) = 3 \ln\left[\overline{\mathbf{X}}_b^* \cdot \left(\overline{\mathcal{C}}_i'\right)^{-1} \overline{\mathbf{X}}_b\right] + \ln|\overline{\mathcal{C}}_i'| - \ln[P_a(i)]. \tag{7.20}$$

Note that the expression (7.20) does not contain a normalization factor. Yueh *et al.* (1988) showed that the classification performance is independent of the normalization scheme.

We can easily see from (7.19) and (7.20) that the optimum distance measures are quadratic functions of $\overline{\mathbf{X}}_b$. Consequently, the decision surfaces implemented by the maximum likelihood classification are quadratic and take the form of parabolas, circles, and ellipses (Richards, 1986).

## 7-4.2 Minimum Distance Classifier

The effectiveness of the maximum likelihood classifier depends on the the accuracy with which the vector $\overline{\mathbf{X}}_b$ and the covariance matrix $\overline{\mathcal{C}}'$ can be estimated. In cases where insufficient training data exist, inaccurate estimates of the covariance matrix may result, leading to reduced classification accuracy. In such cases where the number of training samples per class is limited, a more effective approach may be to resort to a classifier that depends only on the mean positions of the class centers, and not on the covariance (Richards, 1986). The minimum distance classifier uses training data only to determine class means; classification is implemented by placing a pixel in the class with the nearest mean.

Assuming that the polarimetric data are characterized by a multivariate zero-mean Gaussian distribution, the mean value of $\overline{\mathbf{X}}_b$ thus is the zero vector, independent of the class. The minimum distance classifier, as implemented by van Zyl and Burnette (1989), forms the class feature vector from the second-order cross-products of the scattering matrix elements. The distance measure is thus defined as

$$d_i^2(\overline{\mathbf{Y}}_b) = \left|\overline{\mathbf{Y}}_b^* - \overline{\mathbf{Y}}_i\right|^2, \tag{7.21}$$

where

$$\overline{\mathbf{Y}} = \begin{pmatrix} \overline{S_{vv}S_{vv}^*} \\ \overline{S_{vv}S_{hv}^*} \\ \overline{S_{vv}S_{hh}^*} \\ \overline{S_{hv}S_{hv}^*} \\ \overline{S_{hv}S_{hh}^*} \\ \overline{S_{hh}S_{hh}^*} \end{pmatrix}, \tag{7.22}$$

and the subscript $b$ refers to the pixel to be classified and $i$ to the $i$th class. The decision rule in this case is

$$\overline{\mathbf{Y}}_b \in w_i \qquad \text{if } d_i(\overline{\mathbf{Y}}_b) < d_j(\overline{\mathbf{Y}}_b) \qquad \text{for all } j \neq i. \tag{7.23}$$

The minimum distance classifier has been implemented both with unnormalized and normalized $\overline{\mathbf{Y}}_b$ vectors. In the case of the normalized vectors, the elements of $\overline{\mathbf{Y}}_b$ are normalized relative to the total power.

Richards (1986) showed that the maximum likelihood classifier takes $N + 1$ times as long as the minimum distance classifier. Because classification time increases quadratically with the number of components for the maximum likelihood classifier, but only linearly for minimum distance classification (Richards, 1986), for multifrequency polarimetric data, the minimum distance classification thus may be considerably cheaper in terms of computer requirements. However, because the minimum distance classifier does not use covariance data, it is not as flexible as the maximum likelihood classifier. Richards (1986) points out that the maximum likelihood classifier model can account for spreads of data in particular directions. As covariance data is not used in the minimum distance technique, classes are assumed to be symmetric. Elongated classes will therefore not be well modeled.

### 7-4.3 Examples

In this section we present classification results in the areas of land use, forestry, geology and sea ice.

*Land use.* Figure 7.19 shows an L-band total power image of San Francisco, acquired during the summer of 1985. To classify land use in this image, the results of Plate 7.1 were taken into account in choosing the three training areas shown in Fig. 7.19, one each for the ocean, the vegetated areas, and the urban areas. Based on the training data, and assuming equal *a priori* probabilities, the image was classified by using the unnormalized optimum polarimetric classifier. The results are shown in Plate 7.4(a). Even with impure training areas for the vegetated and urban classes, the fully polarimetric classifier correctly classified large areas in the image. Comparing Plates 7.1 and 7.4(a), we notice that in general the colors in Plate 7.4(a) are more pure than those in Plate 7.1. The polarimetric classifier correctly identified tree-lined streets in San Francisco, and even located small lakes within Golden Gate Park, as previously pointed out by Lim *et al.* (1988). The upper right-hand region in Plate 7.4(a) was misclassified as urban and vegetated areas, while the results of Plate 7.1 show this area to be part of San Francisco Bay. The reason for this misclassification is the high return from the water surface for angles near normal incidence. In fact, the classification of the ocean surface in Plate 7.4(a) shows that for the smallest incidence angles with the highest return, the ocean was classified as the brightest training area, the urban area. This is followed by a band that was classified as vegetated area, which is the second brightest training area. Thus, at least for the ocean in the near range, the brightness information seems to dominate any polarization information that may be present in the covariance matrix.

Plate 7.4(b) shows the classification of the San Francisco image, by using the same training sets as in the case of Plate 7.4(a), but using the normalized polarimetric classifier, also assuming the *a priori* probabilities to be equal. Comparison of Plate 7.4(a) and 7.4(b) shows that the upper right-hand part of the image is correctly classified as ocean in Plate 7.4(b) and, as a whole, the classification with the normalized polarimetric classifier

LEGEND:
O = OCEAN
U = URBAN
P = VEGETATED

**Figure 7.19** L-band total power image of San Francisco, showing the training areas used to classify the terrain cover in the image into three classes; urban (U), vegetated (P), and ocean (O).

seems to be more accurate than that with the unnormalized one.

Plate 7.4(c) and (d) shows the classification results using the unnormalized and normalized minimum distance classifiers, respectively. As in the case of the maximum likelihood classifier, the unnormalized classifier misclassifies large portions of the ocean, while the normalized classifier corrects this problem. In general, as expected, the results obtained using the minimum distance classifier are less accurate than those of the maximum likelihood classifiers.

To quantify the performance of the classifiers, Lim *et al.* (1988) used the training regions as absolute truth, and then counted the misclassifications in those regions. The advantage of this method is that all classifiers are treated the same. The results are shown as "confusion tables" similar to those shown by Lim *et al.* (1988). The confusion table not only shows the number of pixels that were misclassified, but also how they were misclassified. Table 7.2 is the confusion table for the unnormalized polarimetric classifier. Reading the first line of the table, the second entry shows the number of the pixels in the urban training area that were actually classified as urban class. The third entry shows the number of pixels in the urban training area that were classified into the vegetated class, and the fourth entry shows the number that were classified into the ocean area class. The quantity in parentheses following the class label is the total number of pixels in the training area of that class.

**Table 7.2** Confusion Table for Unnormalized Maximum Likelihood Classification of San Francisco

| From | To | | |
|---|---|---|---|
| | Urban | Vegetated | Ocean |
| Urban (2948) | 2187 | 761 | 0 |
| Vegetated (2678) | 164 | 2514 | 0 |
| Ocean (3515) | 0 | 1 | 3514 |

Table 7.2 shows that the classification accuracy for the ocean class is very high, with the only confusion occurring between the ocean and vegetated classes. The overall classification accuracy is calculated by taking the total number of correctly classified pixels for all classes, divided by the total number of pixels in the training areas of all classes. The confusion table for the normalized Bayes classifier is shown in Table 7.3, that of the unnormalized minimum distance classifier in Table 7.4, and that of the normalized minimum distance classifier in Table 7.5. Although the individual numbers are different, the general trends shown in the tables are the same for all classifiers. A comparison of the overall accuracies of the four classification schemes is shown in Table 7.6. As expected, the Bayes classifiers give more accurate results than the minimum distance classifiers, and, as is consistent with the better looking results in Plate 7.4, the normalized classifiers show higher classification accuracies than the unnormalized ones. We note here that this result is different from that obtained by Lim *et al.* (1988). In the classifications shown in Plate 7.4, and all the plates that follow, we have used 16-look data, whereas Lim *et al.* (1988) used single-look data. This may explain part of the difference in the results. Another reason for the difference may be that we have used different and smaller training areas than did Lim *et al.* (1988). The final, and we believe the most important, reason for the difference is that the method of quantifying classifier accuracy is seriously flawed. The method assumes that all pixels in a training

area indeed belong to a single class. Obviously, not all pixels in an urban area will show double reflections as the dominating scattering mechanism. Similarly, some pixels in the park area may indeed show double reflections, even if the area is mostly vegetated. Therefore, we believe that the exact quantities shown in Tables 7.2 to 7.6 are less important than the trends indicated by these numbers.

**Table 7.3** Confusion Table for Normalized Maximum Likelihood Classification of San Francisco

| | To | | |
|---|---|---|---|
| From | Urban | Vegetated | Ocean |
| Urban (2948) | 2566 | 382 | 0 |
| Vegetated (2678) | 358 | 2320 | 0 |
| Ocean (3515) | 5 | 24 | 3486 |

**Table 7.4** Confusion Table for Unnormalized Minimum Distance Classification of San Francisco

| | To | | |
|---|---|---|---|
| From | Urban | Vegetated | Ocean |
| Urban (2948) | 1472 | 1464 | 12 |
| Vegetated (2678) | 54 | 1741 | 883 |
| Ocean (3515) | 0 | 30 | 3485 |

**Table 7.5** Confusion Table for Normalized Minimum Distance Classification of San Francisco

| | To | | |
|---|---|---|---|
| From | Urban | Vegetated | Ocean |
| Urban (2948) | 2240 | 682 | 26 |
| Vegetated (2678) | 555 | 2091 | 32 |
| Ocean (3515) | 0 | 0 | 3515 |

**Table 7.6** Comparison of Classification Accuracies for the Four Classifiers Applied to the San Francisco Data

| Classification Method | Classification Accuracy (%) |
|---|---|
| Maximum Likelihood | 89.87 |
| Normalized Maximum Likelihood | 91.59 |
| Minimum Distance | 73.27 |
| Normalized Minimum Distance | 85.83 |

*Forestry.* To investigate the usefulness of multifrequency polarimetric radar data for terrain classification, the images of the forested area near Mount Shasta, California, were classified according to the normalized Bayes classifier, using the training data shown in Fig. 7.20. (We show only the results for the normalized classifier, as this classifier was found to give the best results.) Plate 7.5 shows the results of the classification for the three individual frequencies, as well as that of the full three-frequency data set. As expected from the results of Plate 7.2, the C-band classification accuracy is the most noisy, and the healthy and burned forested areas are almost indistinguishable. The results at L-band are better, but the best single-frequency result is achieved at P-band, consistent with the results of Plate 7.2. The classification map labelled "combination" in Plate 7.5 is the result of assigning a pixel to a particular class based on the full three-frequency polarimetric information. We note from Plate 7.5 that the three-frequency classification is less noisy than that of any of the single-frequency classifications, indicating that multifrequency polarimetric radar data can significantly increase our capability of accurately classifying earth terrain.

*Geology.* Figure 7.21 shows the C-band total power image and a geologic map (after Wise, 1966) of the Pisgah lava flow area in the Mojave Desert in southern California. This image was acquired as part of the Mojave field experiment during the summer of 1988 with the NASA/JPL DC-8 SAR. Plate 7.6 shows the normalized and unnormalized Bayes classification maps of this area based on the training areas outlined in Fig. 7.21, using the C-band data. That the unnormalized classifier results are much more accurate than those of the normalized classifier is immediately clear. This suggests that, for geological scenes, the overall brightness carries much of the information in the polarimetric radar image. Although we show only the C-band results in Plate 7.6, the same results were found using the L-band and P-band data. Plate 7.7 shows the normalized Bayes classification maps of all the frequencies, as well as that of the full three-frequency classification.

**Figure 7.20** L-band total power image of an area near Mount Shasta in northern California. Shown are the training areas used to classify the terrain cover in this image into three classes; healthy forest (H), open water (W), and fire-damaged forest (F).

As in the case of the Mount Shasta image, the C-band classification is the least accurate. The most probable reason for this is that most of the surfaces in this scene have surface roughness comparable to, or larger than, the C-band wavelength. The vertical bands in the P-band images are due to spurious signals from an unknown source on the ground. Notice, however, that the P-band classifications in the lava flow itself are more accurate than those of the other frequencies. As in the case of the Mount Shasta image, the most accurate results are achieved with the three-frequency combination.

*Sea ice.* Three training areas were chosen using the C-band total power image of a part of the Beaufort Sea, Alaska, shown in Fig. 7.22; first-year ice, multiyear ice, and pressure ridges. Plate 7.8 shows the results for the unnormalized Bayes classification for the different frequencies. We notice that in this case the C-band classification discriminates well between the first-year and multiyear ice, but does not seem to be able to discriminate

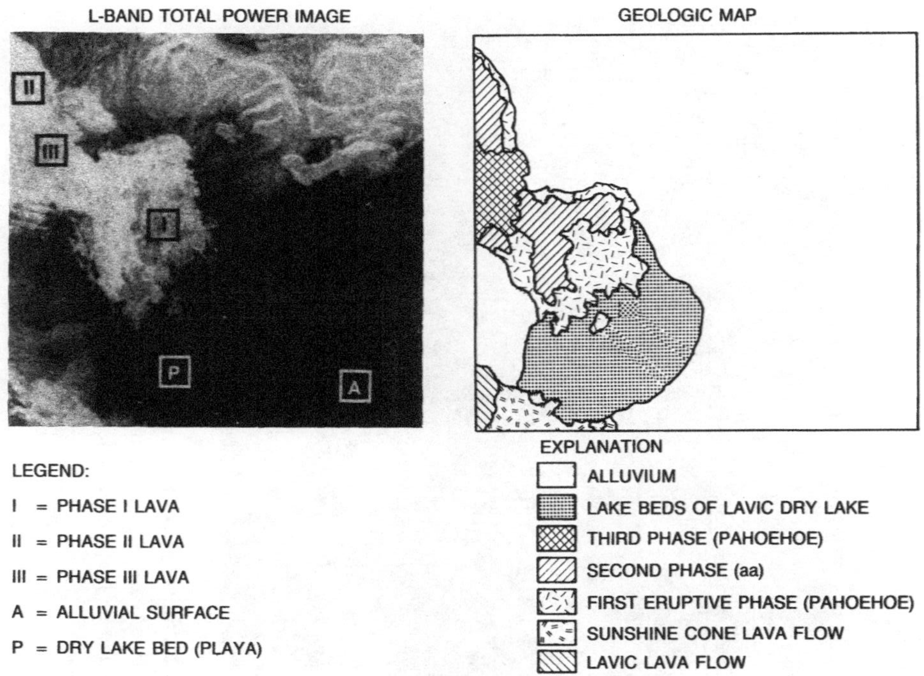


**Figure 7.21** C-band total power image of the Pisgah lava flow area in the Mojave Desert in southern California. Shown are the training areas used to classify the terrain cover in this image into five classes: Phase I, Phase II, and Phase III lava, alluvial surface, and dry lake bed. Also shown is a geological map (after Wise (1966)), indicating the boundaries of the different phases of the lava flows.

well between these ice types and the pressure ridges. The P-band and L-band classifications separate the pressure ridges from the first-year and multiyear ice, but does not separate the ice types as well as the C-band case. As before, however, the most accurate classification is achieved when the information from all three frequencies is used. We show only the unnormalized classifications, because they were found to be far superior to the normalized ones.

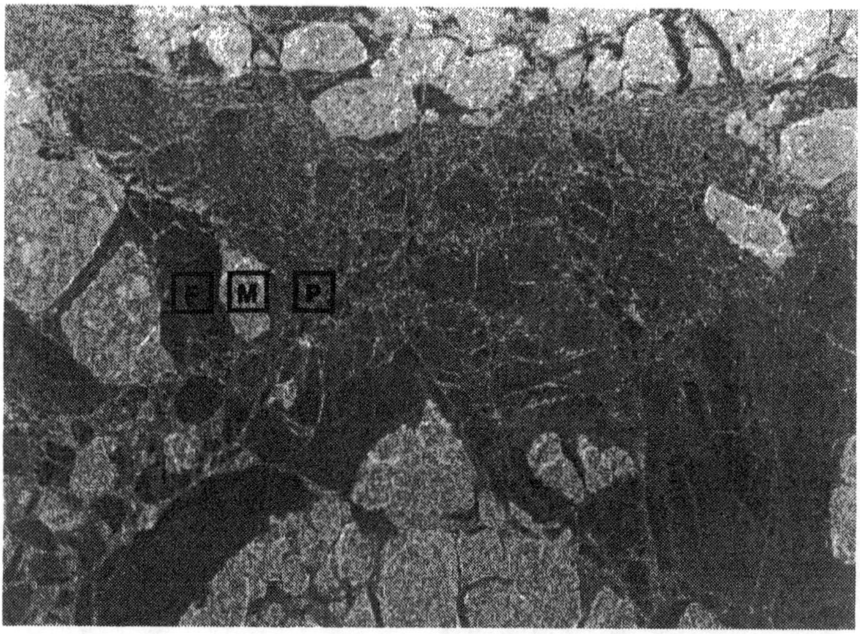


**Figure 7.22** C-band total power image of an area in the Beaufort Sea, Alaska, showing the training areas used to classifiy the image into three classes: first-year ice (F), multiyear ice (M), and pressure ridges (P).

# 7-5 CONCLUSIONS

In this chapter we have discussed several examples of recently developed tools and analysis techniques that permit fuller utilization of polarimetric radar data. The examples show that the polarization responses can be used to study surfaces and terrain cover based on observed scattering behavior. In the case of responses of a forested area, comparison of observed and model derived responses provides information about the dominant scattering mechanisms responsible for the measured radar response. We also discussed the concept of polarization filtering, and derived a useful filter for detecting man-made targets in the presence of clutter. We found that the filter parameters are the same, independent of the type of clutter. When the contrast between areas with similar scattering mechanisms must be maximized, however, we have found that the usefulness of the polarization filter decreases.

Radar data, which provide information about the volumetric properties of vegetation canopies, may increase our ability to isolate the scattering pathway that interacts mostly with tree trunks and stems. This is key to extracting biophysical parameters such as woody biomass and possibly trunk water content. We have shown with some examples how a simple, unsupervised classification scheme of scattering behavior using multifrequency polarimetric radar data can provide some of this information. The examples also show how this algorithm allows mapping of clear-cut areas and wetlands. Dramatically different scattering mechanisms are observed for the same forest under frozen and thawed conditions.

Finally, we presented a number of classification maps of different types of earth terrain cover. These maps were generated by using maximum likelihood as well as minimum distance techniques. We found, as would be consistent with theory, that the maximum likelihood method almost always yielded results that were more accurate than those of the minimum distance method. Unfortunately, in some cases, the unnormalized maximum likelihood method was shown to misclassify large areas in an image. At this stage, there is no recognized consensus on when to use the unnormalized and when to use the normalized classifiers.

Using the techniques presented in this chapter in combination, we can begin to extract information about the biogeophysical parameters that describe different types of earth terrain. Derivation of process-related information and biogeophysical parameters from remote sensing data is one key to our understanding of nature and the significance of contemporary global change. Therefore, future work will undoubtedly emphasize the quantitative determination of the physical state of the surface and cover of the earth through continued development of radar scattering models and model inversion algorithms that can be used to convert radar measurements directly into biogeophysical parameters.

## ACKNOWLEDGEMENT

This work was performed at the Jet Propulsion Laboratory, California Institute of Technology, sponsored by the National Aeronautics and Space Administration. We would like to thank all the people involved in the JPL Radar Program who operated the imaging polarimeter and processed the data used in this manuscript.

# References


Baloga, S., and D. Pieri, "Time-Dependent Profiles of Lava Flows," *Journal of Geophysical Research,* Vol. 91, pp. 9543–9552, 1986.

Basaltic Volcanism Study Project, *Basaltic Volcanism on the Terrestrial Planets,* Pergamon Press, New York, 1286 pp., 1981.

Boerner, W.M., M.B. El-Arini, C.Y. Chan, and P.M. Mastoris, "Polarization Dependence in Electromagnetic Inverse Problems," *IEEE Trans. on Antennas and Propagation,* Vol. AP-29, pp. 262–271, 1981.

Boerner, W.M., B-Y. Foo, and H.J. Eom, "Interpretation of Polarimetric Copolarization Phase Term in the Radar Images Obtained by the JPL Airborne L-band SAR System," *IEEE Trans. on Geoscience and Remote Sensing,* Vol. GE-25, pp. 77–82, 1987.

Cimino, J., A. Brandani, D. Casey, J. Rabassa, and S.D. Wall, "Multiple Incidence Angle SIR-B Experiment over Argentina: Mapping of Forest Units," *IEEE Trans. on Geoscience and Remote Sensing,* Vol. GE-24, pp. 498–509, 1986.

Dubois, P.C., and J.J. van Zyl, "Polarization Filtering of SAR Data," *IGARSS'89 Digest,* Vol. 3, p. 1816–1819, 1989.

Durden, S.L., J.J. van Zyl, and H.A. Zebker, "Modeling and Observations of the Radar Polarization Signatures of Forested Areas," *IEEE Trans. on Geoscience and Remote Sensing,* Vol. GE-27, pp. 290–301, 1989.

Evans, D.L., T.G. Farr, J.P. Ford, T.W. Thompson, and C.L. Werner, "Multipolarization Radar Images for Geologic Mapping and Vegetation Discrimination," *IEEE Trans. on Geoscience and Remote Sensing,* Vol. GE-24, pp. 246–257, 1986.

Evans, D.L., T.G. Farr, J.J. van Zyl, and H.A. Zebker, "Imaging Radar Polarimetry: Analysis Tools and Applications," *IEEE Trans. on Geoscience and Remote Sensing,* Vol. GE-26, pp. 774–789, 1988.

Evans, D.L., and J.J. van Zyl, *Imaging Radar Polarimetry: Analysis Tools and Applications,* Elsevier, Amsterdam, 1989.

Ioannidids, G.A., and D.E. Hammers, "Optimum Antenna Polarizations for Target Discrimination in Clutter," *IEEE Trans. on Antennas and Propagation,* Vol. AP-27, pp. 357–363, 1979.

Kong, J.A., A.A. Swartz, H.A. Yueh, L.M. Novak, and R.T. Shin, "Identification of Terrain Cover using the Optimum Polarimetric Classifier," *Journal of Electromagnetic Waves and Applications,* Vol. 2, pp. 171–194, 1988.

Kuntz, M.A., D.E. Champion, R.H. Lefebvre, and H.R. Covington, "Geologic Map of the Craters of the Moon, Kings Bowl, and Wapi Lava Fields, and the Great Rift Volcanic Rift Zone, South Central Idaho," *U.S. Geological Survey Miscellaneous Investigations Series Map I-1632,* 1988.

Lim, H.H., A.A. Swartz, H.A. Yueh, J.A. Kong, R.T. Shin, and J.J. van Zyl, "Classification of Earth Terrain using Polarimetric Synthetic Aperture Radar Images," *Journal of Geophysical Research,* Vol. 94, 1989.

MacDonald, G.A., "Pahoehoe, A'A and Block Lava," *American Journal of Science,* Vol. 251, pp. 169–191, 1953.

Malin, M.C., "Lengths of Hawaiian Lava Flows," *Geology,* Vol. 8, pp. 306–308, 1980.

Moore, H.J., "Preliminary Estimates of the Geological Properties of the 1984 Mauna Loa lava, *U.S.G.S. Professional Paper 1350,* pp. 1569–1587, 1987.

Peterson, D.W. and R.I. Tilling, "Transition of Basaltic Lava from Pahoehoe to A'A, Kilauea Volcano, Hawaii: Field Observations and Key Factors," *J. Volc. Geoth. Res.,* Vol. 7, pp. 271–293, 1980.

Richards, J.A., *Remote Sensing Digital Image Analysis,* Springer-Verlag, New York, 1986.

Rowland, S.K., and G.P.L. Walker, "Toothpaste Lava: Characteristics and Origin of a Lava Structural Type Transitional between Pahoehoe and A'A," *Volcanology,* Vol. 49, pp. 631–641, 1987.

Stromberg, W.D., and T.G. Farr, "A Fourier-Based Textural Feature Extraction Procedure," *IEEE Trans. on Geoscience and Remote Sensing,* Vol. GE-24, pp. 722–731, 1986.

Swartz, A.A., H.A. Yueh, J.A. Kong, L.M. Novak, and R.T. Shin, "Optimal Polarizations for Achieving Maximum Contrast in Radar Images," *Journal of Geophysical Research,* Vol. 93, No. B12, pp. 15252–15260, 1988.

Ulaby, F.T., R.K. Moore, and A.K. Fung, *Microwave Remote Sensing, Active and Passive,* Volume II, Artech House, Norwood, MA, 1982.

Ulaby, F.T., D.N. Held, M.C. Dobson, K.C. McDonald, and T.B.A. Senior, "Relating Polarization Phase Difference of SAR Signals to Scene Properties," *IEEE Trans. on Geoscience and Remote Sensing,* Vol. GE-25, pp. 83–92, 1987.

Ulaby, F.T., T. Bengal, J. East, M.C. Dobson, J. Garvin, and D. Evans, "Microwave Dielectric Spectrum of Rocks," Report 023817-1-T, Radiation Laboratory, University of Michigan, 1988.

van Zyl, J.J., "On the Importance of Polarization in Radar Scattering Problems," Ph.D. Thesis, Caltech Antenna Laboratory Report No. 120, 152 pp., California Institute of Technology, Pasadena, CA, 1985.

van Zyl, J.J., H.A. Zebker, and C. Elachi, "Imaging Radar Polarization Signatures: Theory and Observation," *Radio Science,* Vol. 22, pp. 529–543, 1987.

van Zyl, J.J., P.C. Dubois, T.G. Farr, and H.A. Zebker, "Inference of Geologic Surface Parameters from Polarimetric Radar Observations and Model Inversion," *Proceedings of IGARSS '88 Symposium,* pp. 51–52, 1988.

van Zyl, J.J., "Unsupervised Classification of Scattering Behavior using Radar Polarimetry Data," *IEEE Trans. on Geoscience and Remote Sensing,* Vol. GE-27, pp. 36–45, 1989.

van Zyl, J.J., "Calibration of Polarimetric Radar Images using only Image Parameters and Tihedral Corner Reflector Responses," *IEEE Trans. on Geoscience and Remote Sensing,* Vol. GE-28, 1990.

Walker, G.P.L., "Lengths of Lava Flows," *Phil. Trans. of the Royal Society of London,* Vol. A274, pp. 107–118, 1973.

Wise, W.S., *Geological Map of the Pisgah and Sunshine Cone Lava Fields,* NASA TL-11, 8 pp., 1966.

Wu, S-T., and S.A. Sader, "Multipolarization SAR Data for Surface Feature Delineation and Forest Vegetation Characterization," *IEEE Trans. on Geoscience and Remote Sensing,* Vol. GE-25, pp. 67–76, 1987.

Yueh, H.A., A.A. Swartz, J.A. Kong, R.T. Shin, and L.M. Novak, "Bayes Classification of Terrain Cover using Normalized Polarimetric Data," *Journal of Geophysical Research,* Vol. 93, No. B12, pp. 15261–15267, 1988.

Zebker, H.A., and L. Norikane, "Radar Polarimeter Measures Orientation of Calibration Corner Reflectors," *Proceedings of the IEEE,* Vol. 75, pp. 1686–1688, 1987.

# Index

# RADAR POLARIMETRY FOR GEOSCIENCE APPLICATIONS

## Errata

Equation (2.43) on page 28 is incorrect. All the elements of the matrix $\overline{\mathcal{W}}$ should be replaced by their complex conjugates. Thus, the correct form is

$$\overline{\mathcal{W}} = \begin{pmatrix} \overline{S^*_{vv}S_{vv}} & \overline{S^*_{vh}S_{vh}} & \overline{S^*_{vh}S_{vv}} & \overline{S^*_{vv}S_{vh}} \\ \overline{S^*_{hv}S_{hv}} & \overline{S^*_{hh}S_{hh}} & \overline{S^*_{hh}S_{hv}} & \overline{S^*_{hv}S_{hh}} \\ \overline{S^*_{hv}S_{vv}} & \overline{S^*_{hh}S_{vh}} & \overline{S^*_{hh}S_{vv}} & \overline{S^*_{hv}S_{vh}} \\ \overline{S^*_{vv}S_{hv}} & \overline{S^*_{vh}S_{hh}} & \overline{S^*_{vh}S_{hv}} & \overline{S^*_{vv}S_{hh}} \end{pmatrix} .$$

www.ingramcontent.com/pod-product-compliance
Lightning Source LLC
LaVergne TN
LVHW101456200926
848483LV00001B/144
* 9 7 8 0 8 9 0 0 6 4 0 6 1 *